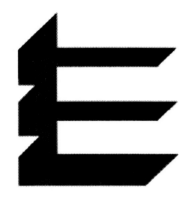

STANDARD ESTIMATING PRACTICE

Sixth Edition

American Society of Professional Estimators

ISBN 1557014817

Copyright ® 2004 by Bni Publications, Inc. All rights reserved. Printed in the United States of America. Except as permitted under the United States Copyright Act of 1976, no part of this publication may be reproduced or distributed in any form or by any means, or stored in a data base or retrieval system, without the prior written permission of the publisher.

While diligent effort is made to provide reliable, accurate and up-to-date information, neither BNi Publications Inc., nor its authors or editors, can place a guarantee on the correctness of the data or information contained in this book. BNi Publications Inc., and its authors and editors, do hereby disclaim any responsibility or liability in connection with the use of this book or of any data or other information contained therein.

2003–2004 Standards Board
The American Society of Professional Estimators

Les Chipman, CPE Teal Construction Company Houston, Texas Houston Chapter	**Don Sellmeyer, CPE** Sellmeyer Construction Consulting, Inc. Tempe, Arizona Arizona Chapter
Michael G. Gioffre, CPE Nason Construction, Inc. Wilmington, Delaware Delaware Chapter	**Doug Warren, CPE** HBG Management Memphis, Tennessee Memphis Chapter
Frank A. Kutilek, Jr., CPE HOK Construction Services St. Louis, Missouri St. Louis Metro Chapter	**Catherine Taylor Yank, FCPE, MBA** Jim Taylor, Inc. Belleville, Illinois St. Louis Metro Chapter

Contributors:

Larry L. Cockrum, CPE
President 2002–2004, American Society of
Professional Estimators
Cockrum & Associates
Ripley, Mississippi

Jeff Coopersmith, CPE
Bancroft Construction Company
Wilmington, Delaware

Edward T. Alexander,/CPE
Bancroft Construction Company
Wilmington, Delaware

The American Society of Professional Estimators wishes to express its gratitude and sincere appreciation to the many individuals, companies, and organizations that contributed to this manual's development.

FOREWORD TO THE SIXTH EDITION - 2004

The American Society of Professional Estimators is proud to present its STANDARD ESTIMATING PRACTICE, SIXTH EDITION. It has been developed and written by the members of the Standards Board of the American Society of Professional Estimators and of the society. It is written in the language of estimators to provide not only basic and fundamental guidance for estimators, but also to define industry recognized standards that are the basis for certification of experienced estimators as Certified Professional Estimators.

This manual is developed for use as an educational tool to be used in the training of new estimators, to provide a foundation for future ASPE publications on estimating, and as a guide to the format and content of construction cost estimates. *The information contained in this work has been obtained by the American Society of Professional Estimators from sources believed to be reliable. However, neither ASPE nor its members guarantee the accuracy or completeness of any information published herein and neither ASPE nor its members shall be responsible for any errors, omissions, or damages arising out of use of this information. This work is published with the understanding that ASPE and its members are supplying information but are not attempting to render engineering or other professional services. If such services are required, the assistance of an appropriate professional should be sought.*

Development of additional standards and the review and upgrading of standards are a continuing responsibility of ASPE.

STANDARD ESTIMATING PRACTICE

CONTENTS

PART ONE -   PRACTICE COMMON TO ALL DISCIPLINES

PART TWO -   PRACTICE COMMON TO SPECIFIC DISCIPLINES

PART THREE - PROFESSIONAL ESTIMATING SERVICE GUIDELINES

PART FOUR -   ETHICS

PART FIVE -   REFERENCE SOURCES

PART ONE

PRACTICE COMMON TO ALL DISCIPLINES

PART ONE
Table of Contents

Section 1 - Basic Standards **5**
    Introduction 5
    Ethics 5
    Integrity 5
    Judgment 5
    Attitude 5
    Thoroughness 5
    Awareness 6
    Uniformity 6
    Consistency 6
    Verification 6
    Documentation 6
    Evaluation 6
    Labor Hours 7
    Value Engineering 7
    Final Summaries 7
    Analysis 7
    Conversion 7
    Change Orders 7

Section 2 - Levels of the Estimate **9**
    Introduction 9
    Level One (1) Order of Magnitude 10
    Schematic Conceptual Design 11
    Design Development 11
    Construction Documents 12
    Bid 13

Section 3 - Scope of Estimate **21**
Section 4 - Estimating Procedures, General **25**
Section 5 - Project Evaluation-Constructors **29**
    Company Policy 29
    Work Scope 30
    Estimate/Bid Due Date 30

Section 6 - Project Evaluation-Value Engineering (VE) **31**
    Introduction 31
    Value Engineering Terminology and Theory 31
    Phases of Value Engineering 31
    Value Engineering Methodology 33
    The Value Engineering Team 34
    Important Elements for Consideration in a Value Engineering Change Proposal (VECP) 34
    Summary 35
    Glossary 35

Section 7 - Bid Documents/Procurement **37**
Section 8 - Checklists/Special Forms **39**
    Master Checklist 41
    Bid Document Inventory Checklist 43
    Site Investigation Checklist 45
    Direct Cost Estimate Checklist 53
    Overhead Estimate Checklist 57
    General Conditions Estimate Checklist 63
    Final Summary 75
    Final Summary Cut & Add Sheet 77

    Account Summary 79
    Estimate Detail 81
    Quantity Survey 83

Section 9 - Specification Review 85
    Invitations to Bid and Instructions to Estimators/Bidders 85
    Relationship to Plans 85
    General and Supplementary Conditions 85
    Determining Pertinent Divisions and Sections 86
    Detailed Review 87
    Interpretation of Intent 87
    Define Scope of Work 88

Section 10 - Plan Review 89
    Develop Familiarity and Format 89

Section 11 - Quantity Survey 91
    Cover Page 91
    Estimate Levels 91
    Estimate Format 91
    General and Supplementary Conditions 91
    Schedule 92
    Defined Areas 92
    Systems Descriptions 92
    Specific Items 92
    Mensuration 92
    Measuring Tools and Devices 93
    Plan Scale Proof 93
    Waste Allowances 93
    Takeoff Procedures 93

Section 12 - Pricing/Summaries 95
    Estimate Detail Form ❑ 95
    Materials 95
    Labor 95
    Equipment 96
    Subcontracts 97
    Account Summary Form ❑ 97
    Final Summary Form ❑ 98

Section 13 - Contingency & Accuracy 103

Section 14 - Bid Day Procedures 107
    Constructors 107

Section 15 - Presentation 121
    Guideline 121
    Example Level 5 Narrative 121

Section 16 - Post-Bid Procedures 125
    Introduction 125
    As-Bid Analysis 125

Section 17 - Estimating Change Orders, Cost or Opportunity? 131
    Recognition 133
    Preparation 134
    Presentation 135
    Change Order Aids 135

Section 18 - Handling Federal Claims and Change Orders 137

Section 19 - Legal Considerations in Construction 151

SECTION 1 – BASIC STANDARDS

Introduction

The practice of construction estimating is a highly technical discipline. It involves certain standards of ethical conduct and moral judgment that go beyond the technical aspects of the discipline.

Estimators are often the persons most familiar with the complete project. They must exercise sound moral judgment when preparing the estimate. Estimators sometime receive pressure from other members of the construction team to make expedient short term decisions that can result in an unsound bid. Resistance to this pressure is a part of the estimator's job.

Examples of expedient behavior litter the history of construction estimating. The fruit of this shows in inaccurate estimates. Deficient estimates also can cause strife between members of the construction team, divisive litigation, and lost profits.

For these reasons, the Society states the following ethical, moral and technical precepts as basic to the practice of estimating.

Ethics

Construction estimators shall follow a high standard of ethical practice as defined by this Society. Members of this Society shall show a commitment to ethical practices by adopting a detailed code of ethics. The Society invites users of this manual and others to join in a discussion of ethical practice in the construction industry.

Integrity

All estimators shall use standards of confidentiality in a manner at least equal to that of other professional societies. The estimator shall keep in strictest confidence information received from outside sources. All members of this society shall follow the Confidentiality Standards. Certified Professional Estimators shall pledge fidelity to this standard. The practice, commonly called "bid peddling", is a breach of ethics and condemned by this society.

Judgment

Judgment is a skill gained by estimators through proper training and extensive experience. Always use sound judgment and common sense when preparing estimates. Proper use of judgment may mean the difference between profit and loss for the company or client.

Attitude

Approach each estimate with a professional attitude and examine in detail all areas of the work. Set aside specific times each day for entry of estimate quantities and data without interruption. Total mental concentration is basic when preparing accurate estimates.

Thoroughness

Allow enough time to research and become familiar with the details of the project and then promptly complete the quantity survey. Review the various aspects of the project with others involved. The estimator with the most knowledge of a project has the competitive advantage.

Awareness

Review the project scope and determine if the company has sufficient financial resources, staff, and plant to complete the project. Consider the time allotted for construction of the project. Examine the general conditions of the contract and determine the effect these requirements have on indirect costs. Consider alternate methods of construction for the project. Conduct an examination of the special conditions that may alter the intent of the general condition. Review all division specifications to gain a better perspective of the total project scope. Estimate the general and special conditions similar to all other divisions of the specifications. Review the deadline for submitting the estimate and determine if there is enough time to prepare the estimate. Review all sections of the plans and find out the degree of coordination between architectural and engineering drawings.

Uniformity

Develop a good system of estimating forms and procedures that exactly meet the requirements of the company. This system should provide the ability to define material, labor hour and equipment hour quantities required for the project. Material, labor and equipment unit costs are then applied to the quantities as developed in the quantity survey. Apply amounts for overhead and profit in the final summaries.

Consistency

Use methods for quantity surveys that are in logical order and consistent with the CSI numbering system. These methods also must meet the specific need of the company or client. Use of consistent methods allows several estimators to complete various parts of the quantity survey. Combine these surveys into the final account summaries. These methods also permit a second estimator to continue the quantity survey from any point where the first estimator stopped.

Verification

The methods and logic employed in the quantity survey must be in a form which can provide a method of proof of the accuracy of any portion of the survey.

Documentation

Document all portions of the estimate in a logical, consistent, and legible manner. Estimators and other personnel may need to review the original estimate when the specific details are vague. The documentation must be clear and logical or it will be of little value to the reader. Such instances may occur in change order preparation, settlement of claims, and review of past estimates as preparation for new estimates on similar projects.

Evaluation

When the estimate involves the use of bids from subcontractors, check the bids for scope and responsiveness to the project. Investigate the past performance records of subcontractors submitting bids. Determine the level of competence and quality of performance.

Labor Hours

The detailed application of labor hours to a quantity survey is primary in governing the accuracy and sufficiency of an estimate. The accuracy of project schedules and work force requirements are dependent on the definition of many hours. The combined costs for worker's compensation, unemployment insurance and social security taxes are significant factors in project costs. The most accurate method for including these costs is to define labor hours and wage rates, then apply percentages to the labor costs.

Value Engineering

Structure the estimate to aid in researching and developing alternative construction methods resulting in cost optimization. Using the same levels of detail in value engineering as in the base estimate is important. This provides a more precise comparison of costs for proposed alternate methods.

Final Summaries

Provide methods for listing and calculating indirect costs. Project scope governs the costs of overhead items such as insurance, home office plant, and administrative personnel. Determine these costs in a manner consistent with quantity survey applications. Consider company work in progress that may have a bearing on projected overhead costs. Each bidder must determine amounts for performance bonding, profits, reserve funds, and shareholder returns.

Analysis

Develop methods for analyzing completed estimates to find out if they are reasonable. When the estimate is beyond the normal range of costs for similar projects, research the detail and determine causes or possible errors. Develop methods of analysis of · post-bid estimates to find the reasons for the lack of success in the bidding process. Calculate the variation of the estimate from the low bid and low average bids, Determine from outside sources if there were subcontract or material bids provided only to other bidders. Determine if the low bidder may have made an omission in the estimate. Properly document this information for future use and guidance.

Conversion

Show estimating procedures that allow conversion of the estimate to field cost systems where management can monitor and control field activities. These procedures include methods of reporting field costs for problem areas. Make reports daily or weekly rather than at some point in time after the project is complete. Field cost reporting, when consistent with estimating procedures, enables estimators to apply the knowledge gained from these historical costs to future estimates. Help train field personnel in labor hour and cost reporting that provide the level of accuracy required.

Change Orders

Apply the highest level of detail from information provided or available to the estimator. State quantities and costs for all material, labor, equipment and subcontract items of work. Define amount for overhead, profit, taxes, and bond. Specific itemization of change order proposals is essential in allowing the client to determine acceptability. Upon approval, use the estimate detail as the definition of the scope of the change order.

SECTION 2 — LEVELS OF THE ESTIMATE

INTRODUCTION

As a project is proposed and then developed, the estimate preparation and information will change based on the needs of the Owner/Client/Designer. These changes will require estimates to be prepared at different levels during the design with varying degrees of information provided. It should also be noted that within each level of estimate preparation, not all portions of the design will be at the same level of completeness. Example could be that architectural is at 80% or 90% complete while the mechanical is only at 50% complete. This is common through the design process, but should be noted in the estimate narrative.

The levels of the estimate correspond to the typical design process and are considered standards within the industry. These levels are as follows:

- Order (Range) of Magnitude
- Schematic/Conceptual Design
- Design Development
- Construction Document
- Bid

These descriptions constitute the levels of the estimate. Estimates within each of these phases may be prepared multiple times during the design process as more information is available or changes made. With each level of the estimate, the estimate will become more detailed as more information is provided and fewer assumptions are being made.

ORDER (RANGE) OF MAGNITUDE

This level is usually prepared to develop a project budget and is based on historical information with adjustments made for specific project conditions. Costs are based on cost per square foot, number of cars/rooms/seats, etc. Allowances must be made for site work and specific project conditions. Information required for this level is a project program and desired quality level from the Owner.

SCHEMATIC/CONCEPTUAL DESIGN

This level is used to price various schemes as the project design develops. It may be used to price various design schemes in order to see which scheme fits the budget best or it may be used to price various materials or methods for comparison. The ultimate goal at the end of schematic design is to have a design scheme, program and estimate that are all within the budget. This estimate is often prepared in UniFormat versus MasterFormat. This allows the design team to easily and quickly evaluate systems and make informed decisions required to progress design. Information required for this level is schematic drawings, sketches, renderings, diagrams, conceptual plans, elevations, sections and preliminary project descriptions.

DESIGN DEVELOPMENT

Estimates prepared at this level are used to verify budget conformance as the scope and design is finalized and final materials are selected. Information required for this level is drawings showing plans, elevations, typical details, engineering design criteria, equipment layouts and outline specifications.

CONSTRUCTION DOCUMENT

This level is used to verify pricing as details are completed, design is modified and completed, and to be aware of and identify "design creep" during the completion of the design. This final construction document estimate can be used to evaluate the subcontract pricing during the bid phase. Information required for this level is detailed drawings showing plans, elevations, sections, details, schedules, specifications and bidding criteria.

BID

The purpose of this level estimate is to develop probable costs in the preparation and submittal of bids for contract with an Owner.

In the typical "design/bid/build" delivery system, this would be with "completed" documents. In other delivery systems becoming widely used (i.e., design/build or GMP), the bid could actually be prepared at a previous level. If this is the case, estimates are prepared as previously described along with progressive estimates as the design is completed. It should be stressed that when preparing a bid at a prior estimate level, it is very important to have a complete and thorough "Scope of Estimate".

It should be noted that it is always good practice to review and evaluate the final cost versus the bid, however this (final cost) is not another level of estimate, and is a cost control issue rather than an estimate.

Level One (1) Order of Magnitude

Prepare this level of estimate with information derived from an outline of the proposed plant. The outline should provide the following information:

- ✓ General description
- ✓ Geographic location
- ✓ Quality
- ✓ Layout
- ✓ Size
- ✓ Intended use

For process areas include the following information:

- ✓ Product capacity
- ✓ Handling requirements
- ✓ Materials
- ✓ Services requirements
- ✓ Raw materials
- ✓ Process layout
- ✓ Utility requirements
- ✓ Storage required
- ✓ Flow diagrams

The purpose of this estimate level is for budgetary and feasibility determinations.

Schematic/Conceptual Design

Prepare this level estimate from Level One information plus outline design criteria with descriptions of the following items:

- ✓ Soil conditions
- ✓ Labor hours by section
- ✓ Foundation requirements
- ✓ Rough sketches
- ✓ Rough utility quantities
- ✓ Construction type/size

For process areas provide the following information:

- ✓ Outline design criteria
- ✓ Electrical one line drawings
- ✓ General arrangement drawings
- ✓ Preliminary motor list/sizes
- ✓ Process identification drawings
- ✓ Preliminary flow sheets/specifications

The purpose of this estimate level is to provide a better defined estimate for budgetary and feasibility determinations.

Design Development

Prepare this level estimate from not less than twenty-five percent complete preliminary design drawings and draft specifications. Information provided should include:

- ✓ General site description
- ✓ Preliminary structural design
- ✓ Site dimensions
- ✓ Elevations
- ✓ Roads
- ✓ Preliminary building equipment plans
- ✓ Impounds & fences
- ✓ Soil bearing condition
- ✓ General arrangements
- ✓ Preliminary plumbing drawings
- ✓ Foundation sketches
- ✓ Preliminary mechanical drawings
- ✓ Architectural construction
- ✓ Preliminary electrical drawings

Use the following preliminary information for the process area estimate.

- ✓ Piping flow sheet
- ✓ Instrument list
- ✓ Equipment list
- ✓ Utility heat balance & flow
- ✓ Insulation requirements
- ✓ Electric substation specifications

The plant and the process area estimate also depends on information provided in the previous estimates.

The purpose of this estimate is to establish probable costs within the range of available information. Continue defining labor hours in this level.

The process area estimate includes information provided in the plant estimate plus the following engineered requirements.

- ✓ Process flow sheet
- ✓ Insulation specifications
- ✓ Equipment specifications
- ✓ Instrumentation list & flow sheet
- ✓ Vessel sheets
- ✓ Heat balance sheets & flow sheets
- ✓ Electrical list/sizes
- ✓ Piping flow sheets
- ✓ Preliminary control wiring specifications
- ✓ Equipment/piping insulation
- ✓ Electrical single line drawings
- ✓

This estimate provides a greater amount of accuracy possible with better definition and detail. Use this level for value engineering applications before the completion of specifications and design drawings.

Construction Documents

Prepare this level estimate from not less than ninety percent complete design drawings and specifications. Use criteria provided for lower estimates with the exception of:

- ✓ Outline design
- ✓ Draft specifications
- ✓ Outline specifications
- ✓ Partial design drawings
- ✓ Preliminary design drawings
- ✓

Additionally, use the following fully developed and engineered data:

- ✓ Site plans
- ✓ Detail drawings
- ✓ Topographical maps
- ✓ Building equipment
- ✓ Plumbing/mechanical/electrical drawings
- ✓ General arrangements
- ✓ Elevations
- ✓ Soil bearing reports

The process area estimate includes the information provided in the plant estimate and the following fully developed and engineered data:

- ✓ Equipment list
- ✓ Piping layout & schedules
- ✓ Electrical distribution specifications
- ✓ Insulation drawings
- ✓ Utility requirements
- ✓ Electrical drawings

Define the plant and the process area labor hours by craft or section. State supervision and general condition labor hours.

This level shows the probable project cost. Use this level for value engineering applications before publication of the project for quotations.

Bid

Prepare this level estimate for both plant and process area from complete design drawings, specifications, and bid documents.

The purpose of this level estimate is to show probable costs in the preparation and submittal of bids for contracts with an owner.

Various types of contracts are:

- ✓ Stipulated sum
- ✓ Lump sum unit price
- ✓ Cost plus a fee
- ✓ Turn key
- ✓ Design - build
- ✓ Cost plus a fee with a guaranteed maximum price

The transfer of estimate information to field cost control systems provides management the opportunity to closely monitor and control construction costs as they occur. Computer estimating and cost control programs, whether industry specific or general spreadsheet type, are especially valuable for rapid and efficient generation of both the estimate and actual construction cost information.

Job Number: ASPE 01

Date: 9/8/97

**Order of Magnitude
Estimate Sample
ESTIMATE SUMMARY**

Description	LABOR HOURS	LABOR $	MATERIAL $	EQUIPMENT $	SUBCONTRCT $	TOTAL DOLLARS
Sitework	0	$0	$0	$0	$1,939,046	$1,939,046
Restaurant	0	$0	$0	$0	$3,906,000	$3,906,000
Parking	0	$0	$0	$0	$4,490,750	$4,490,750
Retail	0	$0	$0	$0	$3,906,000	$3,906,000
Exposition	0	$0	$0	$0	$1,488,000	$1,488,000
Prime Contractor Overheads	0	$0	$0	$0	$1,258,384	$1,258,384
Prime Contractor Profit	0	$0	$0	$0	$849,409	$849,409
Contingency	0	$0	$0	$0	$891,879	$891,879
TOTAL ESTIMATED COST					$18,729,468	$18,729,468

Order of Magnitude
Estimate Sample

Job Number: ASPE 01
Date: 9/8/97

ESTIMATE DETAIL - SITEWORK

SITEWORK			LABOR HOURS		LABOR $		MATERIAL $		EQUIPMENT $		SUBCONTRACT $		TOTAL
Description	Quantity	Unit Meas.	Unit	Total	Unit	Total	Unit	Total	Unit	Total	Unit	Total	DOLLARS
Irrigation system	1	L Sum	0	0	0	$0	0	$0	0	$0	75000	$75,000	$75,000
Site Water	1,500	LF	0	0	0	$0	0	$0	0	$0	40	$60,000	$60,000
Site Electrical	2,000	LF	0	0	0	$0	0	$0	0	$0	65	$130,000	$130,000
Site Sewer	3,000	LF	0	0	0	$0	0	$0	0	$0	40	$120,000	$120,000
Site Storm Drainage	1,500	LF	0	0	0	$0	0	$0	0	$0	40	$60,000	$60,000
Trees and Shrubs (Shrubs 10 times Tree)	346	Each	0	0	0	$0	0	$0	0	$0	525	$181,650	$181,650
Sod Site	16,850	SF	0	0	0	$0	0	$0	0	$0	0.2	$3,370	$3,370
Asphalt Paving	4,334	SY	0	0	0	$0	0	$0	0	$0	19	$82,346	$82,346
Concrete Paving	1,214	SY	0	0	0	$0	0	$0	0	$0	25	$30,350	$30,350
Sidewalks and Stairs	73,050	SF	0	0	0	$0	0	$0	0	$0	2	$146,100	$146,100
Curbs and Gutters	6,115	LF	0	0	0	$0	0	$0	0	$0	2	$12,230	$12,230
Riverside Dock	8,400	SF	0	0	0	$0	0	$0	0	$0	25	$210,000	$210,000
Promenade	16,000	SF	0	0	0	$0	0	$0	0	$0	15	$240,000	$240,000
Excavation and Backfill	8.8	Acres	0	0	0	$0	0	$0	0	$0	10000	$88,000	$88,000
Blasting Allowance	1	L Sum	0	0	0	$0	0	$0	0	$0	500000	$500,000	$5000,000
SITEWORK Estimated Cost				**0**		**$0**		**$0**		**$0**		**$1,939,046**	**$1,939,046**

Section 2 – Levels of the Estimate — Part One

Job Number: ASPE 01

Date: 9/8/97

Order of Magnitude
Estimate Sample

ESTIMATE DETAIL - RESTAURANT

RESTAURANT Description	Quantity	Unit Meas.	LABOR HOURS Unit	LABOR HOURS Total	LABOR $ Unit	LABOR $ Total	MATERIAL $ Unit	MATERIAL $ Total	EQUIPMENT $ Unit	EQUIPMENT $ Total	SUBCONTRACT Unit	SUBCONTRACT Total	TOTAL DOLLARS
Sitework	63,000	SF	0	0	0	$0	0	$0	0	$0	1	$63,000	$63,000
Concrete	63,000	SF	0	0	0	$0	0	$0	0	$0	2	$126,000	$126,000
Masonry	63,000	SF	0	0	0	$0	0	$0	0	$0	2	$126,000	$126,000
Metals	63,000	SF	0	0	0	$0	0	$0	0	$0	1	$63,000	$63,000
Woods and Plastics	63,000	SF	0	0	0	$0	0	$0	0	$0	19	$1,197,000	$1,197,000
Thermal and Moisture Protection	63,000	SF	0	0	0	$0	0	$0	0	$0	5.5	$346,500	$346,500
Doors and Windows	63,000	SF	0	0	0	$0	0	$0	0	$0	9	$567,000	$567,000
Finishes	63,000	SF	0	0	0	$0	0	$0	0	$0	3	$189,000	$189,000
Specialties	63,000	SF	0	0	0	$0	0	$0	0	$0	0.5	$31,500	$31,500
Equipment	63,000	SF	0	0	0	$0	0	$0	0	$0	0	$0	$0
Furnishings	63,000	SF	0	0	0	$0	0	$0	0	$0	0	$0	$0
Special Construction	63,000	SF	0	0	0	$0	0	$0	0	$0	0	$0	$0
Conveying Systems	63,000	SF	0	0	0	$0	0	$0	0	$0	0	$0	$0
Mechanical	63,000	SF	0	0	0	$0	0	$0	0	$0	14	$882,000	$882,000
Electrical	63,000	SF	0	0	0	$0	0	$0	0	$0	5	$315,000	$315,000
Restaurant Estimated Cost				0		$0		$0		$0			$3,906,000

$3,906,000

Part One   **Section 2**- Levels of the Estimate

Job Number: ASPE 01

Date: 9/8/97

Order of Magnitude

Estimate Sample

ESTIMATE DETAIL - PARKING

PARKING Description	Quantity	Unit Meas.	LABOR HOURS Unit	LABOR HOURS Total	LABOR $ Unit	LABOR $ Total	MATERIAL $ Unit	MATERIAL $ Total	EQUIPMENT $ Unit	EQUIPMENT $ Total	SUBCONTRACT $ Unit	SUBCONTRACT $ Total	TOTAL DOLLARS
Sitework	253,000	SF	0	0	0	$0	0	$0	0	$0	3.5	$885,500	$885,500
Concrete	253,000	SF	0	0	0	$0	0	$0	0	$0	12	$3,036,000	$3,036,000
Masonry	253,000	SF	0	0	0	$0	0	$0	0	$0	1	$253,000	$253,000
Metals	253,000	SF	0	0	0	$0	0	$0	0	$0	0.1	$25,300	$25,300
Woods and Plastics	253,000	SF	0	0	0	$0	0	$0	0	$0	0	$0	$0
Thermal and Moisture Protection	253,000	SF	0	0	0	$0	0	$0	0	$0	0.1	$25,300	$25,300
Doors and Windows	253,000	SF	0	0	0	$0	0	$0	0	$0	0.1	$25,300	$25,300
Finishes	253,000	SF	0	0	0	$0	0	$0	0	$0	0.1	$25,300	$25,300
Specialties	253,000	SF	0	0	0	$0	0	$0	0	$0	0	$0	$0
Equipment	253,000	SF	0	0	0	$0	0	$0	0	$0	0.1	$25,300	$25,300
Furnishings	253,000	SF	0	0	0	$0	0	$0	0	$0	0	$0	$0
Special Construction	253,000	SF	0	0	0	$0	0	$0	0	$0	0	$0	$0
Conveying Systems	253,000	SF	0	0	0	$0	0	$0	0	$0	0	$0	$0
Mechanical	253,000	SF	0	0	0	$0	0	$0	0	$0	0.25	$63,250	$63,250
Electrical	253,000	SF	0	0	0	$0	0	$0	0	$0	0.5	$126,500	$126,500
Parking 253,000 SF Estimated Cost				0		$0		$0		$0		$4,490,750	**$4,490,750**

Section 2 – Levels of the Estimate   **Part One**

Job Number: ASPE 01

Date: 9/8/97

Order of Magnitude
Estimate Sample

ESTIMATE DETAIL - RETAIL

RETAIL			LABOR HOURS		LABOR $		MATERIAL $		EQUIPMENT $		SUBCONTRACT $		TOTAL
Description	Quantity	Unit Meas.	Unit	Total	Unit	Total	Unit	Total	Unit	Total	Unit	Total	DOLLARS
Sitework	63,000	SF	0	0	0	$0	0	$0	0	$0	1	$63,000	$63,000
Concrete	63,000	SF	0	0	0	$0	0	$0	0	$0	2	$126,000	$126,000
Masonry	63,000	SF	0	0	0	$0	0	$0	0	$0	2	$126,000	$126,000
Metals	63,000	SF	0	0	0	$0	0	$0	0	$0	1	$63,000	$63,000
Woods and Plastics	63,000	SF	0	0	0	$0	0	$0	0	$0	19	$1,197,000	$1,197,000
Thermal and Moisture Protection	63,000	SF	0	0	0	$0	0	$0	0	$0	5.5	$346,500	$346,500
Doors and Windows	63,000	SF	0	0	0	$0	0	$0	0	$0	9	$567,000	$567,000
Finishes	63,000	SF	0	0	0	$0	0	$0	0	$0	3	$189,000	$189,000
Specialties	63,000	SF	0	0	0	$0	0	$0	0	$0	0.5	$31,500	$31,500
Equipment	63,000	SF	0	0	0	$0	0	$0	0	$0	0	$0	$0
Furnishings	63,000	SF	0	0	0	$0	0	$0	0	$0	0	$0	$0
Special Construction	63,000	SF	0	0	0	$0	0	$0	0	$0	0	$0	$0
Conveying Systems	63,000	SF	0	0	0	$0	0	$0	0	$0	0	$0	$0
Mechanical	63,000	SF	0	0	0	$0	0	$0	0	$0	14	$882,000	$882,000
Electrical	63,000	SF	0	0	0	$0	0	$0	0	$0	5	$315,000	$315,000
RETAIL 63,000 SF Estimated Cost				0		$0		$0		$0		$3,906,000	$3,906,000

Part One Section 2 - Levels of the Estimate

Order of Magnitude
Estimate Sample

ESTIMATE DETAIL - EXPOSITION

Job Number: ASPE 01

Date: 9/8/97

EXPOSITION			LABOR HOURS		LABOR $		MATERIAL $		EQUIPMENT $		SUBCONTRACT $		TOTAL
Description	Quantity	Unit Meas.	Unit	Total	Unit	Total	Unit	Total	Unit	Total	Unit	Total	DOLLARS
Sitework	24,000	SF	0	0	0	$0	0	$0	0	$0	1	$24,000	$24,000
Concrete	24,000	SF	0	0	0	$0	0	$0	0	$0	2	$48,000	$48,000
Masonry	24,000	SF	0	0	0	$0	0	$0	0	$0	2	$48,000	$48,000
Metals	24,000	SF	0	0	0	$0	0	$0	0	$0	1	$24,000	$24,000
Woods and Plastics	24,000	SF	0	0	0	$0	0	$0	0	$0	19	$456,000	$456,000
Thermal and Moisture													
Protection	24,000	SF	0	0	0	$0	0	$0	0	$0	5.5	$132,000	$132,000
Doors and Windows	24,000	SF	0	0	0	$0	0	$0	0	$0	9	$216,000	$216,000
Finishes	24,000	SF	0	0	0	$0	0	$0	0	$0	3	$72,000	$72,000
Specialties	24,000	SF	0	0	0	$0	0	$0	0	$0	0.5	$12,000	$12,000
Equipment	24,000	SF	0	0	0	$0	0	$0	0	$0	0	$0	$0
Furnishings	24,000	SF	0	0	0	$0	0	$0	0	$0	0	$0	$0
Special Construction	24,000	SF	0	0	0	$0	0	$0	0	$0	0	$0	$0
Conveying Systems	24,000	SF	0	0	0	$0	0	$0	0	$0	0	$0	$0
Mechanical	24,000	SF	0	0	0	$0	0	$0	0	$0	14	$336,000	$336,000
Electrical	24,000	SF	0	0	0	$0	0	$0	0	$0	5	$120,000	$120,000
EXPOSITION 24,000 SF Estimated Cost			0	0		$0		$0		$0		$1,488,000	$1,488,000
PRIME CONTRACTOR OVERHEADS	1	L Sum	0	0	0%	$0	0%	$0	0%	$0	8%	$1,258,384	$1,258,384
PRIME CONTRACTOR PROFIT	1	L Sum	0	0	0%	$0	0%	$0	0%	$0	5%	$849,409	$849,409
CONTINGENCY	1	L Sum	0	0	0%	$0	0%	$0	0%	$0	5%	$891,879	$891,879

Standard Estimating Practice

SECTION 3 – SCOPE OF ESTIMATE

Prepare all estimates in an expert and adept manner. This is consistent with standards normally expected of the estimating community.

Estimators are responsible for the quality, accuracy, and timely completion of their product.

Estimators shall follow the Code of Ethics of this Society. Private information given to the estimator shall remain private.

Prepare all estimates with the Construction Specification Institute Numbering System. When estimating defined areas, use the CSI Format in each area. Defined areas are specific areas of the project needing their own estimate. System estimate, work breakdown structure, or special division of work are other names for defined area.

Base estimates on the highest level of detail available from design information. Whenever possible, create estimates on a "quantity times material, labor hours, and equipment cost" format. Use square foot costs or cubic foot costs only when design information is not sufficient to provide a detailed estimate.

Estimators shall provide a narrative of the level and scope of the estimate. Define the level of the estimate according to Part One, Section Two of this manual. Separately, specify sections of work with different levels of design development. This explains the relative percentages of the data available and is valuable for calculating the range of accuracy or estimate contingency.

The narrative should include the following information:

- ✓ Plans & specifications received
- ✓ Project type
- ✓ Project address
- ✓ Addenda issued
- ✓ Legal description
- ✓ Project description
- ✓ Owner name & address
- ✓ Estimate assumptions
- ✓ Estimate due date/time
- ✓ Project quality & size
- ✓ Designer name & address
- ✓ Project labor type

This information helps in the development of historical cost estimate systems.

Section 3 – Scope of Estimate

Prepare estimates in a form that other involved parties can readily understand. The quantity times material and labor hour costs format should contain the following components (read from left to right):

Work breakdown structure	CSI number	Description
Unit measure	Quantity	Material unit
Material costs	Labor hours	Cost per hour
Labor hours	Equipment hours	Cost per hour
Equipment hours	Subcontract	Total

Assemble the information in this form, from the quantity takeoff and extension detail. Direct labor burden is the combined cost for:

- ✓ Workers' compensation insurance
- ✓ Employer's liability insurance
- ✓ Company fringe benefits
- ✓ Union fringe benefits
- ✓ State/federal unemployment insurance
- ✓ Employer-paid social security tax
- ✓ Subsistence
- ✓

Show labor burden as a separate direct field cost in each estimate. Calculate using the following method: Convert hourly costs for benefits or subsistence to a percentage. Add this to the other direct labor costs and determine the total percentage. This applies to each construction craft or trade, recognizing the variation in Worker's Compensation, fringe benefits, and subsistence. Apply this percentage to the labor cost for each division (section) of work. The percentage applied should reflect expected escalation.

General and supplemental conditions are the expected costs for:

- ✓ Construction/physical plant
- ✓ Miscellaneous expenditures

- ✓ Construction equipment
- ✓ Permits, fees
- ✓ Supervision
- ✓ Licenses, testing
- ✓ Supervision service equipment
- ✓ Field engineering

- ✓ Hot/Cold weather protection
- ✓ Other miscellaneous costs
- ✓ Small/hand power tools
- ✓

Show the cost for general conditions as a separate direct field cost in estimate levels three through six. Define each item of general conditions and its cost.

Apply a percentage, based on historical data, of the direct costs for general conditions. This is a judgment based on prior detailed estimates. Show general conditions as a percentage in estimate levels one and two only.

Overhead is the expected contract cost for all items classified other than direct field costs. The overhead cost may include:

- ✓ Home office plant capital cost
- ✓ Project management
- ✓ Home office leasing cost
- ✓ Clerical
- ✓ Accounting
- ✓ Home office plant services
- ✓ General liability & comprehensive insurance
- ✓ Finance expenses
- ✓ Support & pro-rata salaries
- ✓ Outside services

Show the expected cost for overhead separately in each estimate. Calculate these using either of the following methods:

a) Define each item that applies to overhead and the expected cost.

b) Apply a percentage based on historical information and adjusted to reflect contract size and duration.

c) Adjust the percentage for company total contract volume and expected costs for additional staff requirements for a specific project.

Separately show all mark-up amounts assumed for overheads and profits. Also include profits, fees, taxes, contingencies, inflation, escalations, etc.

Total Estimated Contract (construction) Cost (TECC) is the combined total amount for:

- ✓ Direct costs
- ✓ Performance bonding
- ✓ General conditions
- ✓ Escalation
- ✓ Other costs
- ✓ Contingency amounts
- ✓ Overhead & profit
- ✓ Applicable taxes
- ✓ Fixed fees
- ✓ Permits

Total Estimated Project Cost (TEPC) is the combined total amount for:

- ✓ Construction management
- ✓ Design professional costs
- ✓ Inspection
- ✓ Movable furniture, fixtures, equipment
- ✓ Telecommunication equipment
- ✓ Surveys
- ✓ Testing
- ✓ Move-in costs
- ✓ Planning/in-house support
- ✓ Other reimbursables

Add these to the Total Estimated Contract Cost. Include these costs in the estimate at the discretion of and in the owner's format. The estimator should state the nature of the estimate as either Total Estimated Contract (construction) Cost or Total Estimated Project Cost.

Submit cost estimates with back-up material that shows the basis for calculating the estimated cost. The cost estimates prepared under these guidelines are based on the estimator's best judgment using information provided. This is independent of other budgets or estimates furnished for information purposes.

SECTION 4 – ESTIMATING PROCEDURES, GENERAL

Levels of the Estimate

Review the section on Levels of the Estimate to determine and state the level of the estimate. When there are varying degrees of completeness in the division of work, then define these variations. Include the definition of the level or levels in the estimate presentation. Refer to the section on presentation for further information about estimate presentations.

Scope of the Estimate

Refer to Part One Section Three of this manual for further scope definition.

Section 4 – Estimating Procedures, General

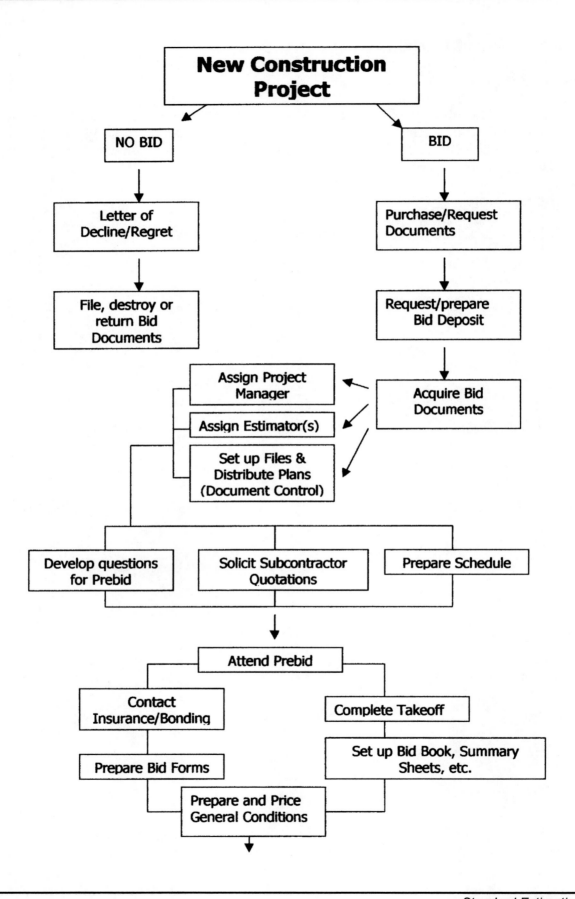

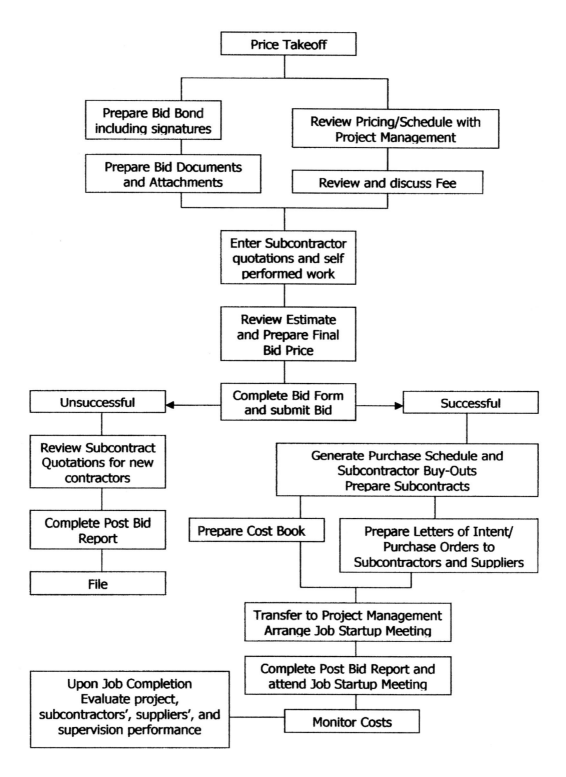

Common Practice

We include Standard Estimating Practices common to other discipline here. Please refer to discipline specific sections for additional Standard Estimating Practice. Part Four, Ethics always applies to all disciplines.

Forms/Checklists

The following forms/checklists are universal in nature and are included in Section 8 Checklists/Special Forms.

No one standard estimating form will meet the needs of the variety of material vendors, specialty contractors, general contractors involved in the construction industry. Additionally, specific types of contracting, e.g. highway and bridges, building, plant and industrial, all have unique formats that preclude a single 'standard' form for quantity surveying or estimating. Included are representative forms that the estimator may use if appropriate to their particular trade or service. Throughout the Standard Estimating Practice are examples to illustrate the specific topics being defined. You may note that there is no consistent format or form used in those illustrations. The format/forms used are those found by the authors to be most appropriate for that particular trade or service. The estimator may use the form shown, change or adapt it to meet his/her need, or use a completely different system. **The key is to find what works best for your situation,**

- Master Checklist
- Bid Document Inventory Checklist
- Site Investigation Checklist
- Direct Cost Estimate Checklist
- Overhead Estimate Checklist
- Estimate Review/Assignment Checklist
- General Conditions Estimate Detail
- Final Summary
- Final Summary Cut and Add Sheet
- Account Summary
- Estimate Detail
- Quantity Survey

Checklists and special forms that are specific to a discipline, if any, are included under each discipline.

SECTION 5 – PROJECT EVALUATION – CONSTRUCTORS

Company Policy

When bidding, use company guidelines established to limit the expense of estimating time and money to those projects that offer the highest potential. Examples of some items which should be given consideration are:

- ✓ Type of construction
- ✓ Project location
- ✓ Construction schedule
- ✓ Time frame
- ✓ Financial ability
- ✓ Bonding requirements

Management should develop estimate/bid policy with guidelines for rating potential estimate/bid projects. Estimate/bid criteria could include:

- ✓ Projects must yield a predetermined minimum profit.
- ✓ There should be an optimum number of bidders.
- ✓ The project is within the geographic target area.
- ✓ Current negotiated work which generates an estimate/bid project.
- ✓ Projects with reasonable opportunities for future negotiated work.
- ✓ Any project that will enhance the company's market position.

Develop estimate/bid policy in concert with the company's business plan of predetermined corporate goals. Positively defined estimate/bid policies reduce emotionalism that may exist in project evaluation. Develop these policies so answers to questions will determine whether it is in the best interest of the company to estimate/ bid the project.

Include these questions when developing estimate/bid policy:

- ✓ Who is the owner?
- ✓ Is their prior experience with owner?
- ✓ Is the project federally funded, thereby, requiring Davis-Bacon wage controls?

- ✓ Is it necessary to employ a project manager or superintendent or is the current staff capable of managing the work?
- ✓ If the project for an existing client?
- ✓ What are the prospects for negotiated repeat business?
- ✓ Can the anticipated estimate/bid project generate enough margin/profit to warrant company involvement?
- ✓ How will the owner fund the project?
- ✓ Is it through appropriations via Congress, state legislature, city council, board of supervisors, fund control or letters of credit?
- ✓ Who is the architect?
- ✓ Is there prior experience with architect?
- ✓ When will the project commence?
- ✓ Are there enough company assets to serve as internal "banker" or is short term financial required?
- ✓ Are performance/payment bonds required? If so, is there sufficient capacity?
- ✓ Are there other bidders? If so, do the other bidders have a similar company make-up, structure, volume, and project management approach? If not, does the difference improve or diminish the chance of being the successful bidder?
- ✓ What is the chance for enhanced margins or profit through scope additions?

Work Scope

Does the scope of work require performance by a prime/general contractor? Or does it require a subcontractor acting as prime contractor, or possibly a subcontractor to a prime/general contractor. The term prime/general contractor means an organization proposing a direct contractual relationship with an owner. This includes general contractors, construction management firms and owner's agents.

Estimate/Bid Due Date

When there is a decision that the project meets the measure of company policy. The estimator's work begins.

SECTION 6 – PROJECT EVALUATION – VALUE ENGINEERING (VE)

Introduction

Value Engineering (VE) is an organized effort to study the function of items that meet user required needs. Maintaining the best cost reduction of initial and life cycle costs and performance of a system is part of the effort. VE is a function-oriented, orderly approach that uses a team concept. The main goal of this team effort is to identify and eliminate unnecessary costs while retaining user required design features.

Accomplishing the effort in the most profitable and effective manner possible is an essential characteristic of VE Cost effectiveness, the tangible result of VE, is a result of both cost avoidance and cost reduction.

To achieve cost effectiveness, the VE process is important to and should involve estimators. VE studies rely on costs and use cost estimates as well as design documents. The professional estimator can provide and check cost figures. The results of VE will be only as reliable as the cost figures applied. Concurrently, VE is most profitable when cost information comes from the efficient contribution of the professional estimator.

Common misconceptions about value engineering include:

- ✓ Value Engineering is a substitute for conventional cost reductions.
- ✓ VE techniques usually result in lowered system performance and design standards.
- ✓ VE is a negative reflection on the professional design engineer.
- ✓ VE studies drain available project funds.

These arguments are invalid when using techniques positively to insure that quality and design favor minimum costs and maximum system effectiveness. Cost benefit ratio studies show that cost savings from VE outweigh the cost of their studies.

Value Engineering Terminology and Theory

The term Value Engineering is often the same as value analysis and value management. Preferred usage is VE, as shown here.

Estimating requires the use of technical theory and estimating language. Similarly, VE utilizes technical standards and terminology of the profession. The estimator involved in VE must understand the terminology and these theories. For help with VE terminology and acronyms, a glossary of definitions is provided at the end of this section.

Phases of Value Engineering

Value Engineering proceeds through specific phases. This established series of steps is also termed a job plan. The job plan helps in the application of VE analyses. It enhances the final analysis achieved as a result of the decision making process. VE approaches include the following basic phases: information, speculative, analytical, proposal, and presentation.

Information or function phase - The goal of the information phase is to provide understanding of the systems undergoing the VE process. There is a review during this phase of the project, including the design, specifications, and cost data. Use function analysis, a basic facet of VE to determine function costs and value standards.

Function is the specific purpose, or intended use, of an item or design that meets the needed requirements of the user. Function is in three principal separate groups. Based function (B), is the primary reason for existence of an item. Secondary function (S), exists to support the basic function. Sometimes there is a third function, Required Secondary (R/S). R/S is a function essential to the performance of the basic function.

Function analysis is a technique for identification and description of the function of an item to separate basic and secondary function. To perform function analysis (describe a work function). VE utilizes a concise technique called the verb/noun approach. The verb answers the question "what does it do?" And the noun answers the question, "what does it do to?" Consider the verb as "the action" and the noun as "the measurable object of the verb." An example of the verb/noun approach is a concrete slab, identified in verb/noun terminology as "support loads." Another example is domestic water piping, identified in verb/noun terms as "send water."

This is a simple approach to the professional unfamiliar with VE The verb/noun approach is an essential part of VE When there is a function description, only one function shows at a time, and the difference between them is clear. Once function analysis is complete, the VE process advances into the speculative phase.

Speculative phase (or creative phase) - The goal of the speculative phase is to generate as many ideas, processes, and methods as possible. Their purpose is to fulfill the function of the design system under study. During the speculative phase, use the Function Analysis System Technique (F.A.S.T.). This phase requires imaginative thinking. Through brainstorming, use alternative design ideas that will perform the required function based on the analyses of the information phase. Also, this identifies irrelevant or redundant design. Rank ideas generated but avoid pre-judgment of ideas. Suspend evaluation of ideas until the analytical phase.

The procedure should provide:

- ✓ An expedient format for recording the study as it progresses.
- ✓ Assurances that all information received attention.
- ✓ A logical resolution of the analysis divided into planned, scheduled, budgeted, and appraised components.

Analytical or judicial phase - Now, review facts and alternatives, and through value recognition, drop design items of needless cost. Rank remaining alternatives based on cost comparisons and the technical feasibility to perform the design functions. Based on rank, select alternatives for further evaluation. View these alternatives for acceptability, performance, and ease of implementation. Estimate initial, overhead and maintenance, and life cycle costs for each selected alternative. Prepare a design cost savings proposal. Initial costs should include direct construction costs plus contractor's markups. Owner overhead and maintenance costs should include the quality and reliability of the design. Also include the costs of maintainability, extended life expectancy, and future procurement. Life cycle costs should include future operating costs that the owner (user) will incur.

Proposal or development and implementation phase - The goal of this phase is to find which alternatives selected represent important savings. From these, develop a constructive proposal. Overcome possible problems of the recommended alternatives, and developed implementation plans. Procedures to accomplish goals may include development of drawings, specifications, and cost estimates to support the alternatives. Justification data accompanies the proposal in the presentation phase.

Presentation or recommendation and follow up phase - Recommend design standards during this phase. Present a plan of action to involved parties. Review a summary of the proposal, and present value standards as back up to support selected alternatives. Rate performance versus cost at this phase, including a review of proposed cost savings data.

Value Engineering Methodology

With the phased approach to value engineering, methodology must achieve maximum effectiveness. This methodology varies depending upon two criteria. First, the level of project development (order of magnitude, conceptual, design development, project control, definitive, bid, or evaluation) will affect the degree of depth of the value engineering study. Pertinent personnel include the cost estimator, architect/engineer, owner, and the contractor.

Always remember that value engineering is proper at any stage of a project. The following describes the stages of a project and how to apply value engineering to each stage.

Order of Magnitude Stage - This is an area for many successful value engineering results. The problem is that this stage contains the least project data. Value analysis at this stage should study the positive factors of the existing project design. Also, ideas studied should focus on design components that are of poor value or worth. Keep in mind the basis of this level is budgetary and feasibility determinations. Any combination of the personnel listed above may participate in analysis at this stage of the project. An estimator should perform cost analysis at this and all other project development stages.

Conceptual Stage - Use enhanced techniques of the analysis process for VE study at this phase. Call more attention to design at this stage with scrutiny of proposed costs for design function. The level of estimate at this stage provides better-defined costs for budgetary and feasibility determination. Thus, value engineering at this stage is more definitive than at the Order of Magnitude Stage.

Design Development (Budget Appropriation) Stage - This stage includes study of construct-ability and feasibility. Suggest alternate approaches and decisions to assure that proposed design and estimated costs reflect the most value. The purpose of the estimate at this stage is to show probable costs within the range of available information. Value engineering team members should include the estimator, design team, and maintenance personnel.

Project Control Stage - This stage contains more detail and accuracy than previous stages. This level of project design and estimate preparation lends itself well to VE application. This is true, though design drawings and specifications are not complete. VE incorporated at this stage includes design proposals and cost concepts. The definitive phase level shows detail design and reasonable project cost. VE at this level consists of a detailed process. This level should use all personnel involved in the design, estimate preparation, procurement, and construction process.

Bid Stage - This stage shows the construction cost of the project based on final design. At this stage, it is possible for the awarded contractor to identify design items that can be successfully value engineered.

Evaluation Stage - Use this stage as an after construction scrutiny of estimated or construction costs. At this level, it is possible for the owner/client or contractor to study value engineering efforts that could improve future projects. This is also known as the lessons learned process. VE studies, at this level, are usually by the client/ owner, estimator, architect/engineering team, or contractor/subcontractor.

The Value Engineering Team

Client/owners usually request Value Engineering (VE) studies from a concern about project constructability and relative costs. A client/owner may engage the services of a VE team to conduct a formal value engineering study. Client/owners usually realize increased capital and life cycle cost savings when using VE AVE study uses the team concept that employs a dynamic group of interdisciplinary problem solvers. Team participants include an appropriate discipline mix of professions experienced in their own fields. Composition of VE teams is usually engineers, maintenance and operating personnel, purchasing agents, research assistants, owner representatives, project managers, estimators.

Team members must have knowledge of past procedures, codes, and union demands. They also must have enough technical training to investigate and rate new products and procedures.

Using a discipline mix is important. Teams usually consist of three to seven persons, with five members being the most effective team. Name one person as the team leader or coordinator to guide the team through the value engineering processes. The team should consist of more than one dominant personality to avoid control of ideas by one individual. Team members should be of equal rank. Isolate VE study team members from their normal work arena during the study. Team members must have experience with the type of project undergoing study. Having open-minded and creative members is a must. Keep team members independent of the project under study. At least 50 percent of the team members should attend a formal value-engineering workshop.

A contractor who performs a Value Engineering study after the award of a contract may do so because of an incentive clause in his construction contract. The contractor identifies a design area where using an alternative method does not compromise design and will achieve cost savings. The contractor, using VE, shares in any cost savings resulting from his proposal. The contractor becomes more competitive for future projects because of these VE practices.

Important Elements for Consideration in a Value Engineering Change Proposal (VECP)

1) Provide a detailed description of the difference between the present design and the proposed design change. List the advantages of each.

2) Describe the justification for the proposed change including the function change and performance change.

3) Describe the contract requirements that may require revision on approval of the VECP. This contractual change may involve drawings, specifications, materials purchased, etc.

4) Estimate construction costs for both the existing and new design. Include applicable economic cost value (labor, material, equipment), overhead and profit, and bond costs.

5) Develop and provide contractor's costs for VECP development, implementation, overhead, and operation and maintenance. Develop costs that are the responsibility of the client/owner as a result of the VECP, including life cycle costs.

6) Develop and provide contractual costs changes respective to the sharing of the proposed cost saving.

7) Prepare and include the schedule required to get the maximum cost benefit for the VECP. The schedule should include the impact of dates upon contract implementation, completion, delivery, and rejection/acceptance considerations.

8) Develop and include the cost impact of the VECP from previously accepted VECP's for the project.

Summary

This standard provides a broad definition of the value engineering process and its need for cost estimators. VE, like cost estimating, includes singular theories. Estimators must learn the terminology and procedures before conducting a VE study.

Glossary

Basic Function - the objective of a design feature to be realized if the item is to work.

Contractor's Implementation Cost - expenses incurred by the contractor in the development, testing, preparation, and submittal of a value engineering change proposal.

Cost - utilization of resources.

Economic Cost Value - the sum of labor, materials, and overhead to produce an item.

FAST (Function Analysis System Technique) - a methodology employed in the value engineering process which uses a beneficial analysis diagram to act as a tool to introduce new ways to perform a basic function.

LCC (Life Cycle Cost) - the economic assessment of a system or facility considering the lifetime ownership costs over an item's economic life.

O&M (Operation and Maintenance) - the costs of an item to perform (e.g., electricity) and the costs to maintain that item in working order (e.g., replace worn belts).

R/S (Required Secondary) Function - a secondary function of value which is essential to the basic function.

S (Secondary) Function - non-essential accessory to the basic value which provides esteem value to an item or system.

Value - the lowest cost to accomplish the required function consistent with the optimum cost of ownership.

Value Standard - worth of critical function.

VE - Value Engineering.

SECTION 7 – BID DOCUMENTS/PROCUREMENT

Sources

For projects of any relative size and complexity, bid documents should come directly from the originating body. This applies to both general and subcontractors. This step will insure receipt of all addenda, communications, etc., promptly and in complete form. This ends problems that occur when the basis for decisions is incomplete and second or third hand information.

Degree of Completeness Required

Estimate accuracy depends on the completeness of the bid documents. Lump sum competitive bids require fully detailed and complete drawings. Negotiated or unit cost type estimates/bids may come from drawings that are incomplete. However, the construction contract must allow for recovery of costs for work determined in the future from complete documents.

Time Available

Adherence to use of bid documents that come directly from the originating body automatically guarantees that estimate/bid documents remain available until publication of estimate/bid results. Always strive to keep all bid documents in hand until there is no longer a need for them. Develop a system of tracking for appropriate refund or credit when returning tendered documents.

Missing or Incomplete Information

Identify these areas of the plans and specifications that affect the cost estimate. Report them to the originating body. Always maintain proper documentation and backup of this information, for possible future requirements, in writing. Missing or incomplete information includes missing plans, missing specifications, ambiguous notes, unclear scope of work, and addenda. Get the missing and deficient information or proceed by written agreement with the originating body. Inform other parties involved of these factors. Adequate communications between involved parties is important.

SECTION 8 – CHECKLISTS/SPECIAL FORMS

Discipline Specific Checklist/Forms

Master Checklist

Bid Document Inventory Checklist

Site Investigation Checklist

Direct Cost Estimate Checklist

Overhead Estimate Checklist

General Conditions Estimate Detail

Final Summary

Final Summary Cut and Add Sheet

Account Summary

Estimate Detail

Quantity Survey

Refer to Part Two for discipline specific checklists/forms.

Part One
Section 8 – *Checklists/Special Forms*

Master Checklist

Project:_____   Estimator:_____

Locations:_____   Estimate #:_____

Checklist		**Date**
Bid Document Inventory	[ ] Yes	_____
Document Review	[ ] Yes	_____
Estimating Assignments	[ ] Yes	_____
Site Investigation	[ ] Yes	_____
Direct Cost Estimate	[ ] Yes	_____
General Conditions Estimate	[ ] Yes	_____
Indirect Cost Estimate	[ ] Yes	_____
Estimate Review	[ ] Yes	_____
Bid Review	[ ] Yes	_____
Profit Determination	[ ] Yes	_____
Bid Adjustments	[ ] Yes	_____
Bid Documents	[ ] Yes	_____
Analysis of Bids	[ ] Yes	_____

Notes:

_____

_____

_____

_____

_____

_____

Part One *Section 8 – Checklists/Special Forms*

Bid Document Inventory Checklist

Project:_____   Estimator:_____
Location:_____   Estimate #:_____
Original Issue:_____   Addenda #'s:_____

Inventory [ ] Completed

Missing Documents:

Document Name	Document Number	Document Name	Document Number
_____	_____	_____	_____
_____	_____	_____	_____
_____	_____	_____	_____
_____	_____	_____	_____
_____	_____	_____	_____

Wrong Revisions:

_____	_____	_____	_____
_____	_____	_____	_____
_____	_____	_____	_____
_____	_____	_____	_____
_____	_____	_____	_____

Illegible Documents:

_____	_____	_____	_____
_____	_____	_____	_____
_____	_____	_____	_____
_____	_____	_____	_____
_____	_____	_____	_____

Acknowledgment Sent   [ ] Yes – Date _____   [ ] No

Comments:
_____
_____

Standard Estimating Practice

Site Investigation Checklist

Project:_____   Performed By:_____
Location:_____   Date Performed:_____

Estimate #:_____

Items Required:

Comments:

[ ] Notepad, Pencils, Straightedge
[ ] Camera and Film
[ ] Pocket Recorder
[ ] List of Questions

_____
_____
_____

En route to Site:

Comments:

Narrow Roads	[ ] Yes [ ] No	
Sharp Turns	[ ] Yes [ ] No	
Load Restrictions	[ ] Yes [ ] No	
Overhead Obstructions	[ ] Yes [ ] No	
Plant Security	[ ] Yes [ ] No	
Rail Access Spur	[ ] Yes [ ] No	
Alternate Routes	[ ] Yes [ ] No	

Signs of Previous Inclement Weather

[ ] Yes [ ] No   Today's Weather_____Temp_____

Site Representatives [Names and Titles]:

_____ _____   _____ _____
_____ _____   _____ _____
_____ _____   _____ _____

Site Utilities:

Power [Descriptipn and distance]:

_____

_____

_____

Votages available:_____Volt_____Amps_____Phase

Is Existing Power Adequate? [ ] Yes   [ ] No

Will overhead power lines restrict our operations? [ ] Yes   [ ] No

Do underground power lines restrict our operation? [ ] Yes   [ ] No

Is there a hookup and/or use charge for power? [ ] Yes   [ ] No

What distribution problems are likely?_____

_____

Comments:_____

_____

_____

Water [Source and capacity]:_____

_____

_____

Is existing water adequate? [ ] Yes   [ ] No

Is there a hookup and/or use charge for non-potable water [ ] Yes   [ ] No

Nearest available source for potable water:_____

Is there a hookup and/or use charge for potable water? [ ] Yes   [ ] No

Comments:_____

_____

_____

Gas [Source and capacity]:_____
_____
_____

Is existing source adequate? [ ] Yes   [ ] No

Is there a hookup and/or use charge for gas supply? [ ] Yes   [ ] No

Are alternative sources available? [ ] Yes   [ ] No

Is there a hookup and/or use charge for temporary gas? [ ] Yes   [ ] No

Comments:_____
_____

Compressed Air

Is compressed are required? [ ] Yes   [ ] No

Is there an existing distribution system available? [ ] Yes   [ ] No
Is this adequate? [ ] Yes   [ ] No
Can single units handle rather than a distributions system? [ ] Yes   [ ] No
Comments:_____
_____
_____

Site conditions
Groundwater
Is the area well drained? [ ] Yes   [ ] No
Will a well point system be required? [ ] Yes   [ ] No
Is the area subject to flooding? [ ] Yes   [ ] No
Comments:_____
_____
_____

Clearing and excavation

Are permits required for clearing or excavation? [ ] Yes [ ] No

Is disposal area available? [ ] Yes [ ] No

Are there ground level or underground obstructions? [ ] Yes [ ] No

Are obstructions located visibly or on drawings? [ ] Yes [ ] No

Disposal area dump charges are:_____

Comments:_____

_____

_____

_____

Construction Plant

Site availability

Is there adequate are for:		Comments:
offices?	[ ] Yes [ ] No	_____
warehousing?	[ ] Yes [ ] No	_____
material yard?	[ ] Yes [ ] No	_____
staff and craft parking	[ ] Yes [ ] No	_____
construction traffic	[ ] Yes [ ] No	_____

Comments:_____

_____

_____

_____

_____

Are improvements or new construction required for:

offices? [ ] Yes [ ] No _____

warehousing? [ ] Yes [ ] No _____

material yard? [ ] Yes [ ] No _____

staff and craft parking [ ] Yes [ ] No _____

construction traffic [ ] Yes [ ] No _____

Comments: _____

_____

_____

_____

Regulatory requirements and permits to obtain:

Building Permit

    Who obtains/pays for it?   [ ] Owner   [ ] General   [ ] Subcontractor

    Whose jurisdiction   [ ] Local   [ ] State   [ ] Federal

                                       [ ] _____

Sanitary Permit

    Who obtains/pays for it?   [ ] Owner   [ ] General   [ ] Subcontractor

    Whose jurisdiction   [ ] Local   [ ] State   [ ] Federal

                                       [ ] _____

Burning Permit

    Who obtains/pays for it?   [ ] Owner   [ ] General   [ ] Subcontractor

    Whose jurisdiction   [ ] Local   [ ] State   [ ] Federal

                                       [ ] _____

Section 8 – *Checklists/Special Forms*   **Part One**

Dumping Permit

    Who obtains/pays for it?   [ ] Owner   [ ] General   [ ] Subcontractor

    Whose jurisdiction   [ ] Local   [ ] State   [ ] Federal

                                              [ ] _____

Other:_____

    Who obtains/pays for it?   [ ] Owner   [ ] General   [ ] Subcontractor

    Whose jurisdiction   [ ] Local   [ ] State   [ ] Federal

                                              [ ] _____

Other:_____

    Who obtains/pays for it?   [ ] Owner   [ ] General   [ ] Subcontractor

    Whose jurisdiction   [ ] Local   [ ] State   [ ] Federal

                                              [ ] _____

Other:_____

    Who obtains/pays for it?   [ ] Owner   [ ] General   [ ] Subcontractor

    Whose jurisdiction   [ ] Local   [ ] State   [ ] Federal

                                              [ ] _____

Other:_____

    Who obtains/pays for it?   [ ] Owner   [ ] General   [ ] Subcontractor

    Whose jurisdiction   [ ] Local   [ ] State   [ ] Federal

                                              [ ] _____

Comments:_____

_____

_____

_____

Part One
Section 8 – Checklists/Special Forms

Local Material, Subcontractor and Labor Supply

Are local suppliers available for: Comments:

small tools and consumables?	[ ] Yes [ ] No	_____
equipment rentals?	[ ] Yes [ ] No	_____
office supplies/furnishings	[ ] Yes [ ] No	_____
construction materials?	[ ] Yes [ ] No	_____
fuel, oil and lubricants?	[ ] Yes [ ] No	_____
_____	[ ] Yes [ ] No	_____
_____	[ ] Yes [ ] No	_____
_____	[ ] Yes [ ] No	_____

Availability of Local Subcontractors

Name-Check (?) = prior work for owner	Trade/Craft	Phone	Union Agreements
_____	( ) _____	_____	[ ] Yes [ ] No
_____	( ) _____	_____	[ ] Yes [ ] No
_____	( ) _____	_____	[ ] Yes [ ] No
_____	( ) _____	_____	[ ] Yes [ ] No
_____	( ) _____	_____	[ ] Yes [ ] No
_____	( ) _____	_____	[ ] Yes [ ] No
_____	( ) _____	_____	[ ] Yes [ ] No
_____	( ) _____	_____	[ ] Yes [ ] No
_____	( ) _____	_____	[ ] Yes [ ] No
_____	( ) _____	_____	[ ] Yes [ ] No
_____	( ) _____	_____	[ ] Yes [ ] No
_____	( ) _____	_____	[ ] Yes [ ] No
_____	( ) _____	_____	[ ] Yes [ ] No
_____	( ) _____	_____	[ ] Yes [ ] No
_____	( ) _____	_____	[ ] Yes [ ] No
_____	( ) _____	_____	[ ] Yes [ ] No
_____	( ) _____	_____	[ ] Yes [ ] No

Comments: _____
_____
_____

Section 8 – Checklists/Special Forms   Part One

Sources for Craft Labor

Name-Check (?) = adequate forces	Trade/Craft	Phone	Union Agreements
_____	( ) _____	_____	[ ] Yes [ ] No
_____	( ) _____	_____	[ ] Yes [ ] No
_____	( ) _____	_____	[ ] Yes [ ] No
_____	( ) _____	_____	[ ] Yes [ ] No
_____	( ) _____	_____	[ ] Yes [ ] No
_____	( ) _____	_____	[ ] Yes [ ] No
_____	( ) _____	_____	[ ] Yes [ ] No
_____	( ) _____	_____	[ ] Yes [ ] No
_____	( ) _____	_____	[ ] Yes [ ] No

Comments:_____
_____
_____
_____

Existing Plant Operation

Will tie-ins be required at existing structures/plant?	[ ] Yes [ ] No
Will tie-ins be requires at existing process lines?	[ ] Yes [ ] No
Will tie-ins be required at electrical distribution?	[ ] Yes [ ] No
Are advance notice/permits required?	[ ] Yes [ ] No
Are shut downs permitted?	[ ] Yes [ ] No
Has time for shut downs been allowed or scheduled?	[ ] Yes [ ] No
Are around-the-clock effort advance notice/ermits required?	[ ] Yes [ ] No

Comments:_____
_____
_____
_____
_____
_____

Direct Cost Estimate Checklist

Project: _____   Bid Date: _____

Location: _____   Estimate Due: _____

Estimate#: _____

Have packages been assembled or contact made for

Subcontractor scope:

_____ [ ] Yes [ ] No		_____ [ ] Yes [ ] No
_____ [ ] Yes [ ] No		_____ [ ] Yes [ ] No
_____ [ ] Yes [ ] No		_____ [ ] Yes [ ] No
_____ [ ] Yes [ ] No		_____ [ ] Yes [ ] No
_____ [ ] Yes [ ] No		_____ [ ] Yes [ ] No
_____ [ ] Yes [ ] No		_____ [ ] Yes [ ] No
_____ [ ] Yes [ ] No		_____ [ ] Yes [ ] No
_____ [ ] Yes [ ] No		_____ [ ] Yes [ ] No

Comments: _____

_____

_____

_____

Material and equipment vendors:

_____ [ ] Yes [ ] No		_____ [ ] Yes [ ] No
_____ [ ] Yes [ ] No		_____ [ ] Yes [ ] No
_____ [ ] Yes [ ] No		_____ [ ] Yes [ ] No
_____ [ ] Yes [ ] No		_____ [ ] Yes [ ] No
_____ [ ] Yes [ ] No		_____ [ ] Yes [ ] No
_____ [ ] Yes [ ] No		_____ [ ] Yes [ ] No
_____ [ ] Yes [ ] No		_____ [ ] Yes [ ] No
_____ [ ] Yes [ ] No		_____ [ ] Yes [ ] No

Section 8 – Checklists/Special Forms Part One

_____
_____
_____

Job site wages, fringes and burdens

Wages:		**Fringes:**	
Base wages	[ ] Yes [ ] No	Health and welfare	[ ] Yes [ ] No
Shift differential	[ ] Yes [ ] No	Vacation and holiday	[ ] Yes [ ] No
High pay	[ ] Yes [ ] No	Pension	[ ] Yes [ ] No
Toxic pay	[ ] Yes [ ] No	Apprentice	[ ] Yes [ ] No
_____	[ ] Yes [ ] No	Legal	[ ] Yes [ ] No
_____	[ ] Yes [ ] No	Industrial	[ ] Yes [ ] No
_____	[ ] Yes [ ] No	Travel	[ ] Yes [ ] No
_____	[ ] Yes [ ] No	Subsistence	[ ] Yes [ ] No
_____	[ ] Yes [ ] No	_____	[ ] Yes [ ] No
_____	[ ] Yes [ ] No	_____	[ ] Yes [ ] No
_____	[ ] Yes [ ] No	_____	[ ] Yes [ ] No

Direct labor burden: **Comments:**

Social security	[ ] Yes [ ] No	_____	
Fed. Unemployment FUTA	[ ] Yes [ ] No	_____	
State unemployment SUI	[ ] Yes [ ] No	_____	
Workers' compensation	[ ] Yes [ ] No	_____	
Liability insurance	[ ] Yes [ ] No	_____	
_____	[ ] Yes [ ] No	_____	[ ] Yes [ ] No
_____	[ ] Yes [ ] No	_____	[ ] Yes [ ] No

Quantity survey

Did this reveal any interference? [ ]Yes [ ] No

Are there any conflicts between the specs and the drawings? [ ]Yes [ ] No

Is there any scope duplication? [ ]Yes [ ] No

Is there any scope omission? [ ]Yes [ ] No

Plan views compared to details/sections? [ ]Yes [ ] No

Any allowances in quantities? [ ]Yes [ ] No

Have any questions been submitted? [ ]Yes [ ] No

Have any questions been answered? [ ]Yes [ ] No

Material listing

Is the listing detailed enough to purchase materials? [ ]Yes [ ] No

Are there confirmed quotations? [ ]Yes [ ] No

Have "plug prices been used? [ ]Yes [ ] No

Have cut and add sheets been prepared for the "plug" prices? [ ]Yes [ ] No

Have extensions and additions been checked? [ ]Yes [ ] No

Have carry-forwards been checked? [ ]Yes [ ] No

Labor units

Development:

Historical units available?	[ ]Yes [ ] No
Compared to similar project?	[ ]Yes [ ] No
Compared to company avg.	[ ]Yes [ ] No
Reviewed by superintendents?	[ ]Yes [ ] No
Developed by crew analysis?	[ ]Yes [ ] No
_____	[ ]Yes [ ] No
_____	[ ]Yes [ ] No
_____	[ ]Yes [ ] No

Application:

Table developed listing units?	[ ]Yes [ ] No
Checks made for correct use?	[ ]Yes [ ] No
Addition/extensions checked?	[ ]Yes [ ] No
Carry-forwards checked?	[ ]Yes [ ] No
_____	[ ]Yes [ ] No
_____	[ ]Yes [ ] No
_____	[ ]Yes [ ] No
_____	[ ]Yes [ ] No

Section 8 – *Checklists/Special Forms* — Part One

Construction Scheduling

Schedule:

		Other:	
Time phased with activities?	[ ]Yes [ ] No	Manpower curves?	[ ]Yes [ ] No
Manpower located?	[ ]Yes [ ] No	Organization chart?	[ ]Yes [ ] No
Major milestones located?	[ ]Yes [ ] No	Resumes of key personnel?	[ ]Yes [ ] No
Multiple shifts shown?	[ ]Yes [ ] No	Work plan?	[ ]Yes [ ] No
Overtime scheduled	[ ]Yes [ ] No	_____	[ ]Yes [ ] No
_____	[ ]Yes [ ] No	_____	[ ]Yes [ ] No

Overhead Estimate Checklist

Project:_____   Bid Date:_____

Location:_____   Estimate Due:_____

Estimate:_____

    The intent of the term overhead, whether it be "General Contractor" or "Subcontractor," is to imply the anticipated contract costs for all items that cannot be classified as direct field costs. The overhead costs include, but are not limited to, home office physical plant leasing or capital investment and maintenance cost, finance expense, home office physical plant services, support and pro rata salaries, project management, clerical accounting, general liability and comprehensive insurance costs, and outside services. The anticipated cost for overhead shall be shown separately in each estimate and may be calculated using either of the following methods: (a) define each item that is applicable to overhead and the anticipated cost, (b) apply a percentage based on historical information and adjusted to reflect contract size and duration, the company total contract volume, and expected costs for additional staff requirements for a specific project.

Home Office Plant

Plant capital investment:

Unimproved real property	[ ]Yes	[ ] No
Improved real property	[ ]Yes	[ ] No
Office buildings	[ ]Yes	[ ] No
Storage buildings	[ ]Yes	[ ] No
Maintenance buildings	[ ]Yes	[ ] No
_____	[ ]Yes	[ ] No
_____	[ ]Yes	[ ] No
_____	[ ]Yes	[ ] No

Plant leasing:

Unimproved real property	[ ]Yes	[ ] No
Improved real property	[ ]Yes	[ ] No
Office buildings	[ ]Yes	[ ] No
Storage buildings	[ ]Yes	[ ] No
Maintenance buildings	[ ]Yes	[ ] No
_____	[ ]Yes	[ ] No
_____	[ ]Yes	[ ] No
_____	[ ]Yes	[ ] No

Plant Maintenance:

_____	[ ]Yes	[ ] No
_____	[ ]Yes	[ ] No
_____	[ ]Yes	[ ] No
_____	[ ]Yes	[ ] No
_____	[ ]Yes	[ ] No

Plant Repair:

_____	[ ]Yes	[ ] No
_____	[ ]Yes	[ ] No
_____	[ ]Yes	[ ] No
_____	[ ]Yes	[ ] No
_____	[ ]Yes	[ ] No

Comments:_____
_____

Home office Staff
Management and engineering

Project manager	[ ]Yes	[ ] No
Assistant project manager	[ ]Yes	[ ] No
Project mgr. Secretarial stall	[ ]Yes	[ ] No
Project structural engineer	[ ]Yes	[ ] No
Asst. structural engineer	[ ]Yes	[ ] No
Project engineer staff	[ ]Yes	[ ] No
Project process engineer	[ ]Yes	[ ] No
Asst. process engineer	[ ]Yes	[ ] No
Process engineer staff	[ ]Yes	[ ] No
Project architect	[ ]Yes	[ ] No
Asst. project architect	[ ]Yes	[ ] No
Project architect staff	[ ]Yes	[ ] No
Chief estimator	[ ]Yes	[ ] No
Senior estimator	[ ]Yes	[ ] No
Estimator	[ ]Yes	[ ] No
Estimator secretarial staff	[ ]Yes	[ ] No
Quality control officer	[ ]Yes	[ ] No
Scheduling personnel	[ ]Yes	[ ] No
Cost engineer	[ ]Yes	[ ] No
Office engineer	[ ]Yes	[ ] No
Safety engineer	[ ]Yes	[ ] No
Aircraft pilot	[ ]Yes	[ ] No
Maintenance engineer	[ ]Yes	[ ] No
Master mechanic	[ ]Yes	[ ] No
Mechanic	[ ]Yes	[ ] No
Heavy truck operators	[ ]Yes	[ ] No
Light truck operators	[ ]Yes	[ ] No
General plant laborers	[ ]Yes	[ ] No
Janitorial personnel	[ ]Yes	[ ] No
Landscape maintenance Personnel	[ ]Yes	[ ] No
Security personnel	[ ]Yes	[ ] No
Survey/layout personnel	[ ]Yes	[ ] No
_____	[ ]Yes	[ ] No

Administrative:

General manager	[ ]Yes	[ ] No
Office manager	[ ]Yes	[ ] No
Managerial staff	[ ]Yes	[ ] No
Contract administrator	[ ]Yes	[ ] No
Comptroller	[ ]Yes	[ ] No
Comptroller staff	[ ]Yes	[ ] No
Payroll clerk	[ ]Yes	[ ] No
Secretarial staff	[ ]Yes	[ ] No
Receptionist	[ ]Yes	[ ] No
_____	[ ]Yes	[ ] No
_____	[ ]Yes	[ ] No
_____	[ ]Yes	[ ] No
_____	[ ]Yes	[ ] No
_____	[ ]Yes	[ ] No
_____	[ ]Yes	[ ] No
_____	[ ]Yes	[ ] No
_____	[ ]Yes	[ ] No
_____	[ ]Yes	[ ] No
_____	[ ]Yes	[ ] No
_____	[ ]Yes	[ ] No
_____	[ ]Yes	[ ] No
_____	[ ]Yes	[ ] No
_____	[ ]Yes	[ ] No
_____	[ ]Yes	[ ] No
_____	[ ]Yes	[ ] No
_____	[ ]Yes	[ ] No
_____	[ ]Yes	[ ] No
_____	[ ]Yes	[ ] No
_____	[ ]Yes	[ ] No
_____	[ ]Yes	[ ] No
_____	[ ]Yes	[ ] No

Part One Section 8 – Checklists/Special Forms

Home office utilities – capital cost:

Waste water systems	[ ]Yes	[ ] No
Domestic water systems	[ ]Yes	[ ] No
HVAC systems	[ ]Yes	[ ] No
Electrical systems	[ ]Yes	[ ] No
Sound systems	[ ]Yes	[ ] No
Alarm systems	[ ]Yes	[ ] No
Security systems	[ ]Yes	[ ] No
_____	[ ]Yes	[ ] No
_____	[ ]Yes	[ ] No
_____	[ ]Yes	[ ] No

Home office utilities – operating cost:

Waste water systems	[ ]Yes	[ ] No
Domestic water systems	[ ]Yes	[ ] No
HVAC systems	[ ]Yes	[ ] No
Electrical systems	[ ]Yes	[ ] No
Sound systems	[ ]Yes	[ ] No
Alarm systems	[ ]Yes	[ ] No
Security systems	[ ]Yes	[ ] No
_____	[ ]Yes	[ ] No
_____	[ ]Yes	[ ] No
_____	[ ]Yes	[ ] No

Comments:_____

_____

_____

Construction Equipment (idle, unassigned to a specific project)

Rental/lease equipment:

Pro rata company rental	[ ]Yes	[ ] No
Pro rata company lease	[ ]Yes	[ ] No
Pro rata outside rental	[ ]Yes	[ ] No
Pro rata outside lease	[ ]Yes	[ ] No
_____	[ ]Yes	[ ] No
_____	[ ]Yes	[ ] No
_____	[ ]Yes	[ ] No

Company owned equipment:

Equipment service vehicles	[ ]Yes	[ ] No
Car, station wagons	[ ]Yes	[ ] No
Vans, pickups	[ ]Yes	[ ] No
Light trucks	[ ]Yes	[ ] No
Heavy trucks	[ ]Yes	[ ] No
_____	[ ]Yes	[ ] No
_____	[ ]Yes	[ ] No

Comments:_____

_____

_____

_____

Standard Estimating Practice

Section 8 – Checklists/Special Forms   Part One

Home office supplies:

		Home office equipment	
Stationery	[ ]Yes [ ] No	Office furniture	[ ]Yes [ ] No
Copier paper	[ ]Yes [ ] No	Copy machine(s)	[ ]Yes [ ] No
Forms – estimating	[ ]Yes [ ] No	Computer	[ ]Yes [ ] No
Forms – administrative	[ ]Yes [ ] No	Fax/printer	[ ]Yes [ ] No
Forms – accounting	[ ]Yes [ ] No	Calculators	[ ]Yes [ ] No
File folders	[ ]Yes [ ] No	Electronic measuring devices	[ ]Yes [ ] No
Printer toner	[ ]Yes [ ] No	Computer software	[ ]Yes [ ] No
Computer disks/peripherals	[ ]Yes [ ] No	Video recorder	[ ]Yes [ ] No
Writing instruments	[ ]Yes [ ] No	Sound recorder	[ ]Yes [ ] No
Graphic supplies	[ ]Yes [ ] No	Postage metering machine	[ ]Yes [ ] No
Postage costs	[ ]Yes [ ] No	File cabinets	[ ]Yes [ ] No
_____	[ ]Yes [ ] No	Plan storage racks	[ ]Yes [ ] No
_____	[ ]Yes [ ] No	Plan storage cabinets	[ ]Yes [ ] No
_____	[ ]Yes [ ] No	Graphics equipment	[ ]Yes [ ] No
_____	[ ]Yes [ ] No	CAD station	[ ]Yes [ ] No
_____	[ ]Yes [ ] No	CAD Plotter	[ ]Yes [ ] No

Comments:_____
_____
_____

Finance Expense

Progress payment delay cost: Interest, debt service charges:

Has the amount been calculated?		Real property	[ ]Yes [ ] No
	[ ]Yes [ ] No	Buildings, structures	[ ]Yes [ ] No
Is it included in the estimate	[ ]Yes [ ] No	Vehicles	[ ]Yes [ ] No
_____	[ ]Yes [ ] No	Equipment	[ ]Yes [ ] No
_____	[ ]Yes [ ] No	_____	[ ]Yes [ ] No
_____	[ ]Yes [ ] No	_____	[ ]Yes [ ] No
_____	[ ]Yes [ ] No	_____	[ ]Yes [ ] No
_____	[ ]Yes [ ] No	_____	[ ]Yes [ ] No

Comments:_____
_____

Part One　　　　　　　　　　　　　　　　　　　　　　**Section 8** – *Checklists/Special Forms*

Outside services:

		Insurance:	
Engineering	[ ]Yes [ ] No	Comprehensive general liability	[ ]Yes [ ] No
Architectural	[ ]Yes [ ] No	Personal injury/prop. damage	[ ]Yes [ ] No
Equipment maintenance/repair	[ ]Yes [ ] No	Business auto	[ ]Yes [ ] No
Building maintenance/repair	[ ]Yes [ ] No	Equipment floaters	[ ]Yes [ ] No
Janitorial	[ ]Yes [ ] No	Builder's risk	[ ]Yes [ ] No
Landscaping	[ ]Yes [ ] No	Fire, extended coverage	[ ]Yes [ ] No
_____	[ ]Yes [ ] No	Insurance exclusions	[ ]Yes [ ] No
_____	[ ]Yes [ ] No	Worker's comp (h.o. staff)	[ ]Yes [ ] No
_____	[ ]Yes [ ] No	Social security (h.o. staff)	[ ]Yes [ ] No
_____	[ ]Yes [ ] No	Fringe benefits (h.o. staff)	[ ]Yes [ ] No
_____	[ ]Yes [ ] No	Earthquake	[ ]Yes [ ] No
_____	[ ]Yes [ ] No	_____	[ ]Yes [ ] No

Comments:_____

_____

_____

Advertising/marketing:

		Taxes/licenses:	
_____	[ ]Yes [ ] No	Real property tax	[ ]Yes [ ] No
_____	[ ]Yes [ ] No	Bus. property/equip tax	[ ]Yes [ ] No
_____	[ ]Yes [ ] No	State sales tax	[ ]Yes [ ] No
_____	[ ]Yes [ ] No	Motor fuel tax	[ ]Yes [ ] No
_____	[ ]Yes [ ] No	Special taxes/assessments	[ ]Yes [ ] No
_____	[ ]Yes [ ] No	Rental equipment tax	[ ]Yes [ ] No
_____	[ ]Yes [ ] No	Contractors license	[ ]Yes [ ] No
_____	[ ]Yes [ ] No	Vehicle licenses/registration	[ ]Yes [ ] No
_____	[ ]Yes [ ] No	_____	[ ]Yes [ ] No
_____	[ ]Yes [ ] No	_____	[ ]Yes [ ] No

Comments:_____

_____

_____

Standard Estimating Practice

Section 8 – Checklists/Special Forms **Part One**

Bonds: **Technical:**

Contractor license	[ ]Yes [ ] No	_____	[ ]Yes [ ] No
Notary public	[ ]Yes [ ] No	_____	[ ]Yes [ ] No
Accountant	[ ]Yes [ ] No	_____	[ ]Yes [ ] No
Employee	[ ]Yes [ ] No	_____	[ ]Yes [ ] No
_____	[ ]Yes [ ] No	_____	[ ]Yes [ ] No
_____	[ ]Yes [ ] No	_____	[ ]Yes [ ] No
_____	[ ]Yes [ ] No	_____	[ ]Yes [ ] No
_____	[ ]Yes [ ] No	_____	[ ]Yes [ ] No

Comments:_____

_____

_____

_____

Home Office Staff equipment

Rental/lease equipment: **Company owned equipment:**

Pro rata company rental	[ ]Yes [ ] No	Equipment service vehicles	[ ]Yes [ ] No
Pro rata company lease	[ ]Yes [ ] No	Cars, station wagons	[ ]Yes [ ] No
Pro rata outside rental	[ ]Yes [ ] No	Vans, pickups	[ ]Yes [ ] No
Pro rata outside lease	[ ]Yes [ ] No	Light trucks	[ ]Yes [ ] No
_____	[ ]Yes [ ] No	Heavy trucks	[ ]Yes [ ] No
_____	[ ]Yes [ ] No	_____	[ ]Yes [ ] No
_____	[ ]Yes [ ] No	_____	[ ]Yes [ ] No

Comments:_____

General Conditions Estimate Checklist

Project:_____   Bid Date:_____

Location:_____   Estimate Due:_____

                                          Estimate #:_____

The intent of the term general conditions is to imply the anticipated costs for division (section) one or project site direct costs, construction physical plant, construction equipment, supervision, supervision service equipment, hot and cold weather protection, small hand and power tools, expendables, permits, fees, licenses, testing, field engineering, and other costs that are required by the general and special conditions and its anticipated cost. A percentage, based on historical information, may be applied to the total direct costs as a judgmental verification of the previously calculated detailed general conditions estimate. General conditions may be shown as a percentage, based on historical information, in estimate levels one and two.

Mobilization

Move-in expense:

		Field supervision:	
Construction equipment	[ ]Yes [ ] No	Project superintendent	[ ]Yes [ ] No
Small tools and consumables	[ ]Yes [ ] No	Architectural superintendent	[ ]Yes [ ] No
Office trailers	[ ]Yes [ ] No	Civil superintendent	[ ]Yes [ ] No
Tool vans & change houses	[ ]Yes [ ] No	Electrical superintendent	[ ]Yes [ ] No
Personnel relocation	[ ]Yes [ ] No	Mechanical superintendent	[ ]Yes [ ] No
_____	[ ]Yes [ ] No	Structural superintendent	[ ]Yes [ ] No
_____	[ ]Yes [ ] No	Equipment superintendent	[ ]Yes [ ] No
_____	[ ]Yes [ ] No	Field engineer(s)	[ ]Yes [ ] No
_____	[ ]Yes [ ] No	Technical aides	[ ]Yes [ ] No
_____	[ ]Yes [ ] No	Secretary/clerk(s)	[ ]Yes [ ] No
_____	[ ]Yes [ ] No	_____	[ ]Yes [ ] No
_____	[ ]Yes [ ] No	_____	[ ]Yes [ ] No
_____	[ ]Yes [ ] No	_____	[ ]Yes [ ] No
_____	[ ]Yes [ ] No	_____	[ ]Yes [ ] No
_____	[ ]Yes [ ] No	_____	[ ]Yes [ ] No
_____	[ ]Yes [ ] No	_____	[ ]Yes [ ] No
_____	[ ]Yes [ ] No	_____	[ ]Yes [ ] No
_____	[ ]Yes [ ] No	_____	[ ]Yes [ ] No

Comments:_____

_____

Section 8 – *Checklists/Special Forms*   **Part One**

Site Staff

The intent of the term site staff is to imply limited participation by home office managerial and administrative personnel on projects when these personnel are assigned to offices and quarters at the site of the project. Refer to Overhead Cost Definitions for projects staffed on site with supervisory personnel only who are fully supported by management and administrative personnel at the home office.

Office management/engineering:

			Administrative:		
Project manager	[ ]Yes	[ ] No	Office manger	[ ]Yes	[ ] No
Assistant project manager	[ ]Yes	[ ] No	Accountant	[ ]Yes	[ ] No
Project engineer	[ ]Yes	[ ] No	Bookkeeper	[ ]Yes	[ ] No
Assistant project engineer	[ ]Yes	[ ] No	Timekeeper	[ ]Yes	[ ] No
Estimator	[ ]Yes	[ ] No	Secretary/receptionist	[ ]Yes	[ ] No
Cost Engineer	[ ]Yes	[ ] No	Clerk/typist(s)	[ ]Yes	[ ] No
Scheduler	[ ]Yes	[ ] No	_____	[ ]Yes	[ ] No
Office engineer	[ ]Yes	[ ] No	_____	[ ]Yes	[ ] No
Safety engineer	[ ]Yes	[ ] No	_____	[ ]Yes	[ ] No
Subcontract engineer	[ ]Yes	[ ] No	_____	[ ]Yes	[ ] No
Technical aides	[ ]Yes	[ ] No	_____	[ ]Yes	[ ] No
Secretary/clerk(s)	[ ]Yes	[ ] No	_____	[ ]Yes	[ ] No

Comments:_____

Temporary buildings:

			Temp utilities/water:		
Site offices – staff	[ ]Yes	[ ] No	Construction	[ ]Yes	[ ] No
Site offices – owner	[ ]Yes	[ ] No	Fire protection	[ ]Yes	[ ] No
Change houses – craft	[ ]Yes	[ ] No	Drinking and ice	[ ]Yes	[ ] No
Tool sheds	[ ]Yes	[ ] No	Dust control	[ ]Yes	[ ] No
Warehouse	[ ]Yes	[ ] No	_____	[ ]Yes	[ ] No
First aid	[ ]Yes	[ ] No	_____	[ ]Yes	[ ] No
Fabrication shop	[ ]Yes	[ ] No	_____	[ ]Yes	[ ] No
Fuel and repair facility	[ ]Yes	[ ] No	_____	[ ]Yes	[ ] No
_____	[ ]Yes	[ ] No	_____	[ ]Yes	[ ] No
_____	[ ]Yes	[ ] No	_____	[ ]Yes	[ ] No
_____	[ ]Yes	[ ] No	_____	[ ]Yes	[ ] No

Comments:_____

Part One Section 8 – Checklists/Special Forms

Temporary constructions:

Barricades and fencing	[ ]Yes [ ] No
Parking/roads-construct/main.	[ ]Yes [ ] No
Material storage & dunnage	[ ]Yes [ ] No
Signs & markers	[ ]Yes [ ] No
Scaffolding and ladders	[ ]Yes [ ] No
Enclosures	[ ]Yes [ ] No
Dewatering system	[ ]Yes [ ] No
_____	[ ]Yes [ ] No
_____	[ ]Yes [ ] No

Comments:_____

Temp utilities/electricity:

Temporary power distribution	[ ]Yes [ ] No
Temporary power services	[ ]Yes [ ] No
Area lighting	[ ]Yes [ ] No
Heating and ventilation	[ ]Yes [ ] No
Systems and maintenance	[ ]Yes [ ] No
_____	[ ]Yes [ ] No
_____	[ ]Yes [ ] No
_____	[ ]Yes [ ] No
_____	[ ]Yes [ ] No

Temporary utilities/other:

Gas systems	[ ]Yes [ ] No
Compressed air systems	[ ]Yes [ ] No
Welding power systems	[ ]Yes [ ] No
Telephone system	[ ]Yes [ ] No
Telecommunication system	[ ]Yes [ ] No
_____	[ ]Yes [ ] No

Comments:_____

Equipment operating expense:

Fuels	[ ]Yes [ ] No
Lubricants	[ ]Yes [ ] No
Repair & maintenance	[ ]Yes [ ] No
Operations labor	[ ]Yes [ ] No
_____	[ ]Yes [ ] No
_____	[ ]Yes [ ] No

Construction Equipment

Rental/lease equipment:

Company rental	[ ]Yes [ ] No
Company lease	[ ]Yes [ ] No
Outside rental	[ ]Yes [ ] No
Outside lease	[ ]Yes [ ] No
_____	[ ]Yes [ ] No
_____	[ ]Yes [ ] No

Comments:_____

Company owned equipment:

Repair & maintenance	[ ]Yes [ ] No
_____	[ ]Yes [ ] No
_____	[ ]Yes [ ] No
_____	[ ]Yes [ ] No
_____	[ ]Yes [ ] No
_____	[ ]Yes [ ] No

Section 8 – *Checklists/Special Forms*   **Part One**

Site office supplies:

Stationary & office supplies	[ ]Yes	[ ] No
Engineering/drafting supplies	[ ]Yes	[ ] No
Outside blueprint costs	[ ]Yes	[ ] No
_____	[ ]Yes	[ ] No
_____	[ ]Yes	[ ] No
_____	[ ]Yes	[ ] No
_____	[ ]Yes	[ ] No
_____	[ ]Yes	[ ] No
_____	[ ]Yes	[ ] No
_____	[ ]Yes	[ ] No
_____	[ ]Yes	[ ] No
_____	[ ]Yes	[ ] No

Site office equipment:

Office furniture	[ ]Yes	[ ] No
Copy machine	[ ]Yes	[ ] No
Fax/printer	[ ]Yes	[ ] No
Blueprint machines	[ ]Yes	[ ] No
Computer	[ ]Yes	[ ] No
Calculators	[ ]Yes	[ ] No
Fire extinguishers	[ ]Yes	[ ] No
_____	[ ]Yes	[ ] No
_____	[ ]Yes	[ ] No
_____	[ ]Yes	[ ] No
_____	[ ]Yes	[ ] No
_____	[ ]Yes	[ ] No

Comments:_____
_____
_____

Site office utilities:

Water hookup & supply	[ ]Yes	[ ] No
Power hookup & supply	[ ]Yes	[ ] No
Gas hookup & supply	[ ]Yes	[ ] No
Telephone hookup & supply	[ ]Yes	[ ] No
_____	[ ]Yes	[ ] No
_____	[ ]Yes	[ ] No
_____	[ ]Yes	[ ] No
_____	[ ]Yes	[ ] No
_____	[ ]Yes	[ ] No
_____	[ ]Yes	[ ] No

Other site office costs:

Photographs	[ ]Yes	[ ] No
Travel expense	[ ]Yes	[ ] No
Permits & licenses	[ ]Yes	[ ] No
_____	[ ]Yes	[ ] No
_____	[ ]Yes	[ ] No
_____	[ ]Yes	[ ] No
_____	[ ]Yes	[ ] No
_____	[ ]Yes	[ ] No
_____	[ ]Yes	[ ] No
_____	[ ]Yes	[ ] No

Comments:_____
_____
_____

Small tools:

Initial purchases	[ ]Yes [ ] No
Replacement purchases	[ ]Yes [ ] No
_____	[ ]Yes [ ] No
_____	[ ]Yes [ ] No
_____	[ ]Yes [ ] No
_____	[ ]Yes [ ] No
_____	[ ]Yes [ ] No
_____	[ ]Yes [ ] No
_____	[ ]Yes [ ] No
_____	[ ]Yes [ ] No
_____	[ ]Yes [ ] No
_____	[ ]Yes [ ] No

Comments:_____

Cold weather protection:

Unit heaters	[ ]Yes [ ] No
Tarps and plastic	[ ]Yes [ ] No
Enclosures	[ ]Yes [ ] No
LP gas/oil	[ ]Yes [ ] No
_____	[ ]Yes [ ] No
_____	[ ]Yes [ ] No
_____	[ ]Yes [ ] No
_____	[ ]Yes [ ] No
_____	[ ]Yes [ ] No
_____	[ ]Yes [ ] No
_____	[ ]Yes [ ] No
_____	[ ]Yes [ ] No

Consumables:

Welding rod	[ ]Yes [ ] No
Welding gas	[ ]Yes [ ] No
_____	[ ]Yes [ ] No
_____	[ ]Yes [ ] No
_____	[ ]Yes [ ] No
_____	[ ]Yes [ ] No
_____	[ ]Yes [ ] No
_____	[ ]Yes [ ] No
_____	[ ]Yes [ ] No

Comments:_____

Hot weather protection;

Fans and ventilators	[ ]Yes [ ] No
Tarps and plastics	[ ]Yes [ ] No
Shelters	[ ]Yes [ ] No
_____	[ ]Yes [ ] No
_____	[ ]Yes [ ] No
_____	[ ]Yes [ ] No
_____	[ ]Yes [ ] No
_____	[ ]Yes [ ] No
_____	[ ]Yes [ ] No

Section 8 – Checklists/Special Forms — Part One

Outside services:

Material testing	[ ]Yes	[ ] No
X-ray and other	[ ]Yes	[ ] No
Authorized inspector	[ ]Yes	[ ] No
Testing and inspection	[ ]Yes	[ ] No
Survey and site layout	[ ]Yes	[ ] No
_____	[ ]Yes	[ ] No
_____	[ ]Yes	[ ] No
_____	[ ]Yes	[ ] No
_____	[ ]Yes	[ ] No
_____	[ ]Yes	[ ] No
_____	[ ]Yes	[ ] No
_____	[ ]Yes	[ ] No

Insurance:

Comprehensive general liability	[ ]Yes	[ ] No
Personal injury/property	[ ]Yes	[ ] No
Business auto	[ ]Yes	[ ] No
Equipment floaters	[ ]Yes	[ ] No
Builder's risk	[ ]Yes	[ ] No
Umbrella excess liability	[ ]Yes	[ ] No
Insurance exclusions	[ ]Yes	[ ] No
Earthquake	[ ]Yes	[ ] No
_____	[ ]Yes	[ ] No
_____	[ ]Yes	[ ] No
_____	[ ]Yes	[ ] No
_____	[ ]Yes	[ ] No

Comments:_____
_____
_____

Other services:

Janitorial	[ ]Yes	[ ] No
Security	[ ]Yes	[ ] No
_____	[ ]Yes	[ ] No
_____	[ ]Yes	[ ] No
_____	[ ]Yes	[ ] No
_____	[ ]Yes	[ ] No
_____	[ ]Yes	[ ] No
_____	[ ]Yes	[ ] No
_____	[ ]Yes	[ ] No
_____	[ ]Yes	[ ] No

Taxes/licenses:

Material sales tax	[ ]Yes	[ ] No
Local sales tax	[ ]Yes	[ ] No
State sales tax	[ ]Yes	[ ] No
Motor fuel tax	[ ]Yes	[ ] No
Property taxes (equipment)	[ ]Yes	[ ] No
Rental equipment tax	[ ]Yes	[ ] No
Vehicle licenses/registrations	[ ]Yes	[ ] No
Gross receipts tax	[ ]Yes	[ ] No
_____	[ ]Yes	[ ] No
_____	[ ]Yes	[ ] No

Comments: Include building permit, plan check and utility connection fees as required

Part One   *Section 8 – Checklists/Special Forms*

_____
_____
_____

Bonds:		**Technical:**	
Bid	[ ]Yes [ ] No	_____	[ ]Yes [ ] No
Payment & performance	[ ]Yes [ ] No	_____	[ ]Yes [ ] No
Material supply	[ ]Yes [ ] No	_____	[ ]Yes [ ] No
Subcontractor	[ ]Yes [ ] No	_____	[ ]Yes [ ] No
Employee	[ ]Yes [ ] No	_____	[ ]Yes [ ] No
Installation warranty	[ ]Yes [ ] No	_____	[ ]Yes [ ] No
Contractor license bond	[ ]Yes [ ] No	_____	[ ]Yes [ ] No
_____	[ ]Yes [ ] No	_____	[ ]Yes [ ] No
_____	[ ]Yes [ ] No	_____	[ ]Yes [ ] No
_____	[ ]Yes [ ] No	_____	[ ]Yes [ ] No

Comments:_____
_____

Home Office Support

The intent of the term Home Office Support is to imply limited participation by home office managerial and administrative personnel on projects when these personnel are assigned to offices and quarters at the site of the project. Refer to Overhead Cost Definitions for projects that are staffed on site with supervisory personnel only who are supported by management and administrative staffs at the home office.

Management:		**Administrative:**	
_____	[ ]Yes [ ] No	_____	[ ]Yes [ ] No
_____	[ ]Yes [ ] No	_____	[ ]Yes [ ] No
_____	[ ]Yes [ ] No	_____	[ ]Yes [ ] No
_____	[ ]Yes [ ] No	_____	[ ]Yes [ ] No
_____	[ ]Yes [ ] No	_____	[ ]Yes [ ] No
_____	[ ]Yes [ ] No	_____	[ ]Yes [ ] No
_____	[ ]Yes [ ] No	_____	[ ]Yes [ ] No
_____	[ ]Yes [ ] No	_____	[ ]Yes [ ] No
_____	[ ]Yes [ ] No	_____	[ ]Yes [ ] No
_____	[ ]Yes [ ] No	_____	[ ]Yes [ ] No

Section 8 – Checklists/Special Forms — Part One

Comments:_____
_____
_____
_____

Demobilization

Move out expense:

Construction equipment	[ ]Yes	[ ] No
Small tools/consumables	[ ]Yes	[ ] No
Office trailers	[ ]Yes	[ ] No
Tool vans & change houses	[ ]Yes	[ ] No
Personnel relocation	[ ]Yes	[ ] No
Punchlists	[ ]Yes	[ ] No
Final job cleanup	[ ]Yes	[ ] No
_____	[ ]Yes	[ ] No
_____	[ ]Yes	[ ] No
_____	[ ]Yes	[ ] No
_____	[ ]Yes	[ ] No
_____	[ ]Yes	[ ] No

Scrap/salvage:

Temporary facilities	[ ]Yes	[ ] No
Temporary construction	[ ]Yes	[ ] No
Freight	[ ]Yes	[ ] No
_____	[ ]Yes	[ ] No
_____	[ ]Yes	[ ] No
_____	[ ]Yes	[ ] No
_____	[ ]Yes	[ ] No
_____	[ ]Yes	[ ] No
_____	[ ]Yes	[ ] No
_____	[ ]Yes	[ ] No
_____	[ ]Yes	[ ] No
_____	[ ]Yes	[ ] No

Comments:_____
_____
_____
_____

Escalation – materials

Permanent	[ ]Yes	[ ] No
Temporary	[ ]Yes	[ ] No
Freight	[ ]Yes	[ ] No
_____	[ ]Yes	[ ] No
_____	[ ]Yes	[ ] No
_____	[ ]Yes	[ ] No
_____	[ ]Yes	[ ] No
_____	[ ]Yes	[ ] No

Escalation – labor:

Wages	[ ]Yes	[ ] No
Payroll	[ ]Yes	[ ] No
Insurance	[ ]Yes	[ ] No
_____	[ ]Yes	[ ] No
_____	[ ]Yes	[ ] No
_____	[ ]Yes	[ ] No
_____	[ ]Yes	[ ] No
_____	[ ]Yes	[ ] No

Comments:_____
_____

Part One   *Section 8 – Checklists/Special Forms*

_____
_____
_____

Finance Expense

Progress payment delay cost: **Other:**

Has the amount bee calculated? [ ]Yes [ ] No   _____ [ ]Yes [ ] No

Is it included in the estimate?   [ ]Yes [ ] No   _____ [ ]Yes [ ] No

_____   [ ]Yes [ ] No   _____ [ ]Yes [ ] No

_____   [ ]Yes [ ] No   _____ [ ]Yes [ ] No

_____   [ ]Yes [ ] No   _____ [ ]Yes [ ] No

_____   [ ]Yes [ ] No   _____ [ ]Yes [ ] No

_____   [ ]Yes [ ] No   _____ [ ]Yes [ ] No

Comments:_____
_____
_____

Contingencies: **Quantity survey:**

Strikes, walkouts, etc.   [ ]Yes [ ] No   Was it done in accordance

_____   [ ]Yes [ ] No   with the plans and specs?   [ ]Yes [ ] No

Abnormal weather   [ ]Yes [ ] No   Are there any conflicting

_____   [ ]Yes [ ] No   areas remaining?   [ ]Yes [ ] No

_____   [ ]Yes [ ] No   _____   [ ]Yes [ ] No

_____   [ ]Yes [ ] No   _____   [ ]Yes [ ] No

_____   [ ]Yes [ ] No   _____   [ ]Yes [ ] No

_____   [ ]Yes [ ] No   _____   [ ]Yes [ ] No

Comments:_____
_____
_____

Section 8 – *Checklists/Special Forms* **Part One**

_____
_____
_____

Material and labor recap:

Has the estimate been assembled so
that all figures are traceable   [ ]Yes [ ] No

Has it been prepared to reflect
how the job will be built?   [ ]Yes [ ] No

Have plug amounts been
identified and adjustments
sheets prepared?   [ ]Yes [ ] No

_____   [ ]Yes [ ] No
_____   [ ]Yes [ ] No
_____   [ ]Yes [ ] No

Estimated labor units:

Is this job similar to past jobs? [ ]Yes [ ] No
Do the labor units reflect the
similarities or differences?   [ ]Yes [ ] No

_____   [ ]Yes [ ] No
_____   [ ]Yes [ ] No
_____   [ ]Yes [ ] No
_____   [ ]Yes [ ] No
_____   [ ]Yes [ ] No
_____   [ ]Yes [ ] No

Comments:_____
_____
_____
_____

General conditions

Do the general conditions correspond to
the length of the project?   [ ]Yes [ ] No

_____   [ ]Yes [ ] No
_____   [ ]Yes [ ] No
_____   [ ]Yes [ ] No
_____   [ ]Yes [ ] No
_____   [ ]Yes [ ] No
_____   [ ]Yes [ ] No

Can the job be properly staffed
with this number of people?   [ ]Yes [ ] No

_____   [ ]Yes [ ] No
_____   [ ]Yes [ ] No
_____   [ ]Yes [ ] No
_____   [ ]Yes [ ] No
_____   [ ]Yes [ ] No
_____   [ ]Yes [ ] No

Comments:_____
_____
_____
_____

REPRODUCIBLE ESTIMATING FORMS FOLLOW

Final Summary

Company Name _____ Estimator _____ Date: _____

Project _____ Pricing _____ Date: _____

Address _____ Extended _____ Date: _____

System/Work Area _____ Estimate # _____ Summary _____ Date: _____

CSI Section _____ Bid Date _____ Verified _____ Date: _____

Labor Hour Summary

	Total
Direct Cost	
Cut & Add Adjustment [+] [-]	
Other Costs	
Adjusted Direct Cost Total	
Overhead as % or $	
Profit as % or $	
Fixed Fee as % or $	
Sales Tax as % or $	
Per. Bond as % or $	
Escalation as % or $	
Contingency as % or $	
Total Estimated Contract Cost =	
Indirect Costs	
Total Estimated Project Cost =	

Cut & Add Summary

	Increase	Decrease
Total Increase		
Total Decrease		
Total Adjustment [=] [-] =		

Standard Estimating Practice

Part One *Section 8 – Checklists/Special Forms*

Final Summary Cut & Add Sheet

Company Name _____ Estimator _____ Date: _____

Project _____ Pricing _____ Date: _____

Address _____ Extended _____ Date: _____

System/Work Area _____ Estimate # __ Summary _____ Date: _____

CSI Section _____ Bid Date ___ Verified _____ Date: _____

	Increase																Decrease	

Total Increase This Page _____   Total Decrease This Page _____

Total Adjustment [+] [-] _____

Standard Estimating Practice — 77

Part One *Section 8 – Checklists/Special Forms*

Account Summary

Company Name _____ Estimator _____ Date: _____

Project _____ Pricing _____ Date: _____

Address _____ Extended _____ Date: _____

System/Work Area _____ Estimate # _____ Summary _____ Date: _____

CSI Section _____ Bid Date _____ Verified _____ Date: _____

Account	Item or Description	Material		Labor		Burden		Equip	Sub	Total
		Cost	Hours	Hours	Cost	X %	Cost	Cost	Cost	
	Total Labor Burden This Page									
	Total Labor Hours This Page									
	Total Direct costs This Page									

Standard Estimating Practice

Estimate Detail

Company Name: _____ Estimator: _____ Date: _____

Project: _____ Pricing: _____ Date: _____

Address: _____ Extended: _____ Date: _____

System/Work Area: _____ Estimate #: _____ Summary: _____ Date: _____

CSI Section: _____ Bid Date: _____ Verified: _____ Date: _____

Item or Description	Material			Labor				Equipment			Sub		Total
Quantity	Unit	Unit $	Cost	Hr/Unit	Hours	Unit $	Cost	Unit $	Cost	Unit $	Cost		

Total Man Hours This Page

Total Direct Costs This Page

Quantity Survey

Company Name _____

Project _____ Estimator _____ Date: _____

Address _____ Pricing _____ Date: _____

System/Work Area _____ Estimate # _____ Extended _____ Date: _____

CSI Section _____ Bid Date _____ Summary _____ Date: _____

Verified _____ Date: _____

Description																			Totals	

SECTION 9 – SPECIFICATION REVIEW

Invitations to Bid and Instructions to Estimators/Bidders

Thoroughly review specifications, to whatever extent complete, during the estimating/bidding process. Conduct this review during periods of uninterrupted time to allow for proper mental concentration. Start at the beginning. Scan the table of contents to find out the included work and the order in which found. The invitation to bid and instructions to bidders provide answers to:

- ✓ Bid date and time
- ✓ Bonding requirements
- ✓ Bid location
- ✓ Availability of extra documents
- ✓ Liquidated damages
- ✓ Bid forms

Estimators should not make the mistake of disregarding this element of the estimate/ bid documents. So doing may result in submitted estimates/bids not being in accord with the requirements and therefore considered valueless though monetarily acceptable. Never assume those receiving your estimates/bids know everything about a particular estimating discipline. More often than not the opposite is true. Estimates/bids should address all addenda, unit prices, and alternates required.

Relationship to Plans

The relationship is such that studying the specifications should be first and then coordinate with the drawings. Specifications have various sections.

- ✓ Each section describes:
- ✓ Standards of quality
- ✓ Installation practices and sequences
- ✓ Material manufacturers or criteria
- ✓ Special project circumstances

The most critical items during the estimating/bidding stage are the specific materials and procedures having a cost or schedule impact.

General and Supplementary Conditions

Read the general and supplementary conditions of the contract and thoroughly understand them. As an example, AIA Forms 101, 201, and 401, have different types and levels of project requirements. Notes on the contract form, insurance requirements, allowances, alternates, and addenda are an integral part of the estimate.

General and supplementary condition items that impact cost may include:

- ✓ Federal prevailing wages
- ✓ Payment date for retention
- ✓ Retention rate
- ✓ Retention rate reductions
- ✓ Alternates
- ✓ Unused material
- ✓ Detailing & shop drawings
- ✓ Date for final payment
- ✓ Date for progress payments
- ✓ As-built drawings
- ✓ Stored material payments
- ✓ Owner personnel training
- ✓ Basis for change orders
- ✓ Startup
- ✓ Operations & maintenance manuals

Identify, list, and appropriately mark items that impact schedule for review by management and others involved in the estimate/bid process. Review each part of the general and supplementary conditions. Maintain a list of items needed for provision and installation, or services necessary for prosecution of the work. Translate this list into questions for subcontractors during the bidding process. Also use this list to cross reference the drawings for coordination. Exercise special care in noting specific items such as:

- ✓ Completion date
- ✓ Partial or beneficial occupancy
- ✓ Liquidated damages/completion incentives
- ✓ Testing requirements
- ✓ Schedules of construction
- ✓ Continuation/interruption of existing facilities
- ✓ Continuation/interruption of existing utilities

Determining Pertinent Divisions and Sections

The particular work and labor force will dictate the exact specification sections application to the work. Be careful in rating the effects of all estimate/bid document elements referenced. Study contractual requirements for general and special conditions in addition to the technical specifications.

Detailed Review

Remove or copy all pertinent element of the specifications and assemble in sequential order. A loose leaf, three-ring binder may be right for this purpose. Read each word from front to back, highlight items that are unusual, contradictory, particularly restrictive or abnormally vague. Be sure to understand and appreciate the cost and schedule implications of references such as:

- ✓ Per manufacturers recommendations
- ✓ Per uniform building code
- ✓ As selected by the architect
- ✓ As approved by owner, etc.

Typical examples of specifications having a cost or schedule impact could include:

- ✓ As-built drawings as a marked-up set of plans
- ✓ As-built drawings as reproducible tracings
- ✓ Shop drawing required
- ✓ Sample submittals
- ✓ Specific material grades and catalog numbers
- ✓ Types of materials allowed
- ✓ Sole sourcing for material or equipment

Also consider site accessibility for the duration of the project. Check unusual requirements for material delivery due to traffic congestion, truck waiting time, and availability of storage areas.

Review all sections of the structural and architectural specifications. Make provision in the estimate/bid for noted items directly related to each specific section. Exercise special care in noting specific items such as:

- ✓ Attachment requirements
- ✓ Mounting requirements
- ✓ Penetration requirements
- ✓ Design criteria

Interpretation of Intent

When reviewing specifications, the estimator should create an analysis and evaluation sheet. This sheet serves to remind the estimator throughout the takeoff process of important impacts on costs and productivity. It also serves to document the estimator's perception of the intent of the documents.

Define Scope of Work

Be specific in defining the scope of work, whether by inclusion or exclusion, in the estimate/bid submittal. For clarity, refer to division, section paragraph, and sub headings by the appropriate numbers and letters contained in the specifications.

Many tables have standard inclusions/exclusions that apply to all projects. A copy of these inclusions/exclusions for trades affecting the estimate/bid should be available during the estimating/bidding procedure. The estimator should note items of direct impact resulting from these standard inclusions/exclusions.

Note inclusions/exclusions unique to a specific project and explain in detail to all trades affected. The estimator must find inclusions/exclusions of other trades that change the specific estimate/bid and incorporate those in the estimate/bid.

SECTION 10 – PLAN REVIEW

Develop Familiarity and Format

Review the plans to interpret the design professional's style and intent. Read all notes and create a list of unusual items. Formulate a method of accounting for various types of work. Analyze details for practicality, intent, and inconsistencies. The preferred method of takeoff is to match the sequential order of actual construction. However, in extremely complex projects, it is usually easier to get started by beginning with the simplest typical systems or components. By the time the bulk of the basic work is complete, the drawings should be easier to understand. This simplifies identification and accounting of the complicated items.

Physical proof that the plan sheets agree with the title sheet index in title and number is critical. Note discrepancies, if any, between the index and the sheets. When there is no title sheet or plan index, prepare a list of the sheet titles and numbers. This is for reference and qualification of the estimate/bid.

Note the drawn to scale signs and prove their accuracy with mechanical or electronic measuring devices used for the quantity survey. Find out if the stated plan dimensions agree with the scaled dimensions noted on the drawing. Prudent observance of these procedures provides proof that the plan copies remain accurate to scale.

Review the plans for each section of work and determine the degree of completeness. State the appropriate levels for each section of the estimate from Part One, Section Two, Levels of the Estimate. Identify the level/levels when submitting the estimate/ bid. This review also should include the scope of work in each section. This is so selection of bidding subcontractors, when applicable, is on an informed basis. Information gleaned from the specification and plan reviews provide the estimator with knowledge required to complete the estimate/bid. This also allows intelligent requests for additional information from the design professional and response to queries from subcontractors and suppliers.

Carefully study the site location plan for items that may hinder access. Review the construction site plan to find out the following conditions:

- ✓ Project locations within property boundaries
- ✓ Security
- ✓ Location of office and storage trailers
- ✓ Distance of these from actual construction

Examine architectural/structural plans in detail for items of work such as:

- ✓ Roof type and roofing material
- ✓ Areas restricting the passage of work

- ✓ Type of exterior wall
- ✓ Areas restricting the specific scope of work
- ✓ Wall and ceiling dimensions
- ✓ Specific design criteria
- ✓ Equipment locations and layout
- ✓ Wall and ceiling materials and finishes

SECTION 11 – QUANTITY SURVEY

Cover Page

On the cover page of each estimate show the following information:

- ✓ Date
- ✓ Address
- ✓ Date of plans
- ✓ Name of governing entity
- ✓ Project name
- ✓ Professional design firm name
- ✓ Name of owner
- ✓

The estimator should date, consecutively number, identify by project name or number, and sign each page of the estimate.

Estimate Levels

Define estimates on the most applicable level of detail equal to design development information available. Prepare estimates on a quantity times material, labor hour, and construction equipment costs. Use other bases, such as square/cubic foot and system costs, only when lack of information and detail prevents a proper quantity takeoff.

Estimate Format

Prepare cost estimates in a uniform format, easily read and understood by other parties with a direct interest or involvement in the project. The quantity times material and labor costs format should contain the following components.

- ✓ CSI section
- ✓ Description
- ✓ Unit measure
- ✓ Labor quantity in hours
- ✓ Subcontract cost (when applicable)
- ✓ Work breakdown structure (when used)
- ✓ Quantity
- ✓ Material cost
- ✓ Labor cost
- ✓ Total cost

Assemble the information in this form from the quantity takeoff and extension detail calculated earlier.

General and Supplementary Conditions

Read and thoroughly understand the general and supplementary conditions of the contract. As an example, AIA forms 101, 201, and 401 entail several kinds of levels of project requirements. Notes on the contract form, insurance requirements, allowances, alternates, and addenda form an integral part of the estimate.

Schedule

Determine the project schedule and estimate the work accordingly. Once the estimate is complete, use it to prove the project schedule.

Defined Areas

Prepare cost estimates, at any level, with Construction Specifications Institute Uniform Numbering System. When specific costs are necessary for defined areas within a project, maintain this format within each defined area. The term defined area is the equivalent of system estimates. Other various equivalents are work breakdown structure and owner's special division of work.

Systems Descriptions

The most effective estimating method for complex projects requiring detailed quantity survey has several names. Some describe the method as build up while others use the word assembly or component. The recommended term is system. By using the system approach, the estimator describes in detail all the components and configuration of a system at one time. Enter the system description on the survey sheet. Then list the first task or component and continue logically with the remaining items in the system. It is usually possible to identify and describe all of the typical systems and many of the non-typical ones before starting the quantity survey. Use recognizable abbreviations to save space. Following is an example of the components in a concrete footing system: excavation, fine grade, formwork, reinforcing steel, concrete, anchor bolts, finish, protect, cure, and backfill.

The application of a systems approach to estimating construction costs is often misapplied. One of the most common errors is the inclusion of different trade components in a given system. While each project will vary in its own format breakdown, prepare the estimate to reflect the specific project requirements. This also requires meeting design development and bidding criteria. Due to the nature of design development and construction methodologies, ASPE adopts and recommends CSI Masterformat as the preferred estimating format.

Specific Items

This term refers to all other work not readily fitting into systems applications. The description of the line items are similar, but consist of only one component or task. Although most specific line items appear towards the end of the quantity survey, it is good practice to write them down when discovered. These discoveries may occur while researching details, sections or elevations for a system assembly. This helps to insure accounting for the items, rather than relying on the estimator's memory. As shown before, it is desirable to organize the estimate sequentially. Accounting for all work, regardless of format, is mandatory.

Mensuration

Survey, identify, and describe a particular portion of work in the headings of the quantity survey sheet. Please refer to Quantity Survey (included in Part One, Section Eight). Measure and record the quantities in their most logical units. Following are some commonly used quantities and their abbreviations.

- ✓ Each - EA
- ✓ Square feet - SF
- ✓ Cubic yard - CY
- ✓ Pound - LB
- ✓ 100 - C
- ✓ Linear feet - LF
- ✓ Cubic feet - CF
- ✓ Gallon - GL
- ✓ Ton - TN
- ✓ 1,000 - M

Measuring Tools and Devices

Manual tools consist of architect/engineer/metric scales, map measurers, and planimeters.

Electronic devices consist of architect/engineer/metric scaled measuring devices, counters, planimeters, digitizers, and light pens.

Plan Scale Proof

Note the drawn to scale signs and use mechanical or electronic measuring devices to provide their accuracy. Determine if the stated plan dimensions agree with the scaled dimensions noted on the drawing. Prudent observance of these procedures provides proof of plan copy accuracy.

Waste Allowances

Round takeoff quantities up to the nearest available unit. As an example, round .83 CY ready mix concrete up to 1 CY or minimum order. As another example, round 14'-4 IN length of 2 × 10-wood joist up to 16'-0".

Takeoff Procedures

Compare plan sheets with sections to find specific components of the structure required for a certain work classification.

Example: the foundation plan will have sections taken through areas that may vary from each other. The plan may indicate where the 24 IN × 12 IN cast in place footings change to 36 IN × 12 IN in certain areas. The section will define the footing size, reinforcing steel requirements, and depth of footing from grade for calculating excavation and backfill. These sections are sometimes keyed to a schedule on the structural plans. Research specifications for concrete design criteria and reinforcing steel grades.

Example: measure the LF and height of each type of interior gypsum board partitions indicated on the plans. This may include partitions to ceiling grid, partitions to underside of deck, number of layers and finish of gypsum board, framing member size, and insulation, if required. Review the specifications for material standards and installation recommendations.

SECTION 12 – PRICING/SUMMARIES

Estimate Detail Form

Transfer information from the quantity survey to this form for pricing and calculating labor costs.

Materials

Bring material descriptions and quantities forward from the quantity survey. Quantities brought forward should include allowances for waste, packaging, and palletization. Provide suppliers these material lists for pricing or use reliable historical prices when applicable. Add sales taxes where applicable, and calculate subtotals.

Labor

Bring forward labor descriptions and quantities from the quantity survey. Stop and review all notes before assigning production rates. Also review specification highlights and everything that is necessary to develop a fully informed understanding of the project. Review the drawings and details of previous similar projects. Most production rate analyses begin with a company or estimator defined standard based on experience. Reflecting earned judgment, the estimator adjusts this rate for favorable or unfavorable conditions and expected degree of difficulty factors. At this point, estimators must keep their own objectivity. Increase or decrease productivity as required but do not allow outside pressures to prevent sound judgment based on experience. Request advice from field personnel in analyzing and determining production rates, where appropriate.

The estimator has the final responsibility for assigning productivity rates after checking input from all sources. The estimator must use experience and ability to carry out this most critical responsibility.

Calculate production rates when expressed as decimals of hours by the following method: Labor hours required (1 hour) divided by units of work completed (200 SF) = production rate (.005 hours per SF). Determine basic production rates and multiply them by the units of work to determine total hours for the work. These total basic hours are very useful when gleaning costs from historical data. On site material/equipment handling by unskilled labor, trainee/apprentice labor, supervision and direct project management hours are examples. Develop crew sizes and production schedules from the completed labor sheet.

Also express production rates by composite crew in the same manner as the previous paragraph. Determine productivity by averaging the hours and costs of the individual components of the composite crew. Example:

	Hours	Rate	Cost
General Foreman	.20	$15.00	$3.00
Foremen	1.00	14.00	14.00
Journeyman	3.00	13.00	36.00
Laborer	2.00	12.00	24.00
Total	**6.20**		**$77.00**

$77.00 = Composite Crew Cost Per Hour

$77.00 Divided by 6.20 hours = $12.42 per hour average labor hour cost.

When the basis of work is linear feet @ 100 LF/HR then 1 divided by 100 = .01 crew hours per linear foot. When the basis of work is square feet @ 500 SF/HR then 1 divided by 500 = .002 crew hours per square foot.

Remember the above example addresses composite crew hours only. To determine total labor hours, multiple the crew hours by the number of workers in the crew.

Develop standard production rates that remain consistent for all estimates. Adjust these productivity standards as needed to compensate for impacts that are unique to each project.

These impacts may include the following:

- Hot and cold weather
- Coordination of trades
- Work in remote areas
- Errors and omissions
- Fatigue
- Lack of supervision
- Shift work

- Morale and attitude
- Overtime work
- Delayed completion
- Crew efficiency
- Beneficial occupancy
- Site access

Equipment

Construction equipment is the mechanical apparatus necessary for the installation and completion of the specified work of a project. It is different from fixed equipment that becomes a permanent portion of the project. This equipment can include motor, electric, electronic, and manually driven machinery.

Construction equipment procurement may use a combination of these methods:

- Company purchased and owned
- Leased by the hour, day, week, month, or extended basis.
- Rented on a specific term basis
- Rented, operated, and maintained on a specific term basis

When construction equipment is under the care, custody, and control of the company as owned or leased, include the following in the estimate/bid.

- ✓ Capital cost
- ✓ Fuel consumption
- ✓ Depreciation
- ✓ Operator training
- ✓ Insurance
- ✓ Operator supervision
- ✓ Maintenance cost
- ✓ Safety requirements
- ✓ Repair cost
- ✓ Operator

Also check the equipment giving special attention to sharing with other trades. Scaffolding is an example of this type of equipment. CSI Division 1, General Requirements, sometimes require the prime contractor to provide certain equipment for use by the subcontractors. Include these costs in the equipment column.

Subcontracts

Note received bids for subcontracts and include in the subcontract column. Without bids, make unit price determinations as required and calculate totals for the page. This method is acceptable only if sufficient historical information exists. Otherwise, develop subcontracts on a material, labor, and equipment basis with overhead, profit, and other costs added to develop an estimated subcontract cost.

General Conditions - General conditions are the expected costs for Division (Section) 1 of the specifications or project site direct costs including the following:

- ✓ Construction and physical plant
- ✓ Expendables
- ✓ Construction equipment
- ✓ Permits
- ✓ Supervision
- ✓ Fees
- ✓ Supervision service equipment
- ✓ Licenses
- ✓ Hot/cold weather protections
- ✓ Testing
- ✓ Small hand and power tools
- ✓ Field engineering

Also include other costs required by the general and special conditions of the contract for construction.

Show the cost for general conditions as a separate direct field cost in estimate levels three through six, list each item that applies and its expected cost. Use of a total direct cost percentage for general conditions is acceptable in Levels One and Two. The percentage is a judgmental verification of other detailed general condition estimates.

Account Summary Form

Transfer information from the Estimate Detail to this form for assigning account codes, if applicable, and calculating direct labor burden.

Direct Labor Burden

- ✓ Direct labor burden is the combined costs required for the following:
- ✓ Workers' compensation insurance
- ✓ Employer's liability insurance
- ✓ Employer paid Social Security taxes
- ✓ State/Federal unemployment insurance
- ✓ Union and/or company fringe benefits
- ✓ Subsistence, if applicable

Be aware that similar statutory required employer contributions also may exists. Include direct labor burden as a direct field cost in each estimate and calculate it using one of the following methods. Convert hourly costs for benefits or subsistence to a percentage. Add this to the combined total of other direct labor burden percentages above to determine the total percentage for each construction craft or trade. Recognize the variation in workers' compensation, fringe benefits, and subsistence, when applicable. Apply this percentage to the labor cost for each division (section) of work. The percentage applied should reflect expected increases or decreases. Calculate direct labor burden in the burden column of the Account Summary. Experienced estimators may include direct labor burden with the hourly wage and fringe benefit costs to give a total labor hour cost.

Final Summary Form

Transfer account summary direct cost total to this form for estimate (cut and add) adjustments. Follow this with applications for overhead, profit, fees, bonding, escalation, and contingency.

Cost estimates shall separately show all mark-up factors determined for subcontractor and general contractor overhead, profit, fees, taxes, contingencies, inflation escalations, etc.

Overhead

Overhead, either for general or subcontractor is the expected contract costs for all items classified as other than direct field costs. The overhead costs may include the following:

- ✓ Home office physical plant leasing
- ✓ Capital investment and maintenance cost
- ✓ Finance expenses
- ✓ Home office physical plant services
- ✓ Support and pro-rata salaries
- ✓ Project management
- ✓ Clerical and accounting

- ✓ General and liability insurance costs
- ✓ Comprehensive insurance costs
- ✓ Outside services

Show the expected cost for overhead separately in each estimate and calculate using either of the following methods:

A) Define each item applicable to overhead and the expected cost.

B) Apply a percentage based on historical information and adjusted to reflect the current estimate. Also consider contract size and duration, the company total contract volume, and expected costs for additional staff requirements for a specific project.

Method B is the less desirable of the two methods.

Profit

Profit is the amount added to the estimate/bid at the general and subcontractor levels for financial gain from the use of capital.

Fixed Fees

A fixed fee is anticipated profit stated as pre-determined amount agreed upon by the parties to a contract.

Sales or Use Taxes

A sales or use tax is a legal assessment by state, county, municipal, and other governmental entities for operating capital. This is usually a material tax, and in some instances, equipment tax. There are certain states that assess this tax on the contract amount. The estimator must be aware of the assessment requirements for the site of the project.

Payment and Performance Bonds

Payment and performance bonds are surety company documents that guarantee (performance) contract completion. They also guarantee payment for all proper claims.

Escalation

Escalation is an increase in cost of goods and services. Treat deflation or de-escalation conditions the same as you would escalation or inflation.

Contingency

Contingency or estimating contingency is additional labor and material cost for items of work possibly required but unclear in the plans and specifications. Reduce contingencies applied to Estimate Levels One and Two as the plans and specifications develop.

Indirect Impact Items

Some of these items are consistent from one estimate/bid to another but most are unique to specific projects. The estimator must rate the indirect impact items to determine their effect on a specific estimate/bid. Listed below are some examples of indirect impact items needing attention during the estimating/bidding procedure. The estimator should create a checklist of these and other items for each project with justification for inclusion/exclusion in the estimate/bid.

1) Dates of all union contract renewals that will occur during the project. Contract negotiations may result in work slow-downs or strikes. Even when the company is signatory to a union contract, slow-downs and strikes of various trades affect the entire project. The estimator should research any labor contract conditions that may occur during the construction and adjust the estimate/bid accordingly.

2) Seasonal weather conditions typical for the locale of the project. Rely on historical data in adjusting this item.

3) Present and forecast economic conditions of the construction industry in the region of the project. This is by far the broadest item of consideration since its effects are extreme and in opposing directions. The law of supply and demand causes variations (upward or downward) of the cost of materials and availability of labor.

4) The volume of construction projects in progress and planned for construction is another consideration. When the volume is small, profit margins decrease due to increased competition for the work, but does insure an available work force. When the volume increases, profit margins increase, but the availability of a qualified work force diminishes.

5) Possible delays in progress payments due to cash flow problems, deliberate or accidental.

6) Proposed federal, state, or local wage determinations that may become effective during construction.

7) Seasonal variations in the available work force and legal holidays that hinder or delay the progress of the project.

8) Other construction projects in progress or starting during construction that are in the immediate vicinity. Determine whether these projects will interfere with accessibility to the project by the work force and for delivery of materials and equipment.

Total Estimated Contract (Construction) Cost (TECC)

TECC is the combined total amount for these costs.

- ✓ Direct costs
- ✓ Performance bonding
- ✓ General conditions
- ✓ Escalation
- ✓ Other costs
- ✓ Contingency amounts
- ✓ Contractor overhead and profit
- ✓ Applicable taxes
- ✓ Fixed fees
- ✓

Total Estimated Project Costs (TEPC)

TEPC is the combined total amount for support costs such as:

- ✓ Construction management
- ✓ Telecommunication equipment
- ✓ Architectural/engineering design
- ✓ Surveys
- ✓ Inspection
- ✓ Testing
- ✓ Movable furniture
- ✓ Move-in costs
- ✓ Fixtures
- ✓ Planning/in-house support
- ✓ Equipment
- ✓ Other expenses

These are in addition to the "Total Estimated Contract (Construction) Cost (TECC)" stated above. Include these costs in the estimate at the discretion of and in the owner's format. The estimator should state the nature of the estimate as total estimated contract (construction) cost or total estimated project cost.

Back-up

Submit cost estimates with adequate back-up materials showing the basis for calculating the estimated cost. This shall never violate confidential information considered as proprietary.

Guidelines

Base cost estimates prepared under these guidelines on the estimator's best judgment using project information provided. These estimates are independent of other budgets or estimates provided for information purposes.

SECTION – 13 CONTINGENCY

There are a wide variety of contingency applications and the following definitions are an attempt to indicate contingency items which are most appropriate to our application. Estimates may be used and reviewed by a wide audience, including owners, developers, accountants, lenders, etc. and each of these entities needs to understand and include contingency for that part of the project for which they are responsible. As estimates are refined and more and finite information becomes available the contingency should be reduced to reflect the completeness of the documents.

Estimating and contracting contingency

This contingency is used for estimating accuracy based on quantities assumed or measured, unanticipated market conditions, scheduling delays and acceleration issues, lack of bidding competition, subcontractor defaults, and interfacing omissions between various work categories. This contingency may be reduced as bids, subcontracts and bonds are received. (This contingency should be for the use of the estimator/project manager and should not be modified or reduced except by the estimator's/project manager's concurrence, not the owner or others.)

The following additional contingencies are not a part of a typical construction estimate but should be discussed and the appropriate parties made aware of the necessity of contingency for the portion of the project for which they are responsible.

Design Contingency

This is used for increased scope due to additional space, added quantities of materials, enhanced quality of materials, code revisions, owner requested items, additional equipment and/or capacity. This contingency is for items during the design process and should be reduced to 0% at construction documents.

Owner's Construction Contingency (Change Order Contingency)

This is used for architect/engineer constructability issues, code interpretations, scheduling delays and acceleration issues, unanticipated conditions underground and owner requested items and is used for items during the construction phase of the project.

Owner's Project Contingency

This is used for additional amounts required for other items in the total budget. This contingency is for items beyond construction related work such as FF&E, moving/relocation allowances for tenants or other non-construction related items.

As you can see each contingency is established and defined for a specific use and purpose and each one is controlled by a different entity. The point is that if a contingency is established for the entire project each time funds need to be reallocated from the contingency getting the funds reallocated can be a divisive, acrimonious, and contentious battle between the entire team.

ACCURACY OF ESTIMATES

A Consultant's View

As a Cost Consultant I am frequently asked by an architect or owner "What degree of accuracy do you guarantee on your estimate?" or "Percentage wise, how close will your estimate be to the bids?" I tell them there is no guarantee; it is my best judgment of cost based on the available information.

I like to tell them about the $103 million estimate where the bids came in at $103.3 million or the $7 million estimate for a renovation/addition project that bid at $7,065,000. There are plenty of success stories to be told but once in a while the bids may be significantly above or below the estimate.

Several years ago I received a bulk mailed advertisement from an estimating service that included a statement that their estimates were guaranteed within two percent. It didn't say what level of estimate this was for nor explain what would happen if the difference were greater. Could they really feel confident to give that guarantee on an estimate from schematic or fifty percent design development plans? It seemed to me that such a statement could be an invitation for more trouble than for marketing advantage.

Thinking about the spread on bids from contractors, even on hard bid projects with completed plans and specifications, I went through the bid tabulations for twenty-five competitively bid general contract projects. Each project had a minimum of four bidders and some as many as ten. On some projects the difference between the low and the high bids was as little as one-half percent (.5%) and on others as much as thirty percent (30%). The average difference for all the projects was seventeen percent (17%).

To confirm the research of several years ago, I went through the same exercise on a dozen projects from the last twelve months. These competitively bid projects had a minimum of three bidders and as many as nine. The lowest spread on these projects was three percent (3%) and the greatest was thirty-five percent (35%). The average difference was sixteen percent (16%).

To me the above surveys say that if general contractors, bidding with the same documents, can't get closer than a 15–20% spread, it is unrealistic to guarantee an estimate to be accurate within a specific, small percentage. A "Murphy's Law"-type axiom for estimating is:

> The same work under the same conditions will be estimated differently by ten different estimators or by one estimator ten times.

Another says:

> It is an *estimate* not an *exactimate*

Estimating is both scientific and an art form. Scientifically we are methodical, meticulous, organized and orderly in our takeoff and pricing of the project as we see it depicted. The art comes in being able to forecast what the economy, market conditions, competitiveness will be in six months or a year when the project is bid. There are times when the estimate is not well received by the architect who wants to give a greater project than the owner can afford, or, the owner wants a monument and suddenly realizes he doesn't have the resources to afford the dream. Tactfulness and thick skin are additional attributes of a cost consultant.

The DPIC's Risk Management Handbook for Architects and Engineers advises that the term "Opinion of Probable Cost" be used rather than "Estimate". The term Estimate carries a connotation of greater degree of accuracy than an Opinion. Of course DPIC's comments are being directed to design professionals whose expertise is not in the field of cost consulting. All too often designers opt not to hire a competent cost consultant for each project, instead relying on historic data and square foot costs to arrive at a projected cost for this unique venture. The result really is an *opinion* of cost, not an estimate.

From a marketing standpoint, Cost Consultants want to stress the expertise contained within the estimating process and the accuracy of previous projects. From a legal standpoint we need to attach disclaimers to the document to protect against liability issues. On the one hand we tell our client how good we are, on the other we say we aren't responsible if there is a cost problem. The key to achieving this balance is through educating your client, whether they are the designer or owner. Impress upon them that the accuracy of an estimate is in direct proportion to the written or verbal descriptions of the project that you are given. Remind them of the risk attached to lengthy time projections. Involve them in the estimating process where possible, define your assumptions and request the client's agreement before you present a price for the project.

Managing the client may be as important as reporting the cost.

Submitted by: Larry L. Cockrum, CPE
Memphis, Chapter 62
February 1999

An opposing view

As professionals, our clients pay us for our knowledge and accuracy. If we don't sell accuracy in our estimates, then what are they buying? At Boyken International, Inc. we set a commitment to every client to be within 3% of the low bid on every project. At that level, the client can make decisions about their project and what courses of action to take. It also places a higher requirement of communication that is necessary between the estimator and the design team. That communication is in the form of information obtained, allowances created and most importantly, the information conveyed in our reports. When our clients understand what is in the numbers, they more often than not, design to the estimate.

Are we successful to meet our commitment every time? No, but over the past nine years we have had over 2,500 project bids and we have been outside our commitment less than 25 times. As a professional predicting the cost of future work, our clients can relyon our accuracy. In fact, that is what all Cost Consultants should be selling, not excuses on how others cannot be right all the time.

BOYKEN INTERNATIONAL, INC.

Donald R. Boyken, CCC
Chairman & CEO

SECTION 14 – BID DAY PROCEDURES

Constructors

The success of a general contractor's business in a competitive bidding situation lies in the completeness of the estimate on bid day. This involves a good degree of risk management and opportunity for the general contractor. The purpose here is to present a procedure developed over a period of years. It is an efficient and effective method for assuring a complete estimate at bid opening. The procedure relies on oral and written communication among the estimating team members. These people are the players who determine the outcome of the bid day procedures.

Today's construction estimating procedures evolved from a rich history of experience in competitive bidding. The traditional trial and error process of estimating has been costly for many contractors. Working late the night before bid day is not the answer either. The amount of time during bid day does not change However, the intensity in which one performs his function increase during various periods of bid day. Estimators must be well organized on bid day to insure a complete estimate with reduced stress and emotional behavior. Each firm is unique in its techniques of operation. However, the basic mechanics of bid preparation are similar among contractors with some slight variation.

An estimating procedure developed over a period of years is an efficient and effective method for insuring a complete estimate at bid opening. It involves the mechanics of bid preparation and the techniques of operation.

All the members of the estimating team should know the mechanics of bid preparation. Each person has a specific function on bid day. Every participant must be aware of the whole picture for the group to perform effectively as an estimating team.

The First Step

The first step is to determine the job functions of each team member. The estimating team consists of the:

- ✓ Receptionist who directs the incoming calls to the bid takers.

- ✓ Bid takers who write down the information on telephone quotation sheets and deposit them into a bid box.

- ✓ Risk manager who analyzes the telephone quotations, enters the proper information on the estimating spreadsheet. The risk manager indexes the telephone quotations for quick reference during the bid process.

- ✓ Estimator who is available to the risk manager for questions, quantity takeoffs, clarification calls to subcontractors and suppliers.

- ✓ Computer operator who enters information into a computerized estimating spreadsheet as directed by the risk manager.

- ✓ Bid runner who takes the signed bid documents to the bid opening. The runner also calls the office to receive and transcribe onto the bid form the final bid amount and any other data. The runner also reports information back to the risk manager.

- ✓ Top management that interfaces directly with the risk manager to decide the project fee.

Combine or expand the various job functions of the estimating team members described above as necessary. For smaller or simpler estimates, the receptionist and the bid-takers still exist. However, the estimator may do some or all of the risk manager's function. For large or more complex estimates, expand the estimating team in the area of bid taking and bid analysis. For example, separate bid takers with discipline-specific knowledge, may handle Divisions 2 (Site Construction), 5 (Metals) 9 (Finishes), 15 (Mechanical), and 16 (Electrical). The bid takers can perform a comprehensive analysis or comparison. The resulting low subcontractor or material supplier bid is then given to the risk manager for final review.

The job functions may vary but there must an adequate number of team members to prevent a chaotic condition. This is usually a decision of the risk manager who is the estimating team captain.

The Second Step

The second step in setting up bid day procedures is making sure the physical set up meets the needs of the team. Items to consider include:

- ✓ adequate size telephone system to handle the rush of incoming calls, faxes, etc., during bid closeout.

- ✓ office layout sized for the estimating team to function effectively. The location of the bid takers, estimators, and the risk manager is important. Other considerations are locations of the telephones, calculators, computers, and each other.

- ✓ a central place to assemble the estimate. This is for completing the estimating spreadsheet. It also is the location of the bid box for receiving the quotations. A coordination station exists for referencing the quotations during the bid process.

The telephone system is an essential ingredient because receipt of most quotations from subcontractors is by telephone. The number of incoming lines is an important judgment call. There may be other business activity during the day. The number of lines may be more than the number of bid takers. There should be a minimum of incoming lines equal to the number of bid takers. Add one for the outgoing calls and one for the fax machine.

Organize the office differently on bid day to help the estimating team. The basic scheme should consist of a dedicated room called the bid room. This is the room in which the estimating team interacts throughout bid day. The bid room must be quiet so a person can think clearly without distraction. Therefore, it is preferable to have the bid takers in separate office spaces nearby.

The bid takers deposit their completed telephone quotation forms into the bid box. The risk manager analyzes each telephone quotation taken from the bid box.

Set up a coordination station for organizing the telephone quotations into the 16 Construction Specification Institute Divisions. Divide these divisions even further into sections of work. For instance, Division 9 consists of various finish trades, such as plaster, drywall, carpet, painting, etc. Segregating these trade groups will aid the risk manager in comparing the telephone quotation forms within each trade group.

This coordination station may consist of a file cabinet or mailbox slots. The purpose of the correlation station is for quick reference to the telephone quotation forms. The risk manager must quickly find the information and file the information as needed throughout bid day.

The Third Step

The third step in setting up successful procedures is the use of a specialized set of preprinted telephone quotation sheets. Every construction company has their own unique format. The basic information such as firm name, telephone number, and date are on the forms. Also listed is the work description. Inclusions, exclusions, and bid amount. The completeness of the telephone quotation depends on the form and how effective the bid takers are at filling in the form.

There should be a checklist of clarification items on the telephone quotations to give the risk manager a better understanding in the analysis. This checklist also can serve as a communication tool to the individual writing the subcontract as well as the people constructing the project. The estimator or risk manager usually develops the checklist. An example of a partial checklist is:

Division 8 - Doors and Frames

Included	Excluded	
_____	_____	Frames/Welded/Knocked Down/Gauge
_____	_____	Doors/Metal/Wood/Plastic Laminate
_____	_____	Prepared/Prehung
_____	_____	Paint/Stained/Prefinished
_____	_____	Fire Rated/20/Min/Hr/Other
_____	_____	Door Vents/Door Hinges
_____	_____	Door Lights/Side Lights
_____	_____	Access Doors
_____	_____	Gate

During the bidding process, estimators must analyze every bid used on the spreadsheet. The bid must pass two tests:

1) Is the bid per plans and specifications without unauthorized substitutions?

2) Is the bid covering their section complete?

Section 14 – Bid Day Procedures

It is necessary to note that some bidders will answer yes to questions. There is the mental reservation that they will supply the items they "normally" cover by the specifications. They fail to state specifically what items from the plans and specifications that are not included.

The fax machine is becoming a much used substitute for phoned bids and the fax bid presents additional problems. The bidder will often list the scope of work but not in an acceptable fashion. They want to list only the items of work they are willing to perform for a stipulated price. General contractors want a section complete bid or a bid with specific exclusions. On each faxed bid ask the questions: "Is this section complete per plans and specifications?" Then attach the question sheet or spreadsheet to their faxed bid and call them to get their answers.

Abstracts and Scope Letters

The purpose of an abstract or scope letter is to inform the contractor of the intent of the bid. This allows the general contractor to be sure of subcontractor bid completeness and the subcontractor has a form in which to educate the general contractor. A good scope letter will include four categories:

1) Items normally covered by this trade.

2) Items specifically covered in this bid.

3) Items normally excluded by this trade.

4) and items specifically excluded in this bid.

A section complete bid will have no exclusions but might have some clarifications such as "no taxes included."

Questions for Bidders

With the information from the abstracts and scope letters, the general contractor should produce a set of questions. Have the subcontractors answer the questions when they phone in their bids. Sort the questions by CSI number for easy access. They should contain a mini scope of work and acceptable exclusions for any section that needs proof. Include questions developed by subcontractors not answered by addenda. Structure the questions for a "yes" or "no" answer thus avoiding ambiguous answers.

Read the questions to the bidder and rewrite on the telephone bid sheet in short form with "yes" or "no" answer. When there are many questions, make copies of the question sheets and attach the sheet with the marked answers to the telephone quote sheet. This procedure will provide the chief estimator with the information necessary for a quick and accurate analysis. The fax machine is an excellent tool for talking to subcontractors even as late as bid day. Use the same question sheet, have the subcontractor fax it back, marked appropriately. Through the questioning process, the estimator receives the information needed to rate the bids of a given trade. The answers may even stimulate further questions. These questions should avoid double coverage or lack of coverage. Also, the process will prompt subcontractors to note inclusions, exclusions and clarifications in their bids to other general contractors. This process will cause the other contractors to improve their cost estimates. This removes any unfair advantage they may have due to ignorance.

A problem may arise after bid day when a subcontractor says that he did not include an item that may be questionable. Consider these questionable items as holes in the estimate when the subcontractor refuses to accept them as part of their scope of work. The checklists on the telephone quotation sheets address many of the questionable items that will clarify any misunderstanding in the scope of the work. Copies of prepared telephone quotation checklists are available by calling. The use of a scope letter will remove this problem.

Once the estimating team learns the mechanics of bid preparation then the group can start to perform effectively. The next step is to learn the techniques of operation to assure a complete estimate at bid opening.

There are essential techniques of operation that the estimating team must be aware of and focus on to assure a complete estimate at bid opening.

The first technique is to have the pre-bid day information ready on bid day. Create a spreadsheet of all cost items of the bid. Prepare the spreadsheet a few days in advance of bid day. The estimator should review the index of the project manual and list each section with its corresponding CSI number. Next, the estimator reads each section to see what is in that section. The estimator may recognize items assigned to a certain specification section not usually done by that particular trade. An example is roofing. The roofers may exclude metal and wood requirements. The estimator must confirm that these items are included by other trades such as sheet metal or roofing specialties. Create a scope of work for each section and list acceptable exclusions. Ask bidders to break out the cost of certain items in their bid. Develop questions for that trade to ask at bid time. Cross reference them into any other trade that might pick up the excluded or break out items.

Step One

Read each section and list any items that are subject to exclusion by subcontractors bidding that section. List these items as separate line items on the spreadsheet. Also prepare a number in case there is no bid for that item on bid day. Write the questions. List the acceptable exclusions. (These extra line items will be scratched at bid time if they are covered on another line.)

Items not listed in specification book: Write a list of items not mentioned in that specification section required for a complete job. Question the subcontractors at bid time about these items.

Step Two

Add items not mentioned in the specification section to the spreadsheet. During the estimating process of company performed work, make notes of special items for this project. Check the specifications for a listing of these special items.

Step Three

Add items to the spreadsheet found in the drawings and not found in the specifications. Next examine the vague areas. Find items on the plans that are either not noted with size or nomenclature or are not in the specifications. However, experience may show these items are necessary to complete the structure. Question the design professional in writing and demand they answer in an addendum.

Step Four

Add items to the spreadsheet that are not on the drawings or found in the specifications necessary for a complete job. Advise the design professional with written questions or the estimator's interpretation of the plans and specifications. Also state this is the basis of the bid if they choose not to answer in an addendum.

Step Five

Add items to the spreadsheet needed for compliance to code. Mention of these codes may be on the plan or in the specifications which satisfies all applicable codes.

There may be a requirement to build a complete functioning project "to code." The design professional may not identify all code requirements. They will not allow an extra and require the contractor to furnish everything to satisfy the codes. This is usually at no increase in contract amount. Again, write questions for the subcontractors and ask the design professional for clarifications of undefined terms.

The following are some suggestions to locate those items which may not be covered:

- ✓ Use professionals in each discipline. Have them analyze the sections to determine if the specified materials will operate as a complete system.

- ✓ Check items missed on previous jobs using historical data. Have the project reviewed by other experienced personnel.

- ✓ Review the drawings and implications of all notes on the plans.

- ✓ Complete an analysis of the construction schedule, take off, and pricing of work by contractor's own forces. Analyze differences in subcontractor scope of work/quotation sheets received in the mail before bid day, and pricing of general condition items.

- ✓ Prepare a bid document form including the signatures and other required information.

- ✓ Develop bid day strategy of the estimating team, their functions, and communication of any important information. The estimator should prepare a brief written set of instructions for bid-takers which includes:

    a) the number of addenda and the subcontractors affected by each addendum.

    b) the unit price or alternate bid information required from each subcontractor.

    c) a list of questions for the material suppliers and subcontractors at bid time.

The estimator usually has the responsibility of preparing the above documentation. However, other company departments may share these tasks. For instance, the prospective project manager and superintendent may provide input in the analysis. The legal and/or accounting department may review the contract documents for bonding, insurance, and liability requirements. These departments must evaluate the risk factors prior to bid day.

The second technique of operation is to communicate effectively early in the bid day process. Some of the key items for maximizing efficiency include developing a system for:

- ✓ the receptionist to channel calls to the next available bid-taker without disruption or overload.

- ✓ continuing the bid process during mealtime. Discuss the meal orders and meal times early in the day.

- ✓ each estimating team player to be available on bid day to concentrate only on the estimating process. It is essential to prioritize one's time on bid day in order to become an effective team player.

- ✓ handling visitors or questions from subcontractors and material suppliers during the bid day process.

The flow of communication starts with the call screening process by controlling disruptions by other business calls. An intercom system helps when there are calls waiting. Get the telephone number of the caller waiting because there may not be another opportunity to get the quotation if the caller hangs up. Handling surprises on bid day, such as the bidders who want to look at the plans, in such a way that the estimating process is not jeopardized.

The third technique of operation is to develop organizational cohesiveness within the estimating team. Some of the guidelines include:

- ✓ Work together as professionals.

- ✓ Communicate openly without unnecessary conversation.

- ✓ Write legibly and communicate clearly.

A positive approach in developing estimating team cohesiveness is for each member to be supportive of each other for the success of the bid. The discipline of the estimating team should be responsive to the needs of the estimator and the risk manager. Written or oral information must be concise and understandable. Unnecessary conversation in the bid room can cause a loss of concentration, jeopardizing the potential success of the entire bid day operation. Working together is being considerate of each other.

The implementation of these techniques of operation will help determine the outcome of a complete bid package. However, there are factors that put these techniques of operation to the test. The estimator has the important task of preparing the estimate before bid day. There may already be an estimating team set up and working on the estimate before bid day. In any event, assemble, review, and have the pre-bid day information ready when bid day arrives. Otherwise there will be more stress on the estimating team.

Bid day is a compressed period in which all the various parts of the estimate come together. The compilation of the estimate becomes a race against time. The fewer the number of members on the estimating team, the more difficult it becomes to complete the bid without holes or overlaps. The intensity of time on bid day is the basic reason for having enough help on that day.

Bid day can be a stressful time when there exists many callers and few bid-takers. It puts a burden on the estimator and the risk manager when callers wait to the closing hours of bid day. One suggestion is to have the bid takers start calling the prospective subcontractors early in the bid day. The bid takers can get each subcontractor's scope of work and possibly an early quotation subject to change later in the bidding process. During the bidding process, estimators must analyze every bid on the spreadsheet to determine the lowest responsive bid. Refer to Step Three of Bid Day Procedures.

The techniques of communication used during the bid process become essential during the final hours of the bid closeout. Good organizational behavior is key to good decision making. The risk manager depends on the full cooperation of the other estimating team members especially during the last closing hour. This involves an organizational cohesiveness within the team built on trust, loyalty, and intestinal fortitude. Otherwise, there is more stress on the estimating team.

The attitude of each team member can reduce or increase stress. Members should be supportive of each other as a group. Listen and take directions from the team captain. Stress will be there on bid day. Recognize the cause of the stress and deal with it effectively before it stifles the estimating process.

After the bidding is over, the estimating team can decompress and discuss the estimate. This post bid period is a good time to re-evaluate the estimate. Confirm the bid analysis, reorganize the telephone quotations, clarify areas of concern, and prepare for the bid results.

There is one last decision that the estimating team makes on bid day. Who will give out the subcontractor bid results? The decision must be consistent with company policy. It is usual to hold all information in confidence until the company receives the final bid results.

The estimating team determines the success of an estimate. A successful estimate is one that is complete. This does not mean that the estimate is the low bid. The success of a general contractor's business in a competitive bidding situation lies in the completeness of the estimate on bid day. The careful implementation of bid day procedures is essential to insuring a complete estimate at bid opening.

American Society of Professional Estimators
Bid Submittal by FAX Machine or E-MAIL

The adherence to a uniform code of practice when submitting and receiving bid proposals transmitted by "Fax Machine" or "E-mail" is required to promote proper communication among the parties involved in "Bid Day."

Prime Contractor Responsibilities

1. Maintain the fax machine & computer in good working condition with proper supplies.
2. Monitor the fax machine & e-mail and promptly deliver proposals to the responsible individuals.
3. Keep all proposals confidential.
4. Notify sub-bidder if the transmission is incomplete or not legible.
5. Request all transmissions be completed at least two hours prior to the time prime bids are due.
6. Do not interfere in any way with another firm's ability to send or receive transmission (such as jamming or keeping their fax line busy).

Subcontract or Vendor Bidder Responsibilities

1. Use standard transmittal forms and filled out bid forms including alternate showing required bid information.
2. Fax or e-mail scope and terms a minimum of 24 hours prior to bid deadline.
3. Transmit only modifications to scope and/or price on bid day (Do not retransmit entire proposal).
4. Verify receipt of transmissions.
5. Consider faxed or e-mailed bids a supplement to, not a substitute for, verbal and written communication.
6. Send a written confirmation of all transmitted bids.
7. Refrain from transmitting scope and/or price modifications during the last two hours prior to bid time. Make all late modifications by verbal communication.
8. Insure that transmitted bids:
   8.1 Show bidder's name, address, phone number, etc.
   8.2 Show project name, bid date & bid time.
   8.3 List technical specification sections and addenda included.
   8.4 Clearly list exclusions and/or exceptions of the bid.
   8.5 State alternate bids as adds or deducts from the base bid and number them consistently with the bid documents.
   8.6 Note terms of proposal (F.O.B. Jobsite, furnished and installed, installation labor only, etc.).
   8.7 Note if bid includes or excludes sales or use taxes.
   8.8 Note if bid includes or excludes performance and payment bond premiums.
   8.9 List any unique contractor status (MBE, SBE, WBE, Contractors License number(s) and bond rate).

Section 14 – Bid Day Procedures

The proper use of fax machines & e-mail can improve the accuracy and efficiency of bid day communications. Whether typed or handwritten, the fax or e-mail must be legible.

For further information, contact:

The American Society of Professional Estimators
11141 Georgia Avenue Suite 412
Wheaton, Maryland 20902
Telephone (301) 929-8848
Telefax (301) 929-0231

Sample Fax Transmittal

BIDDER'S LETTERHEAD W/TELEPHONE # AND MAILING ADDRESS

FROM: FAX #

TO: FAX #

TO:

ATTENTION:

PROPOSAL FOR:
         (TRADE OR MATERIAL)

PROJECT:

TRANSMISSION INCLUDES _____ PAGES (INCLUDING THIS PAGE)

IF ALL PAGES ARE NOT RECEIVED, PLEASE CALL _____ @ (VOICE TELEPNONE #)

REMARKS:

American Society of Professional Estimators

Bid Shopping and Bid Peddling

The American Society of Professional Estimators in its Code of Ethics has stated that bid shopping and bid peddling are unethical and are not to be practiced by members of the Society.

Bid Shopping

Bid shopping, defined in Canon 5 of the Code, occurs "when, after the award of the contract, a contractor contacts several subcontractors of the same discipline in an effort to reduce the previously quoted price."

In other words, if a prime bidder attempts to compel a sub-bidder to lower a previously quoted bid price, that is bid shopping. Bid shopping may occur either on bid day or after bid day; either before or after the award of the contract.

In addition to price information, the status of a sub-bidder's competitive position or technical scope are equally sensitive. Legitimate practice precludes use of this information in haggling, trickery, or coercion of any kind. During contract negotiation, sub-bidders should not be advised, nor should they inquire, of the other bidder's price or scope, nor of any changes that would be required to qualify them as the successful sub-bidder. After sub-bidder commitments are made, or within a reasonable time after prime contract award, sub-bidders should request, and should be advised of their competitive position, both in price and scope.

Owners Participation

Bid shopping is not confined to prime bidders and sub-bidders. Some owners also participate by encouraging prime bidders to bid shop and by bid shopping themselves. Ethical contractors will propose value engineering to lower their bid. They will not engage in bid shopping. Owners will protect the confidentiality of prime bidder's value engineering ideas and all other pre-award submissions.

Bid Peddling

Bid peddling, defined in Canon 7 of the Code of Ethics, occurs when a sub-bidder "approaches a general contractor who has been awarded a project with the intent of voluntarily lowering the original price below the price level established on bid day. This action implies that the subcontractor's original price was either padded or incorrect."

When a sub-bidder lowers a price to get closer to or below the legitimate price, that is bid peddling. Whether or not the low prime bidder has a contract, whether or not the action occurs before or after bid time, it is bid peddling.

Comparable Price

"Price" here means "comparable price," the price which accurately reflects to the prime bidder a scope of work comparable to the other sub-bidders in that trade. It is the prime bidder's responsibility to understand the complete scope of work being bid by the sub-bidder, and to determine the value of adjustments to a sub-bidder's price which must be made to compare it with other prices. In this way, sub-bidder prices are judged "apples to apples."

When negotiating a contract, it is the sub-bidder's responsibility to provide accurate prices for legitimate scope additions and deletions, where necessary,—not to use such pricing as an opportunity to bid peddle.

Why ASPE Prohibits These Activities

The contract (or sub-contract, or purchase order) should go to the qualified prime bidder or sub-bidder determined on bid day at bid time, excluding prime bidders or sub-bidders who shopped or peddled bids prior to bid time. This does not preclude a prime bidder from using a bid higher than the low legitimate bid, but the prime bidder cannot ethically ask the sub-bidder to lower a price as quoted on bid day. This prohibition includes requesting the sub-bidder to perform added work for the price quoted on bid day.

The Ethical Dimension

The ethical basis for this stand is free competition and fair play. The competitive prime bidder assumes the low sub-bidder has carefully quantified the scope of work, has evaluated his risk and pricing options, has included a fee which will justify the risk, offering the best price in confidence. To shop such a price is neither free competition nor fair to the legitimate sub-bidder. For a sub-bidder to peddle his own price is similarly neither free competition nor fair play to the legitimate bidder.

Many construction firms fall prey to the practice of bid shopping or bid peddling in the belief that they will procure contracts not otherwise available to them, or that peddling will allow their firm to maximize profits on the project being bid. In the short run, this may indeed be true. But, in the long run, shoppers and peddlers gain reputations, and soon find it more difficult to obtain legitimate bids. This lack of legitimate bids causes the bidder to "discount" even more, because only those sub-bidders who put "shopping money" in their bids are available. This added risk may be disastrous for the bidder, should he be unable to "sell" the work for his discounted price. Skill and insight are replaced by gambling and often greed. Professionalism is replaced by rolling the dice, and bid shoppers are gradually isolated and change or perish.

The Economic Dimension

In addition to the ethical dimension of bid peddling and bid shopping, there is an economic one. Simply stated, it is this: Bid shopping and bid peddling reduce the total profit available to the construction team. When a bidder cuts a bid below the lowest legitimate bid, the bidder is admittedly taking the contract for less than originally desired and bid. The bidder is, in other words, reducing profit below what is really desired, and is doing so in order to get the job.

No contractor enjoys the prospect of making less profit than desired. Therefore, the shopper has strong incentive to develop ways to recoup that lost profit. One of these ways is to cheapen quality. The shopper may not use the specified material and/or allows workmanship to suffer in order to gain back the profit lost.

Another way is to search for opportunities to increase the amount of one's contract through extras. The bidder is constantly motivated to seek change orders, often pricing them at substantial premiums above the actual cost of the work done. In either of the two scenarios, conflict is sure to result, and legal issues arise. The original fee is lost or reduced by discounting, and the added burden of legal fees to resolve the ensuing conflict is inevitable.

Elimination of bid shopping and bid peddling is essential if the construction industry is to regain its rightful fee structure, and it must begin to eliminate these unethical practices now. For information on how you may help, contact:

American Society of Professional Estimators

11141 Georgia Avenue, Suite 412

Wheaton, Maryland 20902

Telephone: (301) 929-8848

Telefax: (301) 929-0231

Dedicated to serving the construction estimator and the construction industry.

SECTION 15 – PRESENTATION

Guideline

The estimator shall be responsible for providing a Narrative of the Level and Scope of the Estimate/Bid Definition of the Level of the Estimate shall be according to Part One, Section Two of Standard Estimating Practice. Development of the architectural and engineering sections of work may be at various levels. Individually identify the varying levels of these sections in a status evaluation. The evaluation defines the approximate percentages of data available to determine the range of specific or overall contingency for the estimate. The narrative also should include:

- ✓ Date of receipt of plans and specifications
- ✓ Date and time estimate is due
- ✓ Project address, legal description
- ✓ Project description
- ✓ Architect/engineer's name & address
- ✓ Owner's name and address
- ✓ Brief description of the project type
- ✓ Any addenda issued
- ✓ Project quality, size, labor type
- ✓ List of assumptions used

Include this information to better define the project and help in the development of historical records of estimated and bid cost systems.

Example Level 5 Narrative

Estimate Basis

This estimate is a Level 5 - Construction Drawing according to Standard Estimating Practice of the American Society of Professional Estimators. See Part One Section Two and Three, Levels of the Estimate and Scope of the Estimate for additional information.

Per your request, this estimate was prepared to show reasonable expected construction costs for the above project. The expected construction costs reflect the average of the competitive bids for the work. Calculate the average bid (for estimate comparisons only) by using the following procedure: drop the low and high bids and calculate the average or median of the remaining bids.

We base the estimate on information available within the construction documents. The plans are dated _____ and the specifications are dated _____. These documents represent the development of the project to date. The design professional team provided the documents.

The Construction Specification Institute Sixteen Division Format is used to organize estimate information. We will segregate the estimate information into a Work Breakdown Structure (WBS) upon prior request. This provides a special basis for estimating the constructions costs.

Recommendations

The estimate is prepared from the specifications and contract drawings. Additional definition of project requirements will affect the total estimated project costs. EXERCISE CAUTION in using these figures. Design, owner, and regulatory requirement changes can cause costs to vary significantly from those contained in this estimate. Use this estimate to determine bid alternates, if any, in checking bids, preparing schedules or other project control applications.

General Assumptions

Competitive Bidding - This project is subject to competitive bidding with four to six prime contractors submitting bids. Bids outside this range may cause significant differences between estimate amounts and actual bid amounts.

This estimate is based on the assumption that the prime contractor submitting the successful bid will furnish _____ with his own forces. Subcontractors will perform the balance of the work. The anticipated subcontractor overhead and profit margin percentages are included in the division/subdivision schedule.

Quantity Survey - Quantities are from direct takeoff of items, when possible, according to ASPE recommended Standard Estimating Practice. The construction drawings are the basis for determining these quantities.

Materials

Materials are from project area supplier price lists or established reference sources.

Labor

The estimate includes non-union labor rates throughout. Labor rates are per the cost index schedule. Costs for full-time supervision are in the general conditions. Federal government funding requiring the use of prevailing wage labor rates and certified payroll records do not affect this project.

The wage rates for the project are from sources in the area of the project. We add to this, fringe benefits and labor burdens at the subdivision direct cost level for payroll taxes and insurance. We also include estimated craft labor settlements into the labor rates for the overall base wage period.

The labor used within the estimate is our opinion of what it will take to perform the work as shown and described. The size and type of project as it relates to or experience provides the basis for this estimate.

We specifically exclude:

Permits, fees, and taxes.

We specifically include:

Applicable costs, if any, for sewer connection and user fees

Applicable state and city sales or use taxes

Testing costs

Contractor Overhead and Profit

Subcontractor overhead and profit costs are included in each respective section. General conditions costs are a direct cost item as opposed to an overhead cost. For the prime contractor, overhead and profit is in the final account summary report.

Escalation and Contingency

Escalation, or cost growth, and estimating contingency are included in the final account summary report. Escalation is calculated at an annual rate of _____%. Contingency is calculated at _____%.

Other Costs to the Owner

We exclude certain other costs to the owner from the construction estimate. These may include acquisition costs, design fees, finance expenses, administration expenses, furniture, fixtures, and equipment.

Estimate Summary

Refer to Project Total Summary Report for listing of direct costs that include markups. Refer to final account summary report for general contractor direct costs, etc.

Specific Assumptions

Division 1 - GENERAL REQUIREMENTS - The cost for supervision and construction equipment is based on a _____ month construction duration.

ASSUMPTIONS/CLARIFICATIONS for each specific discipline are listed below:

Division 2 - SITE CONSTRUCTION

Division 3 - CONCRETE

Division 4 - MASONRY

Division 5 - METALS

Division 6 - WOOD AND PLASTIC

Division 7 - THERMAL AND MOISTURE PROTECTION

Division 8 - DOORS AND WINDOWS

Division 9 - FINISHES

Division 10 - SPECIALTIES

Division 11 - EQUIPMENT

Division 12 - FURNISHINGS

Division 13 - SPECIAL CONSTRUCTION

Division 14 - CONVEYING SYSTEMS

Division 15 - MECHANICAL

Division 16 - ELECTRICAL

SECTION 16 – POST-BID PROCEDURES

Introduction

Immediately after a competitive bid is submitted, everyone in a general contractor's office will ask the estimator one of the following questions: "How did it go?" "How do you feel about it?" How did we do?" Estimators will have opinions but until the post-bid procedures are complete they cannot be sure that the bid was "good." "Good" does not mean that the bid was low. A low bid may not be in the company's interests if bidding errors erode expected profits. This section discusses post-bid procedure that helps the estimator confirm the effectiveness of bid day procedures.

Bid document requirements and the type of bid opening can affect post-bid procedures. For example, a public bid opening or a requirement for listing of subcontractors may force the estimator to accelerate or change the normal process. The estimator should perform basic post-bid procedures before filing the job away and starting the next one. The following guide addresses post-bid procedures that apply to most situations.

An estimator's first priority after the bid should be to relax. They should take some time to relieve the stress generated by the bidding process. This lets them start the post-bid review with a clear mind.

The estimator should take this time to thank the rest of the estimating team. This includes the receptionist, the bid takers, the bid runner, and top management for their help. Bid day is a team effort and all team members deserve recognition.

As-Bid Analysis

The focus of the period immediately after the bid is on the "as-bid" analysis. The term "as-bid" refers to the estimator's review of the compilation of the project bid. The estimator will recompile the bid to confirm the decisions made on bid day were correct and that the arithmetic was accurate. The lead estimator on the project must take responsibility for this task and personally perform the analysis.

The first step is to put copies of the completed forms, qualifications, summary, and bid bond in a clearly identified file folder or envelope. When the bid summary is on a computer, print it out and place it in the file.

The next step is to collect all the telephone bid forms, scope letters, faxed bids, and add/deduct sheets and organize them by trade. When using a checklist or scope card to analyze subcontractor bids, attach the related paperwork to each checklist.

Place the apparent low bid quotation at the top. Separate bid takers may do the analysis of specific disciplines of work, such as mechanical or electrical. Those individuals must organize the information and clarify their notations before giving the package to the estimator.

The "as-bid" analysis may take a day because the estimators will re-bid the project at their own pace. Carefully review each subcontractor bid to be sure there was a correct analysis on bid day. Estimators may concentrate on the apparent low bidder in each trade, but they also must read the other bids. One reason for this requirement is that occasionally on bid day, a subcontractor qualification or exclusion is misunderstood or missed. Quotations with alternates or extensive breakdowns may not have received proper analysis.

The estimator also may find bids inadequately reviewed at bid time due to time constraints. During the "as-bid" check the arithmetic on each quotation. Compare qualifications or exclusions to the scope the estimator expected for that trade. An unexpected qualification by a third bidder may require review with the apparent low bidder. The estimator must pay particular attention to the analyses performed by the separate bid takers mentioned above. These individuals may be qualified to review bids but depend on the estimator's broader knowledge of the entire project. This may provide a different perspective to the bid analysis.

With the review of each bid, the estimator must show the accurate transfer of information to the bid summary or spreadsheet for the trade. When using a spreadsheet for the analysis of each trade, confirm all entries. Check all arithmetic on the sheet. Upon finding discrepancies, the estimator should note them on a separate piece of paper or on the spreadsheet. Spreadsheet notations should show they occurred after the bid. Enter any bid omissions due to lack of time on bid day, on the spreadsheet with a note about the addition. Erasures or changes to subcontractor bid forms are not appropriate. Do not alter the bid day paper trail. An exception to this rule would be when a telephone bid is not signed or dated by a bid taker. Identify the bid taker and require immediate completion of the form.

Confirm that the initial information on the spreadsheet is correct, then check the accuracy of transfer of numbers to the bid summary. Upon checking the initial transfer, trace all later changes. Use the example of an apparent low price changing twice during the bid. See if the summary matches the spreadsheet. There should be "add" or "deduct" paperwork that tracks the changes.

During the "as-bid," the estimator must pay attention to all notations on the subcontractors' bids. Do they show whether the low subcontractors' bids were per plans and specifications. When estimators knowingly use bids not per plans and specifications, they must recognize the cost exposure.

Maintain a list of suggested substitutions for materials and suppliers. Owner or architect decisions on substitutions can affect subcontractor selection. Ask for cost saving suggestions from subcontractors, since owners may ask for cost saving suggestions after the bid. This is especially true when the bids are over budget. Address value engineering and substitutions as soon as possible to get a decision. The timely completion of the job may depend on the proper processing of these changes at the start.

At the end of the review, the estimator should prepare a new bid summary. Label it "as-bid." This reflects all the potential savings or losses identified by the "as-bid" analysis. A loss may occur when the scope of a subcontractor's bid does not match the scope on the spreadsheet. The general contractor may lose money by having to use a higher, complete bid. The analysis also may identify a "hole," or item not covered by any subcontractor. Savings may occur if the estimator finds a "double-up," or item covered twice.

For example, the HVAC and electrical bidders may both include the same motor starters. The estimator also may find purchasing some scopes of work may differ from bid day assumptions. "Plug numbers" in the bid may eventually result in a saving or a loss. The estimator should try to firm up the plug numbers during the as-bid.

Sometimes a subcontractor other than the one used on bid day actually has a more complete number or lower bid. This may be a result of a hectic bid day. There may be receipt of the number without proper recording or recognition. The estimator should note these issues on the spreadsheet as a reminder to improve receipt and recording of information on bid day.

If a contractor is not the low bidder and the estimators cannot discover why from the analysis, they should try to find the answer. Estimators may determine from industry contacts if there were subcontractor bids not received. They may find that the low bidder omitted an item from the bid. The estimator should do some research because the answers may help on the next bid.

Frequently, late bids are hand-delivered or sent by fax machine. Company policy must address the disposition of late bids. The best policy is to disregard late bids. Late telephone bids should not be acceptable. Some contractors consider late bids as those received after than one to two hours before bid time.

The estimator should carefully document the analysis. This makes the subcontracting process easier for the individuals involved with that process. When a project manager awards the subcontracts, the estimator's notes must clearly identify the status of each trade and his recommendations. Before making an award, the project manager should review the final selection with the estimator.

With the issue of subcontractor selection comes the question of when and how to release information to the subcontractors. Keep subcontractor bid results in confidence until the final general contractor results are available. If there is a public bid opening, the unsuccessful bidders may release results immediately. The successful bidder will want to complete the "as-bid" analysis before releasing information.

Regardless of the status of the bid results, direct all questions from subcontractors to the estimator. Until estimators complete the analysis, they should advise subcontractors that the bid result tabulation is incomplete. The receptionist also should remind the subcontractors to confirm their bids in writing. The estimator's goal should be to have all apparent low bids confirmed in writing before completing the as-bid review. Occasionally, the confirmation alters the information received at bid time. When this happens, the estimator (and bid taker, if applicable) must resolve the discrepancy immediately. The estimator should respond with a letter of record to identify a potential problem.

Company policy must be clear on who releases bid results and when to do it. There is always pressure by subcontractors on the bid takers to provide information. In spite of this, members of the estimating team must direct the calls to the estimator. Simple comments like, "you're in the running," may lead to misunderstandings.

There is always room for personal judgment. However, the best way to let subcontractors know that everyone receives fair treatment is to have clear company policies and procedures.

When a contractor is the successful bidder, the estimator or project manager usually release bid results. This depends on who awards the subcontracts. Notification of subcontractor standings should follow the award of the subcontract. Complete the subcontract awards promptly, usually within thirty days. This is for the contractor's benefit as well as the subcontractors. Subcontractors need the feedback so they can adjust for the next job. When a subcontractor is out of contention, the estimator may choose to convey this information early in the process. The estimator should do this without revealing facts about the other bids. Ideally, advise subcontractors of their competitive position as to price and scope. Sometimes the release of exact figures and subcontractor names may create problems.

General contractors may analyze the scope of bids in different ways. A subcontractor may misinterpret exact figures and names if they receive contradictory information from another general contractor. Another option is to give results to a subcontractor on a percentage variance or standing basis, or both. For example, advise the subcontractor that they were the third bidder and four percent high. Handle the process to prevent or discourage bid peddling and bid shopping in the marketplace.

Release bid results over the telephone, by written notification to bidders or by posting the information in the plan room. The method chosen by the estimator depends on the timing of the release and on the time available to put in a publishable format. The procedure should be consistent from job-to-job within the company. The as-bid review provides other important information to the estimator for future bids and for historical purposes.

For example, a competitive bid will frequently identify new subcontractors for the estimator. During the as-bid, a Dictaphone is a convenient device for developing a summary of the subcontractor bidders by trade and by relative bid order. Listing the subcontractor, the contact, and the telephone number provides important information. Use this to update the telephone numbers and contacts on the subcontractor list. Also use this to identify subcontractors for addition to the permanent list. The relative bid order also may provide an indicator of which subcontractors are most competitive at that period. The subcontractor with the second bid may be the one to watch on the next bid.

The "as-bid" will highlight errors that the estimating team may have made during the bid. For example, if the telephone bid forms are incomplete, review the mistakes with the team at a pre-bid meeting before the next bid. When delivery of critical bids to the bid room is late and not addressed at bid time, review and correct the reason.

The "as-bid" also gives the estimator an opportunity to develop historical information. These include unit prices or cost per square foot figures for various trades. Every piece of information can be helpful on the next project when recorded and accessible. The contractor should have a historical cost data base in place for various building types, systems costs, etc. The estimator then assembles information for the data base during the "as-bid" period. This is to provide the company with up-to-date historical information.

When the contractor was not the successful bidder, the estimator may have additional items to take care of after the bid. Collect bid documents from subcontractors for return to the architect or owner for a deposit refund, when required. When there is a log book of deposit checks, the estimating secretary may have the responsibility for this task. Use a tickler file or log book to record bid bonds and return them as necessary. An apparent low bidder should tell their bonding company they will be receiving the contract. An unsuccessful bidder also should contact the bonding company immediately to maintain the contractor's bonding capacity.

When the contractor is the apparent low bidder, use the post-bid period to address scheduling, bonding, and contract conditions with the proposed subcontractors. Resolution of these issues may affect subcontractor selection. The estimator also must note any supplier or subcontractor qualifications that restrict their bid or material pricing to a limited time. The estimator may give preliminary verbal notification or letters of intent to subcontractors before contract award. Use this to tie down prices for materials and to start the submittal process for long lead items.

During the buy out and job set-up period, the estimator should communicate frequently with the project manager. This will assure the most effective buy out of the job. The estimator may request using specific cost accounts to track important costs during the job. At the completion of job buy out, provide a list of the final subcontractors and suppliers to the estimator.

When estimators complete the post-bid procedures, they will know if the bid meets expectations. "Win or lose," they can put that bid aside and start work on the next one. The post-bid procedures will have provided valuable feedback to make operations more effective.

SECTION 17 – ESTIMATING CHANGE ORDERS, COST OR OPPORTUNITY?

The following is reprinted with the permission of Michael B. Carringer, FCPE of FoxCor, Inc., Roland, Arkansas.

A discussion about change orders should properly begin with some thought about why change orders are so prevalent and what, if anything, can be done to help in bringing about an environment in which they are no longer necessary or, are, at worst, minimized. This assumes a philosophical backdrop of construction in heaven, but it may be instructive.

Certain classes of change orders are necessary and beneficial. These involve owner-induced changes to keep the facility under construction as up to date as possible in a technologically changing environment, and which are anticipated by the owner as a natural part of the development of facilities. Those types are ideally priced by a construction team which is also accustomed to building facilities in which change is natural and expected. Both parties, along with a design team of similar experience and expectation, approach such a project dealing with change orders as an everyday part of the process, not something unusual or adversarial.

Such projects are not the norm, however. Often change orders, or potential change orders, are viewed as an intrusion into the normal building process, the responsibility for which must be laid at the doorstep of one of the members of the construction team, who then is expected to pay for "fixing" whatever is wrong. If there must be a change and there is a cost associated with it, the owner will often take the position that the designer did not properly design the facility, having known what the owner's budget is during design, and therefore the owner will not pay for a change. The designer may take the position that the owner did not properly communicate its intent or that the construction team is not acting in good faith by pricing the change at a multiple of what it is really worth. The contractor may believe that the owner and designer are trying to force their errors upon the construction team, and thereby are acting in bad faith themselves.

Let's back out of this picture and examine the big picture of the construction process today. Let's use as a benchmark of our examination a single concept, not that the situation is so simply analyzed, but that this single one may bring the most light to a heated subject. That concept is competition.

We normally think of competition with regard to construction projects in terms of that between members of the construction team. But this is of course not the only competition present in the process at all.

Begin with the owner and ask, who is the owner in competition with? For a privately or publicly owned company, the answer may be obvious—other firms in the same business. For the governmental or institutional organization the answer may not be as obvious, but a bit of thought may reveal that a particular firm or organization is in competition with others in the same organization, for example, branches of government trying to get funding for their own projects, or institutions of higher learning trying to show themselves leaders in the medical or science or humanities field and so competing with other institutions of like purpose. Or consider different departments of the same institution, say the science and athletics departments of a major university, competing for the same facility construction dollars from this year's budget.

It is important to understand that virtually all owners or users of the facilities the construction industry builds are themselves in competition with some other entity for their construction dollar.

Likewise, it is important to see that design firms are in the same type of competitive environment. This may be easier to understand, but it is not always fully appreciated. Perhaps in days of old owners selected design firms solely on the basis of presentation, concept, and experience. This is rarely the case in the present day. While these facets of the selection process are considered, along with them, and perhaps the most important of them, is the design fee proposed. And just as many contractors argue that margins now are less than they were 20 years ago, so will designers argue that design fees are now less than they were 20 years ago.

Both of the arguments may well be true, but the contractor often fails to consider what lower fees mean to the design team. Many designers, in calculating the cost of design for a project have developed, analogous to the contractor's historical cost records, historical cost records for design. The costs are often expressed in a unit like man-hours/page of documents. A designer will also be able to determine fairly closely how many pages will be required to represent the current project's working drawing design. Total number of pages multiplied by hours/page multiplied by the factor a designer must have on his base wage hour to cover overhead and fee, equals total design fee.

If the designer calculates that this cost to be 7% of the projected cost of the project, and the owner, based upon the designer's competition, is unwilling to pay over 5%, the designer is faced with a dilemma. One answer to that dilemma is to accept the 5% fee, recognize that there are only a certain number of hours that fee will allow, and produce a set of documents that is commensurate with that number of hours. Perhaps the documents will not answer all the construction questions, but it will probably be less costly for the designer to answer the questions the construction team raises than to answer all possible questions in the original documents. This may not be, in the eyes of the contractor, the best perspective for the designer to have, but given the competitive circumstances of the designer, it is understandable.

The competitive pressures on the contractor are well recognized by those in the industry and need no lengthy exposition here. It suffices to say that they are pervasive and unrelenting.

All of the foregoing is prologue to say that changes are, in today's environment, inevitable; that is, unless one happens upon that rare project in which the owner engages the services of the design team to completely design a facility and Compensates the team for a complete design, in which the contractors are allowed, nay encouraged, to do a complete job and make a fair profit, and in which the number of changes, through the cooperation of all the parties, is held to an absolute minimum, identified and resolved quickly and fairly for all parties, and everyone has respect for everyone else. Perhaps we see these once in our construction lifetimes.

How to bring about more of the projects constructed in heaven is a subject for another time, and it is well worth pursuing by all the members of the construction team, because not only does it make for better, more cost effective projects, it is also a lot more fun to work on.

An equally closely related subject, claims (or change orders gone bad) is also one for another time. A claim has a lawyerly ring to it, and is usually the result of a change order, or series of change orders, which could not be resolved by the parties of the first part, that is, those involved every day with the construction of the facility. A claim might properly be called a failed change order.

We want to deal with the recognition, preparation, and presentation of change orders.

Recognition

How does one recognize a condition which gives rise to a change order. While the answers to this question may seem stunningly obvious in theory, they must definitely not be so obvious in practice, because the question of change orders is as notable for the lack of presentation of proper change orders as it is for the presentation of inflated or bogus ones. This may be so because those responsible for representing a company's interest recognize change but decide to forgo pursuing compensation for change for a host or perfectly rational reasons—the tried and true "give and take" of construction. This is the highest and best reason available for not pursuing change orders, a true spirit of cooperation on the jobsite, the cause of which projects finish as well as they do.

In spite of the foregoing, many change orders are not pursued because they are not recognized. Why is this so? Probably the foremost reason is that the company's front line staff, those in the field, are not adequately trained to recognize the change. They may not be as well versed in the construction documents as they ought, not only those which deal directly with their work, but those on the periphery of their work. Many subcontractor's field supervisory personnel have never seen a copy of the contract between their firm and the general contractor, do not have an in-depth knowledge of the project specifications, especially the general conditions portions, and prepare for tomorrow's work after work is complete today. In the grossest terms, how could such employees recognize a change when they do not fully know what their work is?

Related to this lack of knowledge of contractual obligations and scope of work is a more subtle but equally pervasive cause of lack of change recognition—it may be called a lack of interest of management in training supervisory staff in matters other than pushing work. That is to say, the supervisory training which does take place revolves around how to put the widget in place faster, as opposed to the more mundane matters of record-keeping, specifications and contract review and study, etc. A supervisor who cannot put the work in place effectively is short-lived. Those who remain are viewed as valuable because they can put the work in place and short shrift is made of training them in other aspects of their jobs, including the recognition of, not to mention preparation and presentation of change orders. This situation to the contrary, is it not these very staff who are in the best position to recognize changed conditions and react most quickly to them?

The remedy for these conditions is training. By training we mean organized, in-depth, "classroom," ongoing, training. Perhaps it is a one day seminar on changes recognition, record-keeping, etc., conducted by in-house staff or someone brought in; perhaps the training includes project management staff, and emphasizes ongoing, thorough review of possible changes. Perhaps it includes a project start-up meeting where among other things, the specs, plans, contract, etc. are thoroughly reviewed, including the potential for change orders. Certainly, the company's estimator will want to identify potential changes as a part of bid preparation and bidding strategy. Perhaps it involves periodic, in-depth reviews of the job where one of the standard questions addressed is the status of change orders and potential change orders. All of these, as part of a written and effective management plan, will aid immeasurably in recognizing potential changes and thereby potential change orders.

Preparation

Once we recognize a potential change, how do we prepare it for submittal?

The importance of meticulous preparation cannot be overstated. The first impression created by a request for a change order is absolutely critical to the entire process. Shoddily prepared change order requests will almost universally produce the same response, derision and/or anger, leaving the distinct impression that the presenter does not know what it is doing. Perhaps the greatest failing of initial change order requests is that they do not provide nearly adequate documentation, so that a rational person can decide the merit of the claim on the submittal alone. If documentation is provided, it is usually that which is most favorable to the contractor, with none favorable to the other side. The most widely practiced procedure in this area is to adequately describe and price added work while not doing so for deductive work. If the entity reviewing the change order request discovers such omissions, it is extremely difficult for the submitting company to maintain the reviewer's trust, both for the change under consideration and for any that might arise in the future.

Another equally common failure is to use one set of prices for bidding and another set for change orders, the second being substantially higher than the first, without adequate, or indeed, any justification. Many change order proposals are prepared in this way, with the result again of having the reviewer lose trust in the contractor.

A third failure is that the general contractor simply passes through a subcontractor's change request data, not having reviewed it. Again, credibility is lost. It is important for the contractor to both thoroughly review the sub-bidder's prices it receives and to demonstrate that it has done so.

Perhaps the best way to overcome these problems is to do the reverse of the situation descried above. The scope of work should be accurate and complete. It should be described and presented in the same level of detail as the contractor prepared its own detail estimate for the project as a whole. The same pricing structure should be used, and if it is not, variances should be noted, in detail, along with their justification. For example, if the productivity included in the original estimate is more than that provided for the same item of work in the change, a note explaining the variance should be provided. It may be, for example, that the original quantity of a work item was 10,000, all of which is done, while the change will have 1000, which must be done anew, with its concomitant new learning curve, etc. As an aside, it is equally as poor performance to fail to use pricing different from that used in the original bid if the conditions under which the work will be done demand it.

In addition to the above, no change order request which involves the work of a subcontractor should be submitted without careful review and critique. The general contractor representative should be as familiar with this change as it is its own. The critique should include that for the items mentioned above regarding quantity survey, pricing structure, and explanatory notes. One way to demonstrate the level of the general contractor's review is to include a sub-bidder's quote which has been manually changed by the general contractor, and to allow the changes to show on the submittal.

Two items which are often overlooked by those in the construction team are the effect of the change upon general conditions, and upon the project schedule. These are often difficult to determine, but are none the less real. Here the original estimate schedule and general conditions are critical in proving the case, because before expecting to collect more general conditions, the contractor must show what his original bid included. Again, the documentation should clearly set out the original, how it is now changed, and what costs are added because of the change.

Another item often overlooked in change order preparation is the affect the change will have on surrounding work and schedule. Is adjacent work impacted? Does the change entail rescheduling adjacent work so that it is less efficient than before? If so, these impacts must be quantified and priced.

The contractor must not overlook, as well, the cost of preparing the change order. Change orders take time, and if the time or staff is not anticipated in the original bid, it, too, should be quantified and priced.

Presentation

Once the change is properly prepared, it is usually forwarded to the Owner's representative, there to fall into a great black hole, it sometimes seems. It is easy to think that once the change proposal is submitted, the bulk of the work is done; however, in some ways, the process of submittal may be the most important part of the entire process.

The contractor should take a carefully reasoned pro-active approach to change request submittal. A great deal of time, effort, and trouble can be saved thereby. A submittal meeting should be requested, at which the contractor and all parties having an interest in the change order should formally present the change. The presentation will include the scope of the change, how the quantities were derived, the basis for the pricing used, and the rationale for the impacts seen. The presentation should be detailed, as detailed as is the change itself, and should include visual aids to assist the reviewer in understanding its content. The submittal meeting should be viewed as a marketing meeting, helping the contractor to justify the change and therefore "sell" it. Remember, first impressions are important, and that is nowhere as true as in change orders.

The foundation for the entire change order process is, of course, adequate record keeping. Changes involve the original estimate, plans and specifications, contracts, correspondence, meeting minutes, daily reports, and other such documents. Lack of such records has doomed many a legitimate change order request, and their importance cannot be overstated.

A change order is not an opportunity to prospect for gold in the project. Neither is it an opportunity for the owner or designer to put two pounds of project into a one pound bag. Either of those attitudes, when discovered, as they will surely be, will lead to attitudes among other team members which may result in substantial cost increases for everyone involved. Unresolved or inequitable changes lead to claims, and claims often lead to little satisfaction for anyone involved, save the legal profession.

Change Order Aids

There are many resource materials available to assist in educating the construction team in recognizing, preparing, and presenting change orders. One of the best is Contractor's Guide to Change Orders, by Andrew M. Civitello, Jr., published by Prentice-Hall, Inc., Englewood Cliffs, New Jersey. Mr. Civitello takes a very aggressive posture to the subject, and includes not only checklists, but also sample letters to consider for the various situations encountered along the way.

SECTION 18 – HANDLING FEDERAL CLAIMS AND CHANGE ORDERS

The following is reprinted with the permission of Francis J. Pelland, Esquire.

From a contractor's viewpoint, claims are unresolved change orders. In fact, the contractor will make every effort to avoid or mitigate the submission of claims. Change order management and administration of the contract requires implementing systems and procedures to process potential changes. Examples of handling claims and change orders that affect cost and schedule will be discussed.

I. Scope of Changes Clause

    A. General Scope

    B. Cardinal Changes

II. Authority of Government Personnel

    A. Contractor Risk

    B. Protective Measures

III. Types of Changes

    A. Formal Contract Modification

        1. Request for Proposals

        2. Proposal

        3. Acceptance

    B. Bilateral v. Unilateral

    C. Differing Site Conditions

        1. Contract Documents

        2. Subsurface Conditions

    D. Constructive Change

        1. Notice

        2. Timely Submission of Claim

    E. Acceleration

        1. Actual

        2. Constructive

Section 18 – Handling Federal Claims and Change Orders

    F. Delays and Suspensions of Work

    G. Termination

        1. Default

        2. Convenience

        3. Actual v. Constructive

IV. Contract Modifications: Reservation of Impact and Delay Damages

    A. Waiver of Claims

    B. Accord and Satisfaction

    C. "This Modification covers only a performing the changed work. The contractor reserves its right to seek impact and delay damages and time resulting from the change."

V. Proving Responsibility for Delay

    A. Records, Records, Records, and Records

    B. Critical Path Analysis and Narrative

VI. Pricing Certain Changes

    A. Loss of Efficiency

        1. Acceleration

        2. Disruptions

        3. Working Out of Sequence

        4. Overmanning

        5. Unqualified Manpower

        6. Overtime

        7. Productivity Index

    B. Delays and Suspensions

        1. Extended Direct Costs

            (a) Supervision, Trailers, Equipment Rental, etc.

        2. Extended Indirect (Overhead) Costs

            (a) General and Administrative (Eichleay)

    C. Concurrent Delays

    D. Recovering Attorneys' Fees

VII.   Disputes

    A. Settlement

    B. Litigation

    C. Arbitration

Changes and Claims in Construction Contracts

Although this section deals generally with all forms of construction contracts, emphasis is placed on United States governmental procurement and on those private and local government contracts employing AIA 201 general conditions. The principles which have developed over the years, particularly with relation to federal procurement matters, are regularly applied by local courts in resolving private contractual disputes.

I. Scope of Changes Clause - Most contracts, public and private, have a changes clause which empowers the owner to make changes in the work which the contractor is obligated to carry out. Without the clause, there would be no right on the part of the owner to order any changes and the contractor would be free to ignore all such directions.

There is a limit, however, to the changes that can be forced upon a contractor. Only those changes that are within the general scope of the work of the contract are permitted. So-called "cardinal" changes, that is, major modifications which depart from the overall purpose or scheme of the contract, cannot be required. For example, under the guise of a change order, an owner could not modify a single tower office building into a two-tower building, but he could relocate the mechanical equipment from the basement to the roof.

II. Authority of Government Personnel: It is always important to be certain that the owner's representative who issues directions to perform extra work has the requisite authority. This does not often create a loss situation for the contractor, since courts and the Boards of Contract Appeals regularly find, in cases where the government representative did not have the appropriate authority, that his actions were ratified or otherwise accepted by the contracting officer, particularly when the result otherwise would be that the contractor will not be compensated for extra work.

This potential problem becomes real in situations where there is a legitimate dispute between the contractor and the government inspector (who normally has no extra work authority) over whether certain work is or is not extra to the contract; if the contractor performs that work pursuant to verbal or even written direction from the inspector, and if the contracting officer was not aware of the problem and might have taken a different approach had he been informed, the contractor could lose his right to additional compensation by failing to notify the contracting officer before he performs the disputed work. To protect against this potential loss, therefore, the contractor should always notify the contracting officer or his duly authorized representative in writing, before he performs work which he believes to be extra work, that he has been directed to perform that work and will do so under protest unless he receives a stop order from the contracting officer.

III. Types of Changes: There are many ways in which the contract can be changed, either by adding or deleting or modifying particular items of work or by creating conditions where the effort required by the contractor is increased without changing any items of work themselves.

(A) Formal Contract Modification: Where the owner sees a need for a change to overcome a design deficiency or simply in order to upgrade the present design, the owner normally will send to the contractor the scope of work encompassed by the change and request a proposal for an equitable adjustment in the price and time of performance. The proposal is then submitted and the parties come to an agreement, with or without negotiation, on a fixed price and a time extension for the change. If it always worked that way, however, the lawyers who specialize in government contracts would be doing something else for a living. Too often, the Government will demand that the contractor perform the extra work before there is an agreement on the price, thanking the contractor for his proposal and assuring him that they will work out an agreement on the price later. As most contractors know, this practice can cause many problems, all of them dealing with lack of cash flow. It is not unheard of for a contractor to perform $200,000 worth of extra work over a 12-month period without receiving any formal contract modifications, which means that he cannot include the value of those extras in any of his requisitions.

It is my belief that, except in emergencies endangering life or property, a contractor is not obligated, in the absence of a fully executed contract modification, to perform extra work at the direction of the government unless good faith negotiations before the fact have failed to produce an agreed price, and the government has issued a unilateral modification for the amount it deems to be the proper value of the extra. This permits the contractor to draw against that money without prejudice to his right to pursue a claim for the higher amount. Good faith negotiations mean that a contractor lays his proposal on the table, the government puts on the table its analysis of that proposal and its own estimate of the extra work and the parties then negotiate. Under no circumstances is a contractor obligated, whether under a Federal contract or a local public contract or a private contract, to perform what is admittedly extra work without being able to receive some payment therefore on a current basis.

Contracting officers do not like to hear this. Many of them truly believe that the general provisions give them the right to order extra work throughout the life of the contract and later to negotiate prices based upon the time and materials actually expended.

In my view, they are incorrect. I tell those contracting officers that we bid on a lump sum contract, not a T & M contract, that we are not going to permit them to convert major dollar amounts into a T & M situation, that we did not include in our bid the additional administrative effort and expense of trying to keep track of the exact labor and material costs of a particular item of extra work and that even if we wanted to do so, it would be very difficult in view of all of the other work being performed at the same time.

On more than one occasion contractors have stopped all work on Federal contracts because substantial amounts of money had been spent for agreed extra work without formal modifications or because a critical item of proposed extra work needed to be performed before remaining contract work could be completed, and on those occasions the effect was to create a sudden interest on the part of the Government in formalizing those changes so that money could be drawn. We should take this type of drastic action, of course, only when there are serious dollars involved. I have not yet had, and hope I never have, the need to test my theory that the contractor has a right to stop work under those circumstances (which would happen, for example, if the Government default terminated a contractor who withdrew his forces), because in my cases the Government has been so completely in the wrong in failing to settle on changes that it did not elect to challenge our withdrawal.

When a contractor finds himself performing a substantial amount of change order work without being able to requisition for that work because the government has not gotten around to negotiating or finalizing contract modifications, the contractor should put the government on notice as soon as possible that it is not able or is not willing (or both) to continue to finance the work, that it has a right to be paid for the work that it is performing, that his subcontractors are complaining and threatening to pull their men unless they are paid for the changed work they have been performing for so many months, and that unless prompt action is taken to incorporate those changes into the contract, the contractor may have no choice other than to stop work and to assign its forces to jobs that pay. This course of action should not be employed when only small amounts of money are involved or when the government is only slightly delinquent in finalizing changes.

It is important to remember that a contractor does not have to finance the work beyond the normal requisition period. The best way to avoid getting into this problem is for the contractor to make his position known to the contracting officer with the first change order, for while it is common for a contractor who wants to maintain a good working relationship with the contracting officer to go forward and perform extra work in advance of receiving a formal modification - which is certainly acceptable as long as the modification is issued shortly thereafter - it is a sound practice to let the contracting officer know that you are willing to go forward with the work in advance of the modification because you want to help him out, not because you feel obligated to do so. In that way, when the dollars start to accumulate or the change orders become very large, the contractor will have set the precedent for insisting that the price be added to the contract before labor and materials are spent to perform the extra.

(B) Bilateral v. Unilateral: A bilateral modification is one that is executed by both the owner and the contractor, whereas, a unilateral modification is executed only by the owner and constitutes an order to perform the specified work at a stated price without prejudice to the contractor's right to claim additional compensation and/ or time.

In executing a bilateral modification, the contractor should be aware that all claims for additional time or money growing out of that change will be barred unless there is a clear reservation of rights contained in the modification. Appropriate wording to preserve such rights and the typical government resistance to the inclusion of such language in bilateral modifications are discussed later.

(C) Differing Site Conditions: Changes in the scope of the work or in the method or manner of performance which increase the cost or time of performance often arise out of differing or changed site conditions. There are two types of such changes. One category deals with differences between the contract documents and actual concealed conditions at the site. For example, soil borings in the invitation to bid might depict stable subsurface conditions, whereas after excavation is commenced, the contractor might find excessively moist conditions which require well-pointing to dry the site and additional earth handling and time to dry the stockpiled excavated material before backfilling. To support the contractor's entitlement to an equitable adjustment in price and time for this change, the actual subsurface conditions must not be known to the contractor or able to be observed with a reasonable pre-bid site inspection.

The second category of differing site condition changes arises independent of the contract documents. There may be no representation in the contract as to subsurface conditions, but the contractor will still be entitled to an equitable adjustable if the actual subsurface conditions encountered during construction are different from the conditions that were actually known to the contractor at the time of bidding or different from the conditions that could be detected by a reasonable pre-bid site inspection or different from the conditions that normally would be expected to be found in that area. For either type of different site condition change, the contractor must give notice to the contracting officer of the existence of this changed condition before he starts spending money to overcome the condition in order to ensure that he retains his right to an appropriate equitable adjustment.

(D) Constructive Change: A constructive change is an actual change in the work that is not recognized to be a change by the contracting officer. The government may issue what it believes to be a simple clarification of a portion of the specifications, and the contractor may consider that instruction to represent a change in his planned method of performance or an increase in the work that he was required to do under the contract. In that case, the contractor must give written notice to the contracting officer before he undertakes that work that he considers to be a change to the contract and that he will follow the instruction under protest and reserve his right to seek additional money and time, without such advance notice the contractor normally will be barred from any type of equitable adjustments on account of that change. Most contracts contain provisions as to when notice of the change must be given and as to when the claims for additional money or time must be submitted and those time limitations must be carefully observed to avoid losing what otherwise would be a valid extra.

(E) Acceleration: A change occurs when the contractor is ordered by the owner to accelerate his work in order to meet an early completion date. This will normally result in a number of trades over-manning the job simultaneously instead of working with the optimum crew sizes in an orderly sequence; if the contractor pays overtime hours, or, due to the inevitable loss of efficiency, incurs greater labor costs than would otherwise have been spent to perform the same work, he is entitled to an equitable adjustment.

There are two types of acceleration changes. One is an actual acceleration change in which the contracting officer orders the work to be done by a certain date with the full knowledge that date is earlier than the date the contractor is required to meet under the contract. When that occurs, the prudent contractor will simply notify the contracting officer that he intends to submit a claim for his acceleration costs. The other type of acceleration claim is a constructive change, that is, where the contracting officer is not aware of or does not acknowledge that there have been excusable causes of delay and insists that the contract be performed in accordance with the original schedule. When that is the situation, the contractor must notify the contracting officer that his order to complete the work "on time" is considered by the contractor to be an acceleration order because of the presence of excusable delays and that the contractor will reserve his right to submit a claim for acceleration costs. Without that notice, an otherwise legitimate claim often will be barred.

(F) Delays and Suspensions of Work: Delays, disruptions, and suspensions of the work necessarily cost money. If they are the fault or responsibility of the contractor he will be exposed to liquidated damages; if they are the fault or responsibility of the owner, the contractor will be entitled to his direct and indirect increased cost of performance during the overrun period. As in other types of changes the contracting officer can be responsible for both actual delays and constructive delays. These changes are actual when the contracting officer can be responsible for both actual delays and constructive delays. These changes are actual when the contracting officer recognizes that the government has delayed or disrupted or suspended the work, such as when the contracting officer issues a suspension of work order which halts all or some portions of the work while the government considers and decides upon certain design changes. A constructive delay or a constructive suspension occurs when the contracting officer is not aware of or does not acknowledge that his conduct is delaying or disrupting or suspending the work, such as when it permits the work of another contractor on an adjacent site to interfere with the contractor's project or when the government improperly rejects shop drawings on critical material or equipment and thereby extends the time when those items can be ordered and fabricated and installed. As with other constructive changes, notice is vital; as soon as the contractor becomes aware that the government's conduct is or may be disrupting and delaying his progress, he must give the government written notice of that fact in order to be sure that he does not lose his right to additional money and time.

(G) Termination: Most contracts give an owner the right to default terminate the contractor for what the owner perceives to be material breaches of the contract, and most government contracts as well as many private contracts give the owner the right to terminate the contract for the convenience of the owner, that is, a "no fault" termination. All Federal procurement provides that if a contractor is default terminated erroneously, that is, where it is later determined by a Board of Contract Appeals of the Claims Court that the default was improper because the contractor was not in material breach, the remedy is to convert the default termination into a termination for the convenience of the government.

When a contractor is defaulted properly, he (and his performance bond surety) will be liable for the amount by which the government's costs to re-procure and complete the work exceed the balance remaining in the contract. When the government improperly defaults a contractor, it is deemed to be a constructive termination for convenience under which the contractor is entitled to recover his actual out-of-pocket expenses, including payment under the contract for all work done to the date of termination, all restocking charges for material returned to suppliers, all demobilization costs and even the reasonable fees paid to an attorney for the preparation of his convenience termination cost reimbursement claim.

Under the termination for convenience clause, the government is permitted to terminate all or some portion of the remaining work without being guilty of a breach of contract. Without that clause, an owner who decided that it did not want to complete a project and who thereby terminated the remainder of a contract would be guilty of a breach of contract and have to answer to the contractor for the loss of anticipated profit on the terminated portion of the work. In effect, the convenience termination clause converts what would be a breach of contract into a convenience termination by eliminating the loss of anticipated profits as an element of recovery.

IV. Contract Modifications; Reservation of Impact and Delay Damages: As mentioned earlier, a bilateral modification often will constitute a waiver of any other claims that might arise out of the change covered by the modification. "Waiver" is defined as "the voluntary relinquishment of a known right," meaning that if a contractor is aware that he has a right to seek certain damages because of a change to the contract but signs a contract modification giving him time and money but failing to mention such addition rights, he will be held to have waived any additional claim.

There are also changes which cause a direct increase in the cost and/or time of performance where any possible impact costs, such as delaying the performance of other work later in the contract, are not known. When this occurs, and the contractor nevertheless signs a bilateral modification without reserving the possibility of such additional costs, any later claims will be barred under the doctrine of "accord and satisfaction," a legal term meaning that the parties came to a full agreement (accord) on the price of the change and that the price was paid (satisfaction). Many bona fide claims for delay and impact costs have been rejected because they arose out of changes which were incorporated into a bilateral modification.

To protect his right later to seek delay and impact costs arising out of a charge, the contractor must incorporate certain reservation language on the face of the modification. A clause which will work is as follows:

> This modification covers only the direct costs and time of performing the changed work.
>
> The contractor reserves its right to seek impact and delay damages and time resulting from the change.

Contracting officers traditionally resist the inclusion of such a clause into their modifications. It is not unknown for a contracting officer to hold up the issuance of an otherwise settled contract modification until the contractor agrees to eliminate this clause. Do not be pressured into doing so! A contractor has an absolute right to have this language incorporated into the body of the modification document itself. If a contracting officer refuses to recognize that right and sits on the bilateral modification, then the contractor should demand that the contracting officer issue a unilateral modification with the same price and time. Such a modification does not require the agreement of the contractor, and if it contains any clauses which would eliminate any such future claim (as many of them do), the contractor need only delete that language from his copy of the modification and return it to the contracting offer "for his information." Work can then proceed on the change without fear of losing entitlement to delay and impact costs.

The reservation of a right to seek later delay and impact damages, however, should not be used across-the-board. A contractor does not make friends with a contracting officer by uniformly reserving this right on all changes, for the government quite properly does not want to be facing a contract with an unlimited number of potential claims being unresolved. A better practice is to reserve the use of such a clause for changes where the contractor either is certain that there will be a ripple effect or does not know whether or not there will be any later impact.

V. Proving Responsibility for Delay: Just as the three major factors in determining the marketability of a residence are location, location and location, the three most important means of proving responsibility for delays to a project are records, records, and records. There is a fourth item, and that is records. These records include superintendents' logs, superintendents' daily reports, weekly progress meeting reports, architects' field and project reports, logs and reports prepared by the owner's inspectors, and letters, telegrams, and other communications from the contractor to the owner complaining of the various delays at the time they take place. The key ingredient in all of these categories of records is that they are contemporaneous with the events. After-the-fact reconstruction, even with a team of construction consultants and construction-oriented counsel, cannot replace the value of contemporaneous records. If a contractor is being disrupted by having to work his forces out of sequence, that should be noted on the daily report for every day that it occurs, together with an indication of the number of men affected. No costly charts or graphs reconstructing the work will come close to the persuasive power of those daily entries when it comes time to convince the Board of Contract Appeals or a panel of arbitrators that the contractor, in fact, was disrupted and that it resulted in a definite number of wasted labor hours.

These types of essential records form the basis of critical path analysis and narrative that often are necessary in order to show the interrelationship of the various phases of the project and how certain delays affected certain trades and the overall job progress. It takes considerable skill to prepare this analysis so that it does show the cause and effect relationship of delays in a clear manner, but without the underlying records the task is made much more difficult, if not impossible, and the final product will have much less credibility.

VI. Pricing Certain Changes: The art of negotiating acknowledged changes in the work is not a part of this presentation. Forward pricing is done by skilled estimators, and time-and-material pricing is done with good records. There are certain types of changes, however, where the difficulty of establishing the proper value can reduce or even eliminate the recovery if the claim is not presented and handled properly.

(A) Loss of Efficiency: Wasted manpower can be caused by having to accelerate the work by putting more manpower on the job than can efficiently work in the available spaces, by being disrupted or interfered with by the owner or by other trades, by having to work out of sequence and thereby losing the productivity of work crews, by overmanning the work especially at a time when other trades are being forced to do the same, by having to employ unqualified manpower in order to meet the owner's demands to do more work at one time than was planned and by working overtime hours. Any one or combination of these items can result in a contractor spending more money to do the work than he planned to spend or that reasonably should have been spent. Proving the number of wasted dollars due to this inefficiency, however, is often more difficult than establishing the existence and cause of the delays.

Records come into the picture again. The ideal superintendent's daily report will note that half of his six-man crew wasted half of the day because it was forced to hang small portions of its drywall on three floors rather than working uninterruptedly on one floor. Unfortunately, there are very few ideal daily reports (or ideal superintendents), which means that other means must be used to quantify the damages. A comparison of actual labor costs (less change order labor) to the estimated labor costs during the period of disruption is a common method of measuring the damages because often there is no other method, but it is weak in that it assumes that the estimate was not unreasonably low and that the contractor's performance was perfect and did not contribute to the higher actual labor costs. This is often described as the "total cost" approach. Another method is to establish by some means (superintendents' testimony, for example) that on the average a certain portion, such as 25%, of each day's work during the affected period was wasted because of the disruptions. Then it is simply a matter of multiplying that percentage by the actual labor costs expended during that time to arrive at the claim. The infirmity here, of course, is the subjective nature and unreliability of the percentage factor.

There is a method which is persuasive and effective, although there are many situations where it simply is not available. A contractor faced with a period of time during which he is being disrupted can compare the number of labor hours he spends during that period with the number of dollars he is able to bill the owner during that period. That will establish a ratio of labor hours to earnings. If he can then find a reasonably long period of time where the same type of work is being performed during which there were no significant disruptions or delays, he can make the same comparison and come up with the ratio of labor hours to earnings during that "ideal" period. To generate useful figures, the monthly requisitions can be adjusted by eliminating costly pieces of equipment so that the earnings during either period of time are related to labor and are not artificially or unevenly inflated by equipment costs. A comparison of the ratio of each of the two periods will result in a productivity index with which a contractor can accurately measure the amount of his equitable adjustment due to the loss of efficiency. For example, if he can establish that during the disrupted period he was able to earn only $50 for each labor hour of labor, he will have proven that one-half of his actual labor during the disrupted period was wasted.

Given the proper conditions (a project where the work is similar for long stretches of time, sufficiently long periods of both good and disrupted working conditions, a large number of labor hours in each period), this method of measuring damages is quite persuasive. An ideal project to utilize this method is a road building contract where for several months the borrow material to construct the embankment was fast-drying and easy to handle but during another long period of time, it was extremely wet and very difficulty to dry in order to achieve proper compaction.

The payment provisions for that type of contract usually call for the contractor to be paid for each lineal foot of embankment completed and approved by the government inspector, and it then becomes a very easy matter for the contractor to measure the increased cost of performing the work caused by the changed conditions. Most construction contracts, of course, are not as adaptable to a productivity index basis of measuring a loss of efficiency claim, but when it is workable it is quite effective.

(B) Delays and Suspensions: When the work extends beyond the scheduled completion date, two types of cost are generated. The first are the extended direct costs of performance, such as the continuing supervision and the extended use of job trailers, utilities, equipment, etc. These are not particularly difficult to measure as long as adequate cost records have been maintained. The other type of costs are indirect, that is, the extended home office overhead for the overrun period. These are usually measured by the court-approved Eichleay formula under which the contractor determines the amount of his home office overhead attributable to the contract in question, then calculates the daily rate of that overhead and multiplies that rate by the number of days of compensable delay chargeable to the owner. Specifically, the adjusted contract price for the contract in question is divided by the company's total billings for all work, including the subject contract, during the entire period of performance. That fraction is then multiplied by the company's total home office overhead expense during the performance period (after eliminating items disallowed by the applicable Federal acquisition regulations) to arrive at the dollar amount of overhead attributable to the contract in question. That number is then divided by the actual number of calendar days during which the contract was performed to obtain the daily rate of overhead attributable to the contract, and that daily rate then is multiplied by the number of days of owner-caused delay to arrive at the amount of the claim.

As indicated, this formula is an acceptable way of measuring the additional costs of performance arising from government-caused delay as long as the contractor reasonably could not have reduced his overhead costs because of the delays or undertaken more work to fill up that slack and thereby covered his overhead costs during the extended performance period. In most cases, neither of these two conditions is realistic, so that having established the compensable days of delay the contractor can apply this formula in order to arrive at a fairly solid delay damages claim. One word of caution, however, must be noted. This type of claim cannot be prepared carelessly, particularly when the contractor is dealing with the Federal government. The items comprising the home office expense pool which go into this formula usually contain several expenses which are disallowed by the Federal acquisition regulations, and this type of claim will always be audited with the result that if a disallowed item is not removed from the overhead cost pool before the calculations are made, it will cause the contractor trouble, running from delays in negotiating the final price to charges of civil or criminal fraud.

(C) Concurrent Delays: Changes resulting from delays and disruptions also become difficult to price when they overlap with contractor-caused delays. Such concurrent delays will enable the contractor to avoid liquidated damages but they also will prevent him from being paid for the delays. In other words, the government cannot collect liquidated damages and the contractor cannot collect delay damages when both parties are responsible for delays covering the same period of time.

(D) Recovering Attorneys' Fees: The ability to recover attorneys' fees from the Federal government in connection with disputed changes depends upon the legitimacy of the government's defense to the contractor's claims. If the Board of Contract Appeals determines that the contractor was entitled to an equitable adjustment and that the government's defense was frivolous or did not have substantial merit, then the Board is authorized to award the contractor a sum representing his reasonable attorneys' fees expended in prosecuting the claim. In presenting his claim to the Board of Contract Appeals, therefore, the contractor routinely should request attorneys' fees as a part of his relief, for even though it is seldom granted there have been instances where the Board has awarded the contractor his attorneys' fees after finding that the government's position was insupportable.

VII. Disputes: No discussion of changes to a contract which lead to disputes among the parties would be complete without a review of the several methods available to resolve them.

(A) Settlement: Most cases can and should be settled. Many cases which do end up in litigation or arbitration should have been settled. Until a contractor has been through a series of depositions and has experienced the continuing delays of litigation and then has sat through a five day trial, wondering if the jury has any idea what anyone is talking about, he may not appreciate the benefits of settlement. When the two opposing parties are willing to take reasonable positions, with or without the benefit of counsel, and if they make a serious and good faith commitment towards reaching a compromise if at all possible, the chances are excellent that a settlement can be reached. The savings in attorneys' fees and in the contractor's own time will justify substantial financial concessions by thinking people. There are those cases, however, which cannot or do not settle, and there are two basic forums available to adjudicate such disputes.

(B) Litigation: The most widely used mechanism is to go to court. In most jurisdictions, the civil dockets are crowded and delays from one to three years between the filing of a suit and the trial are the rule rather than the exception. Pretrial costs and fees will become quite high if one or both of the parties engage in extensive discovery procedures (depositions, interrogatories, inspection of documents, etc.). Trail dates can be removed from the docket on the morning of trial because the court has been overbooked. And when a complex construction case with dozens of items of claims for extras and back charges which are dependent upon an understanding of the plans and specifications is tried before a jury, the outcome can never be predicted.

(C) Arbitration: The difficulties associated with litigation have prompted many persons to turn to arbitration. Although arbitration has certain advantages, it is not a complete cure for the ills of adjudicating all disputes. There is no such thing as overbooking, since an arbitration hearing is scheduled by the private agreement of the arbitrators and the parties. There is little or no pre-trial discovery permitted, so those expenses are kept to a minimum. On the other hand, the absence of pre-trial discovery will usually cause the hearing to run considerably longer than would have been the case with pre-trial discovery. Further, the American Arbitration Association instructs its arbitrators not to adhere to the rules of evidence which means that the arbitrators are required to listen to almost anyone say almost anything as many times as he chooses, whereas in a courtroom much of that same testimony would not occur because it would be hearsay or otherwise inadmissible.

In addition, when a court trial begins it normally continues day after day without interruption until it is completed, whereas the arbitrators, who do not do that kind of work for a living, may be able to meet for only two days because one or more of them will have other commitments with the result that the proceedings will start and stop and start over a protracted period of time. Once a complicated construction arbitration case before three arbitrators consisted of 25 actual trial dates over a six-month period of time and that was not a rare happening. If that matter had been tried in court, it probably would have taken no more than 10 consecutive trial days. With each of the arbitrators being paid at least $400 for every day of hearing and with the attorneys getting something for their time, arbitration under those circumstances will be far more costly than a courtroom trial. Arbitration has a clean-cut advantage, however, when the issues are relatively compact so that they can be tried in one sitting. When that happens, the arbitration process from beginning to end will take place in a much shorter period of time than litigation.

In comparing the relative merits of arbitration and litigation, it should be kept in mind that arbitrators are not bound to follow the law. If a party has a strong legal defense to a claim which otherwise appears meritorious, that party obviously is better off in a courtroom. Further, when a party's ability to prove his damages, such as in a complex delay situation, are suspect, he is far better off in front of three arbitrators than he is in front of a judge.

In most instances, of course, the choice of the forum has been made long before the dispute arises. There either is or is not an arbitration clause in the contract, and if it is present, the dispute must be resolved through arbitration (unless both parties agree to go to court). Knowing the characteristics and idiosyncrasies of arbitration and litigation, however, can assist the contractor in deciding whether to pursue or defend certain claims, what the costs are likely to be and whether there is a good, an even, or a poor chance of success. It is at this stage that a contractor can benefit by consulting with counsel who is experienced in these matters, for it is at this point that settlement should be discussed as an alternative to a fight.

The foregoing is reprinted with permission of William H. (Harry) Spradlin, FCPE of the Spradlin Corporation.

Overview

Every member of the construction team encounters some form of legal problems during the normal daily conduct of business.

The cost and time delay factors make it impractical to refer each of these occurrences to an attorney for review and recommendations. Therefore, the owners, contractors, architects, and other participating in the design and construction of a project must acquire sufficient basic legal knowledge to orderly conduct their businesses without fear or encountering a catastrophic situation. Proficiency in the skills of developing and understanding the rights and obligations imposed by the construction contract is of extreme importance to the contractor. There is also a need to be knowledgeable in the area of change recognition, claims, supportive records and documentation, delays, default, bonding, insurance, lien rights and warranties.

Through the process of continuing education, contractors and other involved parties can obtain considerable information on these subjects by attending some of the professionally presented seminars by outstanding construction contract lawyers.

In addition, there are numerous legal publications available written by experienced attorneys that provide basic legal information to the industry. These are valuable tools for the contractor and the management team members to use as a general guide in the conduct of business to serve as an awareness of when professional legal assistance is required.

The contractor who proceeds with reasonable caution and conducts business on a fair and reasonable level should enjoy all the success available within the industry.

Is Cooperation Worth the Risk?

As a means of promoting project harmony and developing future business relations with an owner or architect, most contractors display a willingness to cooperate and participate in the resolution of the many problems encountered during the construction of a project. This approach is often of significant benefit to all members of the construction team, and if no complications develop, each member will realize the maximum benefit that only a successful project can produce. In other instances, when the contractor has displayed the willingness to assist in resolving project problems by suggesting alternative methods or materials, providing calculations, drawings and sketches, or has in any way influenced the judgment of the owner and architect in making their final decision, the contractor may have provided the access for a claim by the owner and/or architect should the alternative approach prove unsatisfactory.

An uncooperative contractor refusing to participate in resolving project problems would not be popular and it is unlikely participation in future projects would occur. In order to maintain acceptable business relations and at the same time minimize or avoid the possible claim position, extreme care should be exercised in transmitting information relating to any modification in design or alteration to the plans and specifications for the project.

The design is the responsibility of the owner and architect and it should always be retained by them. If suggestions for modifications are requested of the contractor they should be presented in a manner reflecting cooperation, but silent concerning their capabilities, adaptability or any preference by the contractor. The issue here is to force investigation by the owner and architect so their final decision is based on their determination and not that of the contractor.

It is recognized this is a continuing delicate situation for the contractor and is further complicated by the actions of the field personnel who quite often engage in displaying their knowledge by offering suggestions and alternates in the methods or materials without the knowledge of the contract. To preserve the integrity of the construction company, this subject should be covered in detail in the policy manual with periodic refresher discussions, especially on the projects involving extensive changes or ones where the relationship with the owner and/or architect are less than anticipated.

Should there be evidence claims will be filed against the contractor this entire issue should be presented to the company attorney who may desire to totally eliminate the contractor's role in any future design changes until they are finalized by the architect.

The Search for Mr. Deep Pockets

Throughout the construction industry, there is a search under way — not for talent or profit-producing projects or a means of increasing productivity — the concentration or effort is focused on locating "Mr. Deep Pockets." This is the person or company having the financial stability and capacity to satisfy a claim for damages from a party or parties involved in a construction project. A large number of construction lawsuits have the appearance of a shotgun approach where virtually every party is named in the complaint in an attempt to discover the existence of "Mr. Deep Pockets." Many unsuspecting parties have been judged responsible for acts of other parties and faced with the burden of severe financial costs.

The general and subcontractor experience a dual situation where sufficient financial stability must be maintained to justify bondability and at the same time they do not desire the position of "Deep Pockets." This is a specific problem for each company requiring extensive planning with the company accountant.

There appears to be sufficient evidence to draw the conclusion that many of the lawsuits being filed are the result of the economic instability of the construction industry.

Owners are bringing actions against architects and contractors seeking damages for extra costs or negligence as a delay barrier in making payments to the parties because of their inability to lease buildings in a soft real estate market. General contractors and subcontractors are participating in the lawsuit arena because severe competition forces bid prices to the level of being critical to the profitability of a project should unanticipated situations and costs be incurred as the result of the actions of another party. Making claims against other parties involved in the construction team presents another possible source of additional income to supplement the marginal profit available in the competitive market of construction.

This is not to say if a party has in fact been damaged by the acts of another party there should be any hesitation to make the facts known and aggressively seek compensation equal to the damages suffered. However, when claim filing becomes a standard operating procedure to inflate profits this type of action adversely affects the image of the industry and destroys the harmonious relationship between the members of the construction team leading to more costly projects.

Concentrate on improving contractual relations as a means of improving profitability and avoid the risky adventure of finding "Mr. Deep Pockets."

Lawsuits are Crippling an Industry

Is it possible that the escalation of legal actions within the construction industry can have such a devastating effect on an entire industry? The best answer to this question appears to be a profound "YES".

The industry has historically prospered on the skill, integrity, dedication, and professional performance of all the contributing members. In instances of disagreement and dispute, these were resolved, in general, through the efforts of sincere negotiation and in some instances the parties in dispute were forced to seek remedy through litigation or arbitration. Regardless of the route selected to settle differences, the underlying objective was to develop and complete a construction project within a scheduled time frame, consistent with the design criteria and to reserve a fair and reasonable profit for the services provided.

The filing of a lawsuit is an adversary action dividing the parties into separate and opposed roles, each attempting to protect and preserve their position and in doing so they embark on a mission that extinguishes the elements of cooperation, trust, faith, and normal business relations. With the removal of these elements, it can be anticipated that the entire project and all involved will, to some degree, suffer the consequences of a legal action.

The losses extend far beyond any financial judgment which may be encountered. There will result a reluctance by many of the contributors to the project to become involved again with the litigants. Reputations will be tarnished, personnel can become disillusioned and depart, quality of workmanship and timeliness of project completion can suffer and finally, the first legal action can be the catalyst of many additional actions being filed.

If the same degree of effort was extended to the project as that which would be siphoned off to support a lawsuit, it may be found more worthy to retain the integrity of the project than pursuing some elusive litigation.

As a final reminder, "justice does not have to be fair."

The Legal Burden of Management

The company management team, be it departmentalized or consolidated to the responsibility of one person, is challenged daily to perform a task or respond with action which is or could be of significant legal impact on the contractor.

Basically, most construction personnel lack sufficient legal training to be aware of all the complications which could occur from their actions. Unless problems develop on a project or within the internal organization of a contracting company, management can successfully perform. However, in the situations where legal problems are encountered, all of the past, present, and future actions by management could be subject to minute inspection.

For discovery purposes, all records including financial information and every written note, memo, correspondence or document of every nature can be subpoenaed by the opposing party in a legal action for review, comparison, study, and evaluation. To accumulate and reproduce these records is a monumental task and to then have them exposed to an adversary party can place any or all of the management group at a great disadvantage in the event their presence would be required as a witness. They could encounter situations where seemingly unimportant memoranda or informal notes contain information supporting the position of the opposition or casts reflections on the talent and ability of the manager.

Probably the magnitude of the burden placed on the management team involved in litigation can be measured by the vast amount of time that is devoted to reviewing documents, assembling the facts in an orderly and organized format, reproduction of documents, assisting the attorney with the technical presentation, appearance for depositions, and assisting in interviewing potential witnesses. All of this participation virtually removes the management personnel from their normal assigned duties, thus it can be expected that an over all decrease in efficiency will result in the company operation. Arbitration and litigation are unwelcome events to the knowledgeable construction contractor as it is very questionable if resulting awards are sufficient to offset the losses sustained by the disorganization created by the action.

Contract Knowledge

It is reasonable to assume that most everyone involved in construction contracting has at some time entered into a contract which at a later date proved to be a burden and less attractive than had been anticipated. There are many reasons these situations occur and the contractors can remove most of these unpleasant experiences by expanding their knowledge of the construction contract and establishing company policy for the information and guidance of employees engaged in developing or monitoring construction contracts.

This presentation is divided into three basic categories: (1) Pre-Bid; (2) Bidding Process, and (3) Contract Performance Period.

Pre-Bid

This is the period when a project is selected for bidding and the bid documents are obtained. At this point a contract does not exist, but there is the intention through the bidding process to obtain a contact and to accomplish this ultimate objective. It is mandatory that each and every condition set forth in the bidding documents be responded to so the contractor will be considered responsive when submitting the bid and at the same time be better informed on details having a later impact on the project.

Bidding Process

During the Phase 2 bidding process the general contractor faces a multitude of situations involving the legal effect on their decisions and actions.

The contract documents are not expected to reflect perfection but when taken in concert, the plans and specifications should be completed in sufficient detail to permit an experienced contractor to develop an accurate estimate and construct the project for the purpose it was intended.

The prudent contractor will give written notice to the architect and/or owner of any errors, omissions, conflicts or ambiguities discovered in the plans and specifications at the earliest possible date, even if it is not a contract requirement, so clarification can be issued to all prospective bidders. Any contractor relying on an ambiguity to gain a competitive advantage is at great risk because the interpretation applied by the contractor must stand the test of being "reasonable under all circumstances". A specification clause cannot be considered ambiguous by applying a hypothetical or unreasonable interpretation to the disadvantage of the other party.

During bidding, considerable interrelation exists with the specialty subcontractors, especially during the final hours of bid preparation. In an effort to achieve a favored position the negotiations between the parties can be misinterpreted which frequently results in disputes between the parties.

As a deterrent to such problems, the general contractor receiving notice of award should promptly notify each selected subcontractor, in writing, of the acceptance of their bid and the conditions under which it was accepted.

During or immediately after the general contractor bidding procedure, there can be cause for a "bid protest" or "request for permission to correct the bid" or "request to withdraw bid".

In the event a competitor's bid failed to meet all of the bidding requirements, a protest should be filed. In public openings, this initially could be made orally and at the earliest possible time confirmed in writing giving all of the specifics. Timeliness and the reasons for the protest are most important. When requesting permission "to correct a mistake and receive the award" this is complicated by policy and regulations controlling such situations in the public sector. However, the private sector would probably be more receptive. Early notification is extremely important and it should provide detailed information concerning the nature of the error. Notification of the willingness to furnish documentation supporting the existence of the error will enhance the position of the contractor.

The mechanics of a notice to "withdraw a bid" are similar to the procedures for "permission to correct a bid" except the intent is to withdraw and not encourage acceptance after the correction is applied.

It is suggested that all notices involving protests and/or bid mistakes be submitted early as mailgrams and then confirmed by certified or registered mail.

Contract Performance

This period begins with the receipt of the general contract from the owner and concludes when the project is totally finalized, including periods of warranties.

The contractor, before embarking upon the task of preparing subcontracts, should be aware of some of the elements and risks associated with this important undertaking.

Contracts are offered and accepted under almost ideal circumstances without projecting how this relationship will survive the numerous problems to be encountered during the extended period required for construction. It could be a first venture for both parties in a joint effort with limited knowledge of the other parties' financial, technical, and managerial abilities.

Awarding portions of a project to specialty contractors does not insulate the general contractors from their failures or defects. Actually the contractor inherits all of the strengths and weaknesses of each subcontractor and unfortunately, the contractor has limited control over the performance of subcontractors.

In many instances, subcontractors also experience unsatisfactory relations with the general contractor for a variety of reasons.

By observing there are three (3) basic elements to the contract which are offered consideration - acceptance - and drafting - the contract in clear, understandable language so there is mutual agreement of the entire content the parties will have constructed a document destined to serve their best interests for the duration of the contract.

Hastily developed contracts containing ambiguous clauses, inclusion of conditions not previously agreed to or incomplete information become the fertile breeding ground of future conflict and dispute.

Monitoring the project during the contract performance period should include the following activities:

- ✓ Written notification of schedule slippage
- ✓ Maintain correct status on submittals
- ✓ Monthly review of progress payments
- ✓ Establish progress meetings. Keep agenda and minutes.
- ✓ Give prompt notice of any defect in performance
- ✓ Projections on the progress required on a trade basis to accommodate completion schedule
- ✓ Review prior month daily reports

These activities will produce a realistic position report on a project and identify areas of possible future problems. In addition, they improve the management skills of a company and the resulting record keeping may serve to insulate the contractor from a future claim.

Administer Every Contract as if it will End in Litigation

Recognizing the proliferation of litigation within the construction industry demands that the general and subcontractors, in order to protect their position, accept this policy as a standard operating procedure on all contracts.

It is not envisioned as another paper war, but more as a refinement of the policy under which most construction companies are currently operating.

The following suggestions are offered to be observed as normal and regular course of business activities:

- ✓ Review general and subcontracts for accuracy and clarity
- ✓ Develop accurate progress schedules
- ✓ Develop cash flow charts for each project
- ✓ Establish and maintain daily job report records
- ✓ Confirm oral directions or agreements in writing

- ✓ Give prompt written notice as problems arise
- ✓ Monthly review and written report on project status
- ✓ Give written notices required by the contract
- ✓ Project problems recorded by photographs, laboratory or consultant investigation and daily reports
- ✓ Educate field personnel on company policy

Should disputes or claims have to be resolved by arbitration/litigation the quality of the above records will determine the degree of risk in proving the contractor's position.

Ethical Performance Prevails

Adherence to ethical business practice by owners, architects, general contractors and subcontractors would significantly contribute to improving contract relations between the parties. Trust would be restored, cooperation would improve, negotiation of problems would reappear, projects would be completed more timely and arbitration/ litigation would certainly diminish.

In general, the entire construction industry would benefit through the collective effort of each member of the construction team. This would be a return to viewing the completion of a project as a time of personal accomplishment to be shared with so many others who contributed their talent and skill to achieve the desired end product, rather than filing lawsuits.

The choice rests with the team members. Are they willing to place ethics above greed?

Litigation vs. Arbitration

The selection of the arena to resolve difficult disputes which cannot be disposed of through negotiation is one of personal choice. In instances where legal issues are the basis of a dispute, it would appear the court system would be the logical choice. However, arbitration would more likely be appropriate in resolving the disputes involving technical problems within the construction industry.

Among attorneys there exists divided opinion on this matter, but most accept the less formal atmosphere and flexibility of scheduling as positive advantages. Arbitration is considered as an acceptable alternate to litigation for the following reasons:

- ✓ Panel members are industry oriented
- ✓ Less formal atmosphere allows better presentation
- ✓ Scheduling is faster
- ✓ Hearings are in closed sessions
- ✓ Less costly than litigation
- ✓ Panel decisions are more timely and considered final in nature

In general, the arbitration process produces a better decision than a jury trial and at least equal in quality to a trial before a judge.

Value of Communication

The value of effective oral and written communication cannot be overestimated in the construction industry. The complexity of the industry requires the services and input of a large number of people into a composite effort to accomplish a desired end result.

Each segment is specifically educated, experienced and motivated to its particular discipline and it is through the channels of communication that coordination of individual achievement can be focused into a joint or team action The ability to communicate orally or in written form is a key to career success. Effective communication is part of management responsibility simply because it is necessary to provide information, directives, reports and notices to business associates and employees in a clear, understandable format so the content of the communication is obvious and presented in an orderly and logical manner.

The skill of communication can also become an art form if it is developed to the degree where oral and written efforts produce positive results without offending any other party.

Effective communication can be achieved through home study courses or attendance at any of the available seminars on this subject. Universities and community colleges also offer credit and non-credit courses to improve proficiency in communication.

Sample letters

The legal profession frequently criticize contractors for their failure to document their actions so there will exist an identifiable paper trail to record the events encountered in the administration of a construction contract.

As an assistance to contractors, the following sample letters are offered as a general guide:

Confirmation of Subcontractor's Bid:

Ladies/Gentlemen:

This letter will confirm the following:

    1. Your base bid on the subject project was in the amount of $_____. Alternate Bid 1 deducted from the base bid in the sum of $_____.

    2. Your bid was inclusive of all requirements under specification section _____.

    3. In the event we receive an award on the prime contract, your bid as recorded, will be accepted and you will execute our standard subcontract form for the stipulated work in strict compliance with requirements of the plans and specifications.

    4. In the event Alternate Bid 1 is accepted, your base bid will be reduced by the stated amount.

Bid Protest:

Ladies/Gentlemen:

This letter confirms our verbal and telegraphic protest of this date concerning the bid of Hippy Contractors which is considered non-responsive for the following reasons:

    1. Bid bond not included with the bid.

We hereby demand Hippy Contractors be considered non-responsive and the contract be awarded to this firm as the lowest responsive bidder.

Request to Correct Bid Mistake and Receive Award:

Ladies/Gentlemen:

This letter will confirm telegraphic notice of a bid error on Project No. _____, bid this date.

Review of our estimate reflected an extension error in calculating the excavation work. Our bid reflected the excavation as 40,000 CU YDS at a unit cost of $2.80 per CU YD, for a total cost of $11,200.00. The correct extension is $112,000.00 or $100,800.00 more than the amount included in the estimate.

It is respectfully requested we be given the opportunity to correct the mistake by adding the $100,800.00 to our bid price and to receive the contract award as this company would still be $16,240.00 below the recorded second bidder.

Upon your request we will make available all supporting documentation and the appearance of our chief estimator to verify the existence of the mistake. In the event you deny our request to correct the bid, you are to consider our bid withdrawn.

Request for Withdrawal of Bid:

Ladies/Gentlemen:

This letter confirms our verbal notice of this date of the existence of a mistake in our bid for Project No. _____.

We have discovered a mistake in computing the total estimated price for excavation. The correct amount would be 40,000 CU YDS at the unit price of $2.80 per CU YD, for a total of $112,000.00 Our estimate reflected a total cost of $11,200.00 for the excavation or $100,800.00 less than the correct cost.

In view of the magnitude of this error, we request permission to withdraw our bid. Upon notification, we will provide any requested documentation to verify the existence of the mistake.

Notice of Ambiguity:

Ladies/Gentlemen:

We observe on Drawing A-12 the building indicates a structure consisting of a basement area and five floors above ground. There is also a note stating "no interior work to be performed on the fifth floor." In order to make this building operational certain electrical and mechanical work will be required on the interior of the fifth floor. The electrical and mechanical drawings show a complete installation of both systems.

It is our interpretation the only interior work required on the fifth floor is the basic electrical and mechanical work necessary to make the project functional on the remaining floor areas. It is requested you clarify this situation by issuing an addendum prior to the bidding date.

In the absence of further direction by you, we will prepare our bid consistent with our interpretation as stated.

Differing Site Conditions:

Ladies/Gentlemen:

During the excavation being performed this date on "B" Wing of Building 10, we encountered at elevation 106.00 unsuitable material in the form of rubbish and debris. The existence of such material was not shown on the drawings or present in the soil borings.

A preliminary investigation indicates the unsuitable material extends to at least elevation 96.00 in the area of approximately 20,000 square feet,

We have suspended work in this area and request you investigate and advise how this matter is to be resolved.

Work in "B" Wing is critical to job progress and should we suffer time delay, increased costs or impact costs, we reserve all rights to file a claim at a later date.

Time Extension for Weather:

Ladies/Gentlemen:

Notice is given that we have experienced unusual weather conditions during the month of December.

Climatological records indicate freezing weather occurs on an average of four (4) days. However, we have encountered extreme freezing conditions for twenty-eight days which has severely delayed the project. We hereby request a time extension in the amount of twenty-four days (24) representing the period of unusual conditions.

Deferred Submittal of Cost Proposal:

Ladies/Gentlemen:

The additional work required by Change Order No. 6 is being performed at the present time in the interest of job progress.

Due to the involvement of multiple trades we are not in a position to submit a proposal for the increased costs at this time. We anticipate filing our claim for additional time, costs, impact and extended overhead within 30 days.

Protest Reduction of Pay Estimate:

Ladies/Gentlemen:

Our November pay request in the amount of $160,000.00 has been reduced by you, without explanation, to $90,000.00 and we have this date received a check in that amount. We protest your wrongful action and demand immediate payment of the $70,000.00 withheld.

We have deposited your check with the understanding we have not waived our rights to the remaining sum due and reserve all rights to collect the full amount of the November requisition.

Protecting Impact Costs:

Ladies/Gentlemen:

We have received Change Order No 12 and note you have deleted the charges for impact costs.

We will proceed with this additional work with the understanding all rights are reserved to file a claim for the impact costs as part of this change order within 30 days from date hereon.

PART TWO

PRACTICE COMMON TO SPECIFIC DISCIPLINES

PART TWO
TABLE OF CONTENTS

Discipline Specific Estimating Procedures 169
Scope of the Estimate 169
Specific Practice 169
Terminology 169
Forms 169
Project Evaluation through Plan Review 169

Division 1 General Requirements 171
Introduction 171
General Conditions 171
Project Site Indirect Costs 171
Services to the Owner, Architect, Engineer 172
Recap Summary 172
Project Site Staffing 172
Management and Engineering 173
Supervision 173
Administrative 173
Temporary Plant/Facilities 174
Temporary Utilities 174
Project Office Expense 175
Environmental Protection 175
Construction Equipment 175
Mobilization 176
Personnel Mobilization 177
Plant/Facility Mobilization 177
Construction Equipment Mobilization 177
Small Tools and Consumables Mobilization 177
Demobilization 177
Miscellaneous 177
Testing/Inspection and Outside Services 178
Job Site 179
Small Tools and Consumables 181
Weather Protection 181
Company Office Support 181
Management Support 182
Escalation 182
Finance Expenses 182
Cash flow Analysis 183
Contingencies 183
Summary 183

02260 – Excavation Support and Protection – Soil Nailing for Bank Stabilization 185
Types and Methods of Measurement 185
Factors Affecting Takeoff and Pricing 185
Overview of Labor, Material, Equipment and Indirect Costs 186

02310 - Grading 199
Introduction 199
Type of Measurement Used 199
Factors Affecting Takeoff and Pricing.. 199
Calculating Quantities 200
Pricing 201

02315 - Excavation and Fill - Trenching... 203
Introduction 203
Types and Methods of Measurement 203
Factors Affecting Takeoff and Pricing.. 203
Overview of Labor, Material, Equipment and Quantity 203
Sample Estimate 204
Conclusion 205

02465 - Bored Piles -- Drilled Caissons 207
Introduction 207
Types and Methods of Measurement 207
Factors Affecting Takeoff and Pricing.. 208
Overview of Labor, Material, Equipment and Indirect Costs... 208
Special Risk Consideration 209

02465 - Bored Piles -- Auger Cast Grout Piles 213
Introduction 213
Types of Measurement 213
Factors Affecting Takeoff and Pricing.. 213
Drilling and Grouting Pile Process 214
Overview of Labor, Material and Equipment 215
Conclusion 218
Glossary 219
Reference.. 219

02530 – Sanitary Sewerage.. 221
Introduction 221
Types and Methods of Measurement Used... 221
Factors Affecting Takeoff and Pricing.. 222
Overview of Labor, Material, Equipment 222

02750 – Rigid Pavement.. 237
Introduction 237
Types and Methods of Measurement 237
Specific Factors Affecting Estimate 237
Organization of the Estimate 238
Overview of Labor, Material, Equipment 239

02770 – Concrete Curb and Gutters 241
Introduction 241
Types and Methods of Measurement 241
Specific Factors Affecting Estimate 241
Organization of the Estimate/Quantity Survey and Calculation Quantities...................... 241
Overview of Labor, Material, Equipment 242
Conclusion 242

02775 – Sidewalks 245
Introduction/Description 245
Types and Methods of Measurement 245
Specific Factors Affecting Estimate.. 245
Organization of the Estimate/Quantity Survey and Calculation Quantities.... 245
Overview of Labor, Material, Equipment 246

03050 - Basic Concrete Materials and Methods 249
Hot & Cold Weather Procedures.. 249

03100 - Concrete Forms and Accessories 251
Introduction 251
Analyzing the Job... 251
Plans and Specifications.. 252
Horizontal Forming 253
Vertical Forming 253
Considerations... 255
Major Building Groups 255
Project Characteristics... 257
Accessories.. 258
Other Concerns.. 260
Takeoff Procedures 261
Walls 261
Columns 263
Beams... 264
Slabs 267

03200-Reinforcing Steel 271
Reinforcing Steel Placement 273

03300-Cast-in-Place Concrete 285
Takeoff Procedures 285
Waste Allowances 285
Concrete Footings 285
Concrete Walls 286
Columns 290
Beams... 291
Slabs 294

03410-Plant-Precast Structural Concrete 297
Introduction 297
Types of Measurements 297
Specific Factors Affecting the Estimate.. 297
Organization of the Estimate... 297
Overview of Labor, Material and Equipment 298
Estimate 298

03470-Tilt-up Precast Concrete 303
Introduction 303
Types and Methods of Measurements.. 303
Organization & Specific Factors Affecting the Estimate.............................. 303
Overview of Labor, Material and Equipment 304

Estimate	308
04050-Basic Masonry Materials and Methods	**311**
Introduction	311
General Requirements	311
04060-Masonry Mortar	**313**
Introduction	313
Quantity Survey	313
04090-Masonry Accessories	**315**
Quantity Survey	315
04210-Clay Masonry Units-Masonry	**317**
Quantity Survey	317
Site Improvements	318
General	318
04220 – Reinforced Unit Masonry Assemblies	**321**
Quantity Survey	321
05120 – Structural Steel	**323**
Introduction	323
Types and Methods of Measurement	323
Specific Factors Affecting the Estimate	323
Organization of the Estimate/Quantity Survey and Calculation Quantities	324
Overview of Labor, Material, Equipment and Indirect Costs	325
Estimate	327
06100-Rough Carpentry	**331**
Introduction	331
Types and Methods of Measurement	331
Factors Affecting Takeoff and Pricing	332
Overview of Labor, Material, Equipment, Indirect Costs, Approach and Mark-ups	333
06170 – Prefabricated Structural Wood	**339**
Introduction	339
Types and Methods of Measurement	339
Factors Affecting Takeoff and Pricing	339
Truss Roof Systems	340
Sample Estimate	341
06200-Finish Carpentry	**351**
Introduction	351
Types and Methods of Measurement	351
Factors Affecting Takeoff and Pricing	351
Overview of Labor, Material, Equipment, Indirect Costs, Approach and Mark-ups	352
06410-Custom Cabinets	**357**
Introduction	357
Types of Measurement	358
Factors Affecting Takeoff and Pricing	358
Special Risk Considerations	360
Post Project Analysis and Recording of Historical Data	360
07100-Dampproofing and Waterproofing	**361**
Introduction	361

| Quantity Survey | 361 |

07190-Water Repellents ... 363
Introduction	363
Quantity Survey	363
Multipliers for Developed Surface Area of Concrete Masonry Units	364
Multipliers for Developed Surface Area of Brick Masonry	364
Multipliers for Formed Concrete Walls	364
Extension of Quantities	365

07240 – Exterior Insulation and Finish Systems ... 367

07500 – Membrane Roofing ... 375
Introduction	375
Determining Material Cost	376
Assessing the Scope of the Job	376

07590-Roof Maintenance and Repairs ... 379
Introduction - An Ounce of Prevention	379
Quantity Survey	379

07920-Joint Sealants ... 381
Introduction	381
Quantity Survey	381
Extension of Quantities	381
Determining Lineal Feet Per Gallon	381

08100 – Metal Doors and Frames ... 383
Introduction	383
Types and Methods of Measurement	383
Facts That May Affect Quantity Takeoff	386
Overview of Labor, Material, Equipment and Indirect Costs	387
Risk Considerations	388
Ratio and Analysis Tools	388

08710 – Door Hardware ... 395
Introduction	395
Types and Methods of Measurement used	395
Specific Factors Which Affect the Takeoff, Pricing, Etc.	397
Overview of Labor, Material, Equipment and Markups	398
Ratios, Rules of thumb and Analysis Tools	398

09510 – Acoustical Ceilings ... 415
Introduction	415
Types and Methods of Measurement	416
Factors Affecting Takeoff and Pricing	416
Performing the Takeoff	417
The Spreadsheet and Calculating the Cost	418

09720 - Wall Covering ... 425
Introduction	425
Quantity Survey	425

09900 - Paints and Coatings ... 427
Introduction	427
Quantity Survey	427

Standards of Measurement 428
GROUP I: Walls, Ceilings and Floors 428
GROUP II: Doors and Windows 429
GROUP III: Millwork and Interior Wood 431
GROUP IV: Metals 432
GROUP V: Exterior Walls and Soffits 434
GROUP VI: Exterior Wood 435
Use of Spreadsheet Database Sorting 435
Waste Allowances 436

09970 - Coatings for Steel 439
Introduction 439
Quantity Survey 439
Extension of Quantities 440

10800 - Toilet, Bath, and Laundry Accessories 443
Types of Measurement 444
Factors Affecting Takeoff and Pricing.. 444
Overview of Labor, Equipment and Material 444
Ratios and Analysis Tools Used 446
Method and Approach to Quick Budget Estimates.. 446
Post Project Analysis and Recording Historical Data.. 447
Miscellaneous Pertinent Information 447
Glossary 447

13280 - Hazardous Material Remediation 449
Indirect Cost Items Affecting the Cost of Work 449
Direct Cost Items... 450
Mine Tailings as a Hazardous Material 450
Estimating the Disposal of Polychlorinated Biphenyls.. 451
Estimating the Removal of Asbestos Contaminated Material 452

The Anatomy of Mechanical Estimate 458

DISCIPLINE SPECIFIC ESTIMATING PROCEDURES

Scope of the Estimate

Refer to Part One Section 3 of this manual for further scope definition.

Specific Practice

The purpose of all sections of the standards is to provide detailed information on discipline specific estimating. It is not intended to include items common to all disciplines which is included in Part One. This includes general conditions and special conditions.

Terminology

Terminology common to all estimating is in Part Five of this manual. Terminology common to a specific discipline is at the end of each Discipline Specific Section.

Forms

The following forms are universal in nature and are in Part One Section 8. Checklists and special forms that are specific to a discipline, if any, are at the end of each section.

- ✓ Master Checklist
- ✓ Bid Document Inventory Checklist
- ✓ Site Investigation Checklist
- ✓ Direct Cost Estimate Checklist
- ✓ Overhead Estimate Checklist
- ✓ General Conditions Estimate Checklist
- ✓ Final Summary
- ✓ Final Summary Cut & Add Sheet
- ✓ Account Summary
- ✓ Estimate Detail
- ✓ Quantity Survey

Project Evaluation through Plan Review

These Sections match similar Sections in Part One. Please refer to Part One for this information.

The section numbering system will be according to the CSI MasterFormat, the Master List of Numbers and Titles for the Construction Industry, November, 1996.

DIVISION 1 GENERAL REQUIREMENTS

Introduction

This introduction provides a definition of the term "general conditions." It also includes a condensed summary of costs associated with this group of project related expenses.

General Conditions

General Conditions include any cost required to complete the project other than costs associated with the technical specifications sections. These costs are not normally a permanent part of the work. The portion of the estimate addressing specified portions of the permanent work is the direct cost part of the estimate. Estimate the "General Conditions" using the same methods as for the direct costs. The items may be quantified, and priced. General condition items are dependent on time constraints for completion of each element of work. Never estimate general conditions as a percentage of the direct costs. This percentage factor or rationale has merit in judgmental value only.

Group general conditions costs as follows:

1. Project site indirect costs, general or subcontractors.
2. Home office support and services, general contractor.
3. Services to the owner, architect, and engineer that are specified or eminently required.

The intent now is to provide a broad-based listing of those functions and costs that usually occur for each of the above groups. This listing, however, is not in any order of priority, importance, or magnitude.

Project Site Indirect Costs

The specific items of general conditions costs that usually occur in this group are:

- ✓ Personnel
- ✓ Committed temporary plant/facilities
- ✓ Environmental protection
- ✓ Construction equipment, owned or leased
- ✓ Site and home
- ✓ Temporary utilities
- ✓ Miscellaneous

A brief description, with checklist type headings follow elsewhere in this section. Format examples of each specific heading are in the appendix. For this portion of the Standards, the term "temporary plant/facilities" means such items required to accomplish the permanent work. The contract documents may not show these items in detail. They may include:

- ✓ Access roads
- ✓ Drainage swales
- ✓ Berms, pits or sumps
- ✓ Excavation shields
- ✓ Temporary enclosures
- ✓ Home office services

The items of home office support and services required for an individual project or estimate vary widely and are unique to individual firms. Estimators must address individual specific requirements within the guidelines established by management.

Subcontractors usually include home office support and services costs in their quotations for work.

Services to the Owner, Architect, Engineer

These general conditions costs are usually specified as furnished by the general contractor and may include:

- ✓ Owner/architect site office
- ✓ Janitorial services
- ✓ First aid/emergency treatment facilities
- ✓ Progress photographs
- ✓ As-built drawings, microfilm records
- ✓ Testing laboratory or facilities
- ✓ Quality control
- ✓ Consulting and testing fees
- ✓ Office supplies/blueprinting

Recap Summary

These general conditions procedures are not in the order of quantification or in the priority of estimating. They are checklist oriented. Other information sources required to generate the general conditions estimate are:

- ✓ Direct cost estimate
- ✓ Contract documents
- ✓ Company historical costs
- ✓ Personnel records

The direct cost estimate contains information on quantities, labor hours by craft, schedule and sequence requirements. Some projects require submittal of the progress schedule, including precedence diagram or CPM format with the proposal.

Project Site Staffing

Project site costs are a function of project size and duration, subcontractor work scopes, contractual requirements, and craft labor levels. The costs associated with the project site staffing involves the salaries and fringe benefits associated with the staff employees on the project. Costs on large projects may include management and engineering supervision and administrative personnel. Smaller projects may only have a part-time superintendent to watch the work. Use the schedule developed from the specifications and the direct cost estimate information to determine the staff size, duration, and when to mobilize or demobilize. Using the direct cost estimate activities and specifications requirements, it is possible to determine when each person must be at the project site.

Remember, the descriptions shown in the checklists do not mean there must be one person for each position. Construction bidding is a competitive process. Determine the staffing requirements and prove the estimated staffing of the project to the staff/ craft ratio on similar projects.

Management and Engineering

Management and Engineering personnel may include:

- ✓ Project manager
- ✓ Assistant project manager(s)
- ✓ Project engineer(s)
- ✓ Assistant project engineer(s)
- ✓ Estimator(s)
- ✓ Scheduler(s)
- ✓ Cost engineer(s)
- ✓ Safety engineer(s)
- ✓ Subcontract administrator(s)
- ✓ Clerk(s)
- ✓ Secretary(s)
- ✓

The managers, engineers, and estimators shown above may also serve at various levels such as chief and senior. At contract award, employment of the project manager, estimators, cost engineers, schedulers, and general superintendent is common practice before site mobilization.

An often overlooked group is quality assurance and quality control personnel. This area may require a QA/QC manager, QA/QC assistant manager, and QA/QC inspectors and technicians. The QA/QC staff also requires support personnel such as clerks and secretaries.

Supervision

Supervision may include personnel such as:

- ✓ General superintendent
- ✓ Typist(s)
- ✓ Secretary(s)
- ✓ Field engineer(s)
- ✓ Discipline specific superintendent(s)

Administrative

Project administrative personnel may include:

- ✓ Office Manager
- ✓ Accountant(s)
- ✓ Bookkeeper(s)
- ✓ Timekeeper(s)
- ✓ Secretary(s)
- ✓ Receptionist(s)
- ✓ Clerk(s)

Temporary Plant/Facilities

Temporary plant and facility costs are a function of project size and duration. These plants and facilities may include:

- ✓ Contractor's staff and owner offices
- ✓ Change houses for the crafts
- ✓ Material storage yards
- ✓ Emergency medical treatment
- ✓ Tool rooms
- ✓ Warehousing
- ✓ Fuel and maintenance areas

When there are requirements for considerable amounts of field fabrication, providing field fabrication shops may be required. In addition to the physical plant, there also may be requirements for barricades, fencing, signs and markers, and maintenance of traffic areas. Consider special shoring, trench shields and bracing, scaffolding, ladders, and temporary shelters for material storage or special warehousing. Size staff and owner offices by number and position. The project manager(s) and owner's representative(s) may require individual offices. Offices for the support staff probably would be satisfactory with an open or landscape design. Determine trades change houses or rooms by the number of trades and their size, climatic conditions, and common practice in the area.

Tool and material control distribution is a prime factor in the determination of the proper type of storage area for them. Storage of major equipment items may require blocking, bracing, shoring, and preparation and maintenance of storage areas. It is good business practice, and mandatory, to provide emergency medical facilities at the project site. The size and type of these areas is obviously dependent on the number of employees. In certain instances, a first aid kit in an office will suffice. Other projects may require separate offices, ambulances, and trained medical technicians. Site conditions and operations may require barricades, fencing, and signs. Specifications may require the maintenance of traffic patterns, especially in highway construction where there is considerable responsibility for protection of the public.

Temporary Utilities

Temporary utilities are requirements that may include:

- ✓ Construction water
- ✓ Dust control
- ✓ Sanitary systems
- ✓ Fire protection
- ✓ Electrical construction power
- ✓ Heating or cooling systems
- ✓ Potable water
- ✓ Lighting

Also consider the advantages and disadvantages of independent gas, air, and welding power systems for use in construction operations. The construction water source may range from a simple hydrant tap to a complex distribution system.

Fire protection water may involve storage tanks, the distribution of water in barrels, or having it available at the site through other means. Fire protection of electrical devices and electronic equipment requires systems other than water. The project may require fire protection curtains, blankets, and fire watch personnel. Give consideration to adequate lightning protection.

Project Office Expense

Project office expenses are those costs associated with the expense of maintaining and operating a project office. Determine these costs based on the scope and duration of the project. Included are the costs for supplies such as stationery, general office supplies, and engineering/drafting material. Office equipment may include desks, tables, chairs, filing cabinets, copy machines, blueprint machines, calculators, computers, and electronic data systems. In this section include utility hookups and monthly costs for water, power, gas, and telephone. Also remember applicable permits and licenses.

Environmental Protection

Information about this increasingly important area is under development for future publication as a supplement to this section.

Construction Equipment

Include construction equipment necessary to perform the work in this section of the general conditions estimate. Identify each area of the direct cost estimate that requires equipment for its assembly or installation. The costs for equipment usually may include:

- ✓ Rental
- ✓ Fuels, lubricants, repairs
- ✓ Depreciation
- ✓ Operating labor

Include equipment costs for items such as dust control and warehousing in this section.

Investigate the advisability of using company-owned equipment compared to leased (rented) equipment. Types of equipment for a specific project may include:

- ✓ Automotive, trucking
- ✓ Excavation
- ✓ Welding
- ✓ Boring and tunneling
- ✓ Cranes and rigging
- ✓ Pumps and compressors
- ✓ Drilling and blasting
- ✓ Instrumentation and testing

Automotive equipment includes sedans and pickup trucks.

Trucking includes:

Tractor units	Trailers
Water trucks	Fuel and maintenance trucks
Farm type tractors	Mining and heavy hauling off road trucks

Cranes and rigging equipment may include crawler, self-propelled and truck mounted cranes, pile driving and other lifting equipment, hoists, man lifts and derricks.

Standard Estimating Practice

Excavation equipment may include:

- ✓ Scrapers and elevating scrapers
- ✓ Boring machines
- ✓ Compactors
- ✓ Loaders
- ✓ Backhoes and trenchers

Pumps and compressors may include test pumps, dewatering pumps, and air compressors.

Regardless of the group used, each requires an ownership cost, whether external or internal rental plus operating costs. Geographical areas, climatic conditions, and availability have an impact on equipment selection and usage. The estimator should thoroughly investigate these items before selecting particular equipment. This is true especially if the project site is in a region unfamiliar to the estimator. Three basic situations apply to determine the location of equipment for estimating mobilization costs. These are:

The company has its own equipment.

The company will rent the equipment needed.

The company will purchase the equipment.

Using a combination of these three cases is not unusual when estimating a project. Locate owned pieces of equipment and identify the distance for transportation to the project site. Does the transportation equipment require permits for over width, overweight, or overweight? Determine if there are requirements for activities by internal forces, by a transport company, or by a combination of both. Moving equipment may require up to five general operations. These are:

- ✓ Disassembly
- ✓ Transporting
- ✓ Reassembly
- ✓ Loading
- ✓ Unloading

Larger equipment such as cranes and large excavating equipment, may require disassembly. While most types of equipment require some form of loading, this operation usually consumes only a little time. Accurately define transport of the equipment to get accurate quotations from a transportation company. Estimate internal transport cost by identifying hauling equipment required, crew size, and distance.

Offloading is the opposite of the loading operation. Reassembly is also the opposite of the disassembly operation but will require more time than the disassembly operation. When purchasing used equipment, review purchase agreements and determine if direct site delivery is an advantage. A quotation for rented equipment should indicate the appropriate hourly rate plus delivery charges to the project site.

Mobilization

Mobilization cost is a function of the job size, type of work, distance and time. It is costs incurred when starting a project. Items to consider for mobilization include:

- ✓ Construction equipment
- ✓ Small tools and consumables
- ✓ Personnel
- ✓ Construction plant

Personnel Mobilization

For longer duration projects located remotely from other company operations, personnel relocation may be necessary. This relocation may include rental of an apartment and furniture for the employee, moving their entire household, or a combination of both. Determine the relocation duration from the project start and completion dates and major schedule milestones. When estimating relocation costs, examine the conditions to determine which method has an economical advantage.

Plant/Facility Mobilization

Consider using a specialized company for moving project office trailers and similar equipment. The scope of this subcontract may include transportation and setup of the unit, usually on blocks or timbers. The contract will not include utility connections (electric, gas, sewer, and telephone). Provide for these connections in a separate section.

Construction Equipment Mobilization

Construction equipment is the major equipment required for the performance of the work. After determining the equipment needed for the project, consider costs for mobilizing the equipment. These costs include location of the current equipment, its size and weight.

Small Tools and Consumables Mobilization

Small tools and consumables usually present few transportation problems. Ship current inventory of tools to the project site in portable vans and bins for initial or permanent setup at the project site. It is simple to estimate transportation charges for portable units approved for highway use or adapted to fit on a flat bed truck.

If the project requires purchasing of new tools and consumables, arrange delivery to the project site with the supplier to lessen associated labor costs.

Demobilization

Demobilization is removing to another location those items brought to the project under mobilization. Include funds for final project cleanup, removal of surplus materials, punch list, and final inspection in the estimate. Investigate temporary enclosures to determine if there may be a credit for scrap or salvage material.

Miscellaneous

Include expected costs for insurance, taxes (other than direct labor burden), and bonds in this section. As defined in Part One, Section Three, Scope of the Estimate of this manual, the term direct labor burden means the combined costs for:

- ✓ Required worker's compensation insurance
- ✓ Subsistence
- ✓ Union and company fringe benefits
- ✓ State and federal unemployment insurance
- ✓ Employer paid Social Security taxes

Show direct labor burden as separate direct field costs in each estimate. Refer to that section of the manual for the balance of the definition.

Some of the insurance considerations include:

- ✓ General liability
- ✓ Builder's risk
- ✓ Equipment floaters
- ✓ Business vehicle
- ✓ Umbrella excess liability
- ✓ Other required special policies

Taxes may include:

- ✓ State and local sales or use
- ✓ Local, state and federal taxes on motor fuels
- ✓ Property and equipment
- ✓ Vehicle registration and licensing

Certain states require payment of sales taxes on the total contract amount. Identify the tax structure governing at the site of the project before completing the estimate. The specifications will define various bond requirements. These include:

- ✓ Bid
- ✓ Material supply
- ✓ Payment
- ✓ Installation Bonds
- ✓ Performance

These requirements for bonds may include both the general contractor and subcontractors. Payment and performance bonds are good business practice, whether required or not. Each project is subject to the domain of various levels of government that may require purchase of permits. Usually, county and city departments adopt the permit and plan check fee schedule of the uniform building code. Find the governing agency and determine its policy concerning permits and inspections.

Permits also may include demolition, shoring, and access.

Testing/Inspection and Outside Services

This section includes:

- ✓ Material and soil testing
- ✓ Authorized inspectors
- ✓ Personnel testing/clearance

Include testing and inspections performed by company forces in the direct cost estimate. These costs may include:

- ✓ Craft labor
- ✓ Supervision for the tests
- ✓ Testing equipment/materials

It is good practice to set a separate section within the direct cost estimate for this work. The specifications define tests and inspections required for a project. These tests may include:

1. Destructive and non-destructive examinations of materials

2. Testing of installed materials by x-ray or other means

3. Soil and concrete testing

The performance of these tests may require the use of an authorized inspector. Some material quotations may include testing costs in the price of the materials. Identification of the testing requirements to the supplier is essential to get probing prices. Use specialized, licensed firms when x-ray or similar testing is a requirement. These firms will usually provide unit price or hourly quotations for their services. The degree of estimate reliability for the costs for these services depends on accurately identifying their scope.

Soil and concrete testing may be the responsibility of either the contractor or the owner. Consult the specifications to determine where the responsibility rests. When required, specialized testing firms usually do the tests and quotations are on a unit price or hourly basis.

An authorized inspector is usually an independent individual or firm qualified or licensed to perform the inspections on the owner's behalf. Quotations for these costs may be on a lump sum or hourly basis.

Job Site Security

Security of the job site should be considered and an estimated budget established. Security is needed not only for the protection of material and products; but also the protection of the job site workers and the general public. Access to the site by those not authorized may result in the theft loss or injury to persons, such as curious children. The following items may be considered for a security budget:

- ✓ Temporary fencing

  - Perimeter with or without barbed wire or concertina wire

  - Job site office yard

  - Material storage / lay down area

  - Parking areas

  - Gates for access and egress control

  - Personnel flow control

- ✓ Security guards

  - Armed or unarmed

  - Patrols with or without vehicles

- Public law enforcement or private company
- Communications by radio or cell phone
- Written log report for period of surveillance
- Written reports of any activity after hours
- How many shifts and periods of
- Camera to photograph any incidents

✓ Security systems

- Motion detectors
- Video Cameras
- Alarm horns
- Monitored alarm system with auto dial to security company
- Fire alarms

✓ Physical barriers and miscellaneous accessories

- Burglar bars for doors and windows
- Temporary lock sets and construction keys
- Temporary doors
- Storage trailers or lock boxes with add on lock covers
- Knox key storage boxes for gate control authorized access for key personnel or fire department
- Worker badge system
- Gate guards during working hours for authorizing access and material loss control

✓ National security alerts

Each project and its location should be evaluated in regards to the effects of the different levels of the national security system. This could limit personnel access and material deliver to the project. Particular projects could require increased security personnel and systems if the national security level is increased.

Progress and Final Cleaning

Progress cleaning is maintaining the building and site in a neat and safe condition during construction. This may include general debris, material packaging and material waste. The estimate for this should include gathering, loading in a container and haul off from the job site. The proper disposal of waste may include fees from the dump site. After the trash has been removed, there normally is additional floor or surface sweeping required. In addition to the on site cleaning, street or other off site cleaning may be required. Renovations of businesses that remain operational will have special on going cleaning requirements.

The final clean is a thorough detailed cleaning of all exposed and unexposed surfaces such as: windows, store front, interior glass, walls, flooring, doors, counters, plumbing fixtures, mirrors, cabinet and drawer interiors. This is the final cleaning prior to turning the project over to the owner. This includes removal of all stickers, decals, labels and tags.

Small Tools and Consumables

Consider small tools and consumables necessary for the installation of the work as having a useful life only for the project duration. Two common methods for inclusion of this section in the estimate is either on a dollar or percentage basis. Use of the percentage method is common. Recommended practice is to identify and price them. The cost of the small tools is usually a function of the labor and type of project. Basing the amount of small tools and consumables required for the project on labor hours is more reliable than labor costs or total project cost. Refer to historical records for judgmental proof.

Weather Protection

This section covers the provision for protection against the elements. Extreme heat, cold, and moisture occur on most projects. Adequate weather protection may range from simplistic procedures to elaborate and expensive systems. Typically, materials and labor required to install and remove them are:

- ✓ Unit heaters and fans
- ✓ Plastic enclosures
- ✓ Tarps and curing blankets
- ✓ Temporary shelters
- ✓ Evaporative coolers and ice

Company Office Support

The costs for company office support will be minimal it the project has its own support staff. Maximum costs will be incurred if there is no project support staff or facilities and the project is being administered from the company office.

If the project is to be administered from the company office, these costs may be regarded as overhead and not included in the body of the estimate. If the project is to have its own support staff and facilities, these costs should be itemized in the general conditions section of the estimate.

Before deciding about the method of company office support, carefully check the advantages and disadvantages of both methods. This evaluation should include the requirements of the project and the overall company scope of work.

Management Support

A majority of management support will be jobsite or from the company office depending on the decisions made about the economics of these two scenarios. Technical support is a function of the contractual requirements and the complexity of the project. Administrative support from the company office should be tailored to assist the field operations staff. When there is a full time project staff, company office administrative support may take the form of aiding in the development of payroll, maintaining files for the project, change order administration, etc.

Escalation

Escalation is an increase in prices for labor, materials, and equipment for the duration of the project. When estimating and bidding work, provide for escalation.

Use the project schedule to determine wage rate changes on affected hours. With materials purchased at the start of the project, escalation is minor. Items with long lead times priced at the time of delivery may affect material costs escalation. In addition to the potential for escalation on permanent materials and wages, temporary materials, freight charges, and insurance may escalate more rapidly than other items.

Finance Expenses

"Progress payment delay cycle" will cause each project to incur finance expenses to build. Chart the progress payment delay cycle to determine the cash requirements for expenditures exceeding the income. There will always be an outflow of cash at the beginning of a project. By knowing the amount used from borrowed or reserve funds, calculations may determine the finance expense. To calculate this cycle, the specifications will ordinarily define:

- ✓ Time for submitting applications for payment
- ✓ Owner review and approval time
- ✓ What is allowable in the request
- ✓ Retainage applicable to the payment request

Usually, the contractor will work for nearly thirty days before submitting a progress payment request. During this time, the contractor will incur expenses for:

- ✓ Mobilizing the project
- ✓ Equipment rentals
- ✓ Utility hookups
- ✓ Staff salaries and craft wages
- ✓ Operating costs

Cash flow Analysis

Consider these possible delays in receiving payments. Submittal preparation may take up to five days to prepare from the cutoff date. The owner may take as much as two to three weeks to review and confirm that the request is in order. After this review, it may take up to thirty days to process the payment to the contractor. There usually is a retainage withheld. Give serious consideration to determine the funds required from project start until receiving the first payment. Consider retainages withheld until project completion separately and include the length of time withheld.

Contingencies

To many, the use of contingencies is an acceptable inclusion in the estimate. Contingency is the amount required to adjust the estimate to the project's proper valuation. Others use the contingency as a risk item for unknowns. For estimates other than competitive bids, this is acceptable and illustrates the use of contingencies. Contractors, in a competitive bidding situation, usually do not include contingencies or risk factors in their proposals.

The use of historical labor and equipment costs, accurate material quotations, and bids from responsible subcontractors decreases contingency requirements. Proper evaluation of the contract documents, getting document clarifications, taking exception to items in question further reduces the use of contingencies. Should there be major items of concern remaining at bid time, price them out or withdrawal from the bidding procedures may be advisable. In essence, costs included would be for those items identified as risk producing areas, defined as to scope and priced accordingly.

Summary

Define the major areas of project general conditions costs, then identify and price the details within each area accordingly. The most common error is the misidentification of the scope of the general conditions, whether under or overpriced. Conduct a thorough evaluation of each general conditions item to insure it is required. Be aware of the required scope of work, requirements of the specifications, and company operations.

02260 EXCAVATION SUPPORT AND PROTECTION — SOIL NAILING FOR BANK STABILIZATION

Soil nailing is a method of retaining the side of an excavation or stabilizing a slope by reinforcing the existing soil in place.[1] The term "Soil Nailing" actually refers to a complete earth support system, of which the soil nail is an integral part. The system involves reinforcing the existing soils by placing grouted steel tendons (soil nails) along the bank as the excavation proceeds downwards. A predetermined quantity of soil nails is installed in a pattern that essentially reinforces the ground into a stable block. Typically, an eight inch, reinforced, shotcrete wall completes the system. The finished product will have similar structural characteristics of a retaining wall (refer to Figure 02260-1). Soil nailing walls can be faced with precast panels, cast-in-place concrete, or an additional layer of shotcrete sculpted for aesthetics.

Construction of a soil nailing system for bank stabilization begins with a shallow cut along the top edge of the slope to be stabilized. The height of this cut will be determined by the capacities of the existing soils to stand vertically. Once this first level of excavation is completed, end threaded steel tendons are inserted into the augured holes. The tendons are held to the centerline of the auger hole by premoulded centralizers. The steel tendons are pre-cut to match the depth of the auger hole and will leave approximately 12 IN exposed to receive reinforcing and shotcrete.

Upon completing the installation of the steel tendon and grouting operation, reinforcing and shotcrete will be installed. Once the shotcrete has reached design strength, anchor plates and tension nuts will be installed to the complete the system.

Soil nailing is commonly referred to as a top down method of construction.[2] With the steps completed as described above, another level of excavation can now begin and the soil nailing process for stabilizing the next section of exposed bank can be repeated.

Types and Methods of Measurement

Knowledge of three major divisions, horizontal auguring, steel reinforcing, and concrete including shotcrete and grout, is needed when estimating soil nailing systems. Horizontal auguring is measured in lineal feet of depth and cubic yards of augured material.

Soil nailing sitemaps for bank stabilization typically have two types of reinforcing; steel reinforcing bar and wire mesh reinforcement. Steel reinforcing bar is typically grade 60 with a threaded end added to one end of the bar. The bars are cut to the design depth of the bore holes, are commonly referred to as tendons and measured in pounds. The wire mesh is calculated and priced by the square foot and allowances must be made for lap splices and waste.

The placement of shotcrete and grout for the soil nailing system is measured by the cubic yard. Some degree of finishing is required for the shotcrete but the larger costs are associated with the materials and placement. Soil nail systems do not require any formwork when using shotcrete, thus making the system more economical when compared to traditional cast-in-place concrete walls where two-sided formwork is often required.

---

[1] Soil Nailing in Varied Geological Formations, David E. Ferworn

[2] An Overview of Soil Nailing, H.P. Ludwig

Factors Affecting Takeoff and Pricing

Since soil nailing systems must be completed by sequencing the work around the excavator's bulk excavation operation, smaller projects (with cuts less than fifteen feet deep) tend to have slower production rates and thus higher costs.[3] Increased mobilizations and demobilizations drive up set-up costs. Extensive ground water conditions require countermeasures, which increases the overall project cost.

The type of soil conditions largely determines the viability of soil nailing systems. In areas with cohesionless sands (Mid-Atlantic States), expensive countermeasures are required making soil nailing systems less economically feasible. In areas where heavy freeze-thaw conditions occur for extended periods of time, denser nail patterns, additional reinforcing and thicker shotcrete wall profiles are common.[4]

Overview of Labor, Material, Equipment and Indirect Costs

An accurate quantity takeoff, organization by major categories, pricing based on site specific productivity, and verification by historical data are the basis for a complete estimate. The sample estimate is based on a site, which is a two hundred foot bank stabilization project. The depth of the vertical cut is twenty feet. The engineered soil nail pattern is three feet on center horizontally and four feet on center vertically (refer to Figure 02260-2). Both the nail spacing and depth will vary for each site and are directly related to the existing soil capacities. Augured holes of four inches in diameter are required and will be grouted with the tendon in place. One layer of welded wire reinforcing (wire mesh) is required for an eight inch thick, 4000 PSI shotcrete wall.

Auguring for Soil Nails

Carefully review the geotechnical reports and boring logs in order to determine auger production rates and the length of time augured holes can remain open. If the reports indicate a rock ledge ten feet below the surface of a proposed twenty foot deep nail wall, adjust the productivity rates of the auger operation within the rock layer. Typical auger production rates of four inch horizontal borings in clays and gravels can be more than two hundred lineal feet per hour. However, rates for certain rock formations can be as low as fifty lineal feet per hour.

Once soil conditions have been analyzed, determine the quantities of the auger operation. The bank to be stabilized is 200 LF long and 20 LF deep with soil nails 20 LF long. A nail pattern of 4 LF on center vertically and 3 LF on center horizontally will require 68 auger holes horizontally and 5 auger holes vertically (assuming soil nails would be required at the top of the bank). Multiplying the horizontal count by the vertical count will provide the total quantity of augured holes.

On sites with multiple layers of varying soil strata, it is convenient to calculate the quantity of auger holes by each level. This allows the estimator to alter production rates at each level as the soil strata changes (refer to Sample Estimate Sheet 02260-1). For example, the first level of a proposed excavation may be sands and gravels and the next level may be a rock ledge. The production rates for the second level (rock) will be less than the preceding level. Consider the quantity of spoil or excess material created by the auger

---

[3] U.S. Department of Transportation
[4] Reinforcement of Earth Slopes & Embankments, J.K. Mitchell

operation. Once the total length of auger holes is determined, the spoil calculation (in cubic yards) is the total auger hole length multiplied by the area of the hole converted into cubic yards. Multiply the amount by a swell factor to account for the natural compaction of the soils and the over excavation of the auger. Include an additional expansion factor to the spoils quantity. Costs for auguring equipment, spoils removal, fuel and maintenance are applies in determining the labor costs for the auguring operation. Equipment costs are included in the total cost of the auger operation.

Installing the Tendon

Determine the steel tendon or nail in lineal feet. The total length is calculated by multiplying the quantity of tendons by the design depth of the auger hole, providing for the thickness of the shotcrete wall. Convert the total length to weight. Typical tendon sizes range between No. 7 rebar to No. 9 rebar. For tendons exceeding manageable lengths, threaded couplings are used to splice the bars together. To insure that tendons remain along the center line of the auger hole, premoulded centralizers are installed at 6 LF intervals along the tendon (refer to Sample Estimate Sheet 02260-3).

Grouting the Tendon

Typically the grout is delivered as a ready-mix product to the site. However, remote sites may require the grout to be mixed on site. Caution should be used when on site mixing is used, to insure that proper water-cement ratios are maintained. The grout is then introduced into a mechanical pump.

It is critical to the design of soil nail systems that grout be uniformly introduced into the auger hole. The most common method of grouting the tendon is by low pressure tremmie. The tremmie allows for grouting to begin at or near the bottom of the auger hole with the tendon in place. Grout is uniformly installed as the tremmie is removed from the auger hole. The grout is calculated by the cubic yard. Compare the grout quantity against the auger spoil quantity for accuracy.

Material, labor costs and equipment rates, including anticipated pump productivity rates are illustrated in Sample Estimate Sheet 02260-5. Quantity calculations are shown on Sample Estimate Sheet 02260-4.

Shotcrete Wall Reinforcing

Wire mesh reinforcing is typically used as the reinforcing for the shotcrete wall. Calculations for labor and material costs are based on the square feet of contact area (SFCA).

Typical lap splices of 12 IN are common and must be factored into the cost analysis with a waste factor applied to the overall material quantity. Refer to Sample Estimate Sheet 00260-5 for detailed calculations for the mesh reinforcing.

Shotcrete Facing Wall

Shotcrete by definition refers to a mechanically driven concrete mix. It may be delivered as a ready-mix product or mixed at the site. A special pump is used to deliver the material through a series of hoses to the work area. The shotcrete is applied to the wall face after the tendons have been installed and grouted, and the wire mesh is in place. Installation for walls exceeding four inches will require two lifts, with no finishing or cure-time required between lifts.

After completing the final lift of shotcrete, the wall face will be roughly finished. Since the wall face can be covered with a finished surface (such as precast panels) or left unfinished, the estimates for labor to finish the wall surface varies greatly (refer to Sample Estimate Sheet 02260-6).

Subsurface conditions should determine the waste factor of shotcrete at each site. For example, higher waste factors would be anticipated for soils with low cohesion. These "soft" soils will result in over excavation by the site contractor and consequently additional shotcrete will be required to maintain the surface elevation of the wall.

Vertical expansion joints are sometimes installed during the shotcrete process. When used, vertical expansion joints are normally spaced at forty foot centers. Additional vertical, hand tooled crack control joints are installed at twenty foot centers. The cost for installation of control joints is typically spread across the shotcrete wall costs.

Miscellaneous Accessories

Once the grout for the tendon and the shotcrete wall have cured, the steel face plates and tension nuts are installed to complete the soil nail system. Today's high strength, fast curing grouts and shotcrete, typically allow this work to occur before moving to the next level of excavation.

Listed below is a typical relationship of major cost elements for a soil nail system used for bank stabilization.

Historical data should be used for comparison of similar projects.

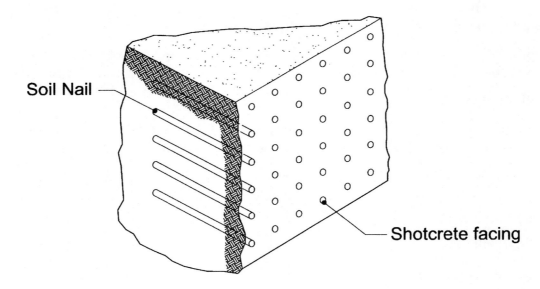

SK-1 SCHEMATIC DIAGRAM

Figure 02260-1

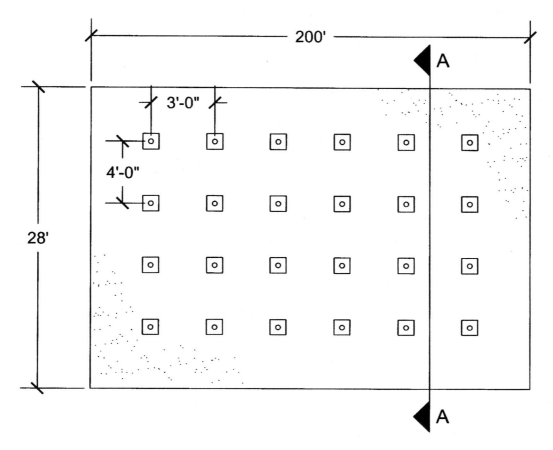

SK-2 ELEVATION - PARTIAL NAIL WALL FACE
NOT TO SCALE

Figure 02260-2

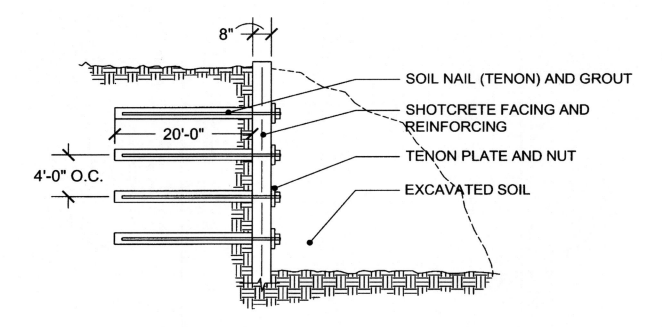

SK-3 SECTION A-A
NOT TO SCALE

Figure 0226-03

Sheet SE – 1

QUANTITY SHEET						
Project	ASPE - SAMPLE			**Date**		5/99
Description	BANK STABILIZATION				**Estimator**	
Architect					**Checked**	
Description	4" ⊙ AUGER HOLES					
	Description	Quantity		Auger Hole Qty.	Length	Total Auger Length
1	LEVEL 1 0-5'd SAND/GRAVEL					
2	200' LEGNTH @ 3' O.C. = 68 EA					
3	5' DEPTH @ 4' O.C. = 2 EA					
4	TOTAL = 68 H x 2 V = 136 EA			136	20	2,720 LF
5						
6	LEVEL 2 5'-10' d SAND/GRAVEL					
7	200' LEGNTH @ 3' O.C. = 68 EA					
8	5' DEPTH @ 4' O.C. = 1 EA					
9	TOTAL = 68 H x 2 V = 68			68	20	1,360 LF
10						
11	LEVEL 3 10' - 15' d SAND/GRAVEL					
12	200' LEGNTH @ 3' O.C. = 68 EA					
13	5' DEPTH @ 4' O.C. = 1 EA					
14	TOTAL = 68 H x 1 V = 68			68	20	1,360 LF
15						
16	LEVEL 4 15' - 20' d SAND/GRAVEL					
17	200' LEGNTH @ 3' O.C. = 68 EA					
18	5' DEPTH @ 4' O.C. = 1 EA					
19	TOTAL = 68 H x 1 V = 68			68	20	1,360 LF
20						
21	**TOTALS**			340		6,800 LF
22						
23						
24						

Sample Estimate Sheet 02260-1

Sheet SE-1A

QUANTITY SHEET							
Project	ASPE - SAMPLE				**Date**		5/99
Description	BANK STABILIZATION				**Estimator**		
Architect					**Checked**		
Description	Spoils Removal							
	Description	Quantity	Hole Diameter	Area Constant $\pi \cdot r^2$	Depth	Total Cu. Ft.	
1							
2	4" Auger	340 ea	4" (r)	.087	20'	591.6 cu. ft.	
3							
4	Add 20% swell factor					119.32 cu. ft.	
5							
6						709.92	
7							
8						26.29 c.y.	
9							
10							
11							
12							
13							
14							
15							
16							
17							
18							
19							
20							
21							
22							
23							
24							

Sample Estimate Sheet 02260-2

Sheet SE-2

	QUANTITY SHEET					
Project	ASPE - SAMPLE			**Date**	5/99	
Description	BANK STABILIZATION			**Estimator**		
Architect				**Checked**		
Description	Steel Tendon - No. 8 Rebar					
	Description	Quantity	Length	Subtotal	Weight	Total Weight
1						
2	LEVEL 1	136 EA	21'	2,856'	2.67#/lf.	7,625.52 #
3						
4	LEVEL 2	68 EA	21'	1,428'	2.67#/lf.	3,812.76
5						
6	LEVEL 3	68 EA	21'	1,428'	2.67#/lf.	3,812.76
7						
8	LEVEL 4	68 EA	21'	1,428'	2.67#/lf.	3,812.76
9						
10						
11						19,064 #
12	CENTRALIZERS	340 EA	20'	21/6'oc.	= 4 x	1,360 EA.
13						
14						
15						
16						
17						
18						
19						
20						
21						
22						
23						
24						

Sample Estimate Sheet 02260-3

Sheet SE – 3

QUANTITY SHEET

Project	ASPE - SAMPLE	Date	5/99
Description	BANK STABILIZATION	**Estimator**	
Architect		**Checked**	

Description 4,000 psi Grout - 4" ⊙ Auger

	Description	Quantity	Area Constant $\pi \cdot r^2$	Length	Total Cu. Ft.	
1						
2	LEVEL 1	136	.087	20'	236.64	
3						
4	LEVEL 2	68	.087	20'	118.32	
5						
6	LEVEL 3	68	.087	20'	118.32	
7						
8	LEVEL 4	68	.087	20'	118.32	
9						
10						
11					591.60	
12	Add 10% Waste	Factor			59.16	
13					650.76	cu. ft.
14						
15					24.1	cu. yd.
16						
17						
18						
19						
20						
21						
22						
23						
24						

Sample Estimate Sheet 02260-4

Sheet – SE – 4

	QUANTITY SHEET					
Project ASPE - SAMPLE				**Date** 5/99		
Description				**Estimator**		
Architect				**Checked**		
Description Wire Mesh Reinforcing - WWF 6.1 x 6.1 w/1' LAP						
	Description	Quantity	Width / Height	Vert. LAP	Horoz. LAP	Total S.F. Area
1						
2	LEVEL 1	200'	5'	---	---	1,000 s.f.
3	Vert. Lap	8 EA	5'	1' (w)	---	40 s.f.
4	Horoz. Lap	0	0	---	---	0 s.f.
5						
6	LEVEL 2	200'	5'	---	---	1,000 s.f.
7	Vert. Lap	8 EA	5'	1' (w)	---	40 s.f.
8	Horoz. Lap	200'	1'	---	1' (h)	200 s.f.
9						
10	LEVEL 3	200'	5'	---	---	1,000 s.f.
11	Vert. Lap	8 EA	5'	1' (w)	---	40 s.f.
12	Horoz. Lap	200'	1'	---	1' (h)	200 s.f.
13						
14	LEVEL 4	200'	5'	---	---	1,000 s.f.
15	Vert. Lap	8 EA	5'	1' (w)	---	40 s.f.
16	Horoz. Lap	200'	1'	---	1' (h)	200 s.f.
17						
18					TOTAL	4,760 s.f.
19						
20						
21						
22						
23						
24						

Sample Estimate Sheet 02260-5

02260 Excavation Support and Protection

Sheet – SE – 5

	QUANTITY SHEET				
Project ASPE - SAMPLE				**Date** 5/99	
Description BANK STABILIZATION				**Estimator**	
Architect				**Checked**	
Description 8" Shotcrete Wall					
	Description	Quantity	Height	Thickness	Total cu. ft.
1					
2	LEVEL 1	200' (L)	5'	.67	670 cu. ft.
3					
4	LEVEL 2	200' (L)	5'	.67	670 cu. ft.
5					
6	LEVEL 3	200' (L)	5'	.67	670 cu. ft.
7					
8	LEVEL 4	200' (L)	5'	.67	670 cu. ft.
9					
10				Subtotal	2,680 cu. ft.
11					
12				10% Waste	269 cu.ft.
13					
14				TOTAL	2,948 cu.ft.
15					
16					109.2 cu. yd.
17					
18					
19					
20					
21					
22					
23					
24					

Sample Estimate Sheet 02260-6

Sheet SE – 6

	QUANTITY SHEET				
Project ASPE - SAMPLE			**Date**	5/99	
Description			**Estimator**		
Architect			**Checked**		
Description Hardware - Tension Plate & Nuts					
	Description	Quantity			
1					
2	LEVEL 1	136 EA			
3					
4	LEVEL 2	68 EA			
5					
6	LEVEL 3	68 EA			
7					
8	LEVEL 4	68 EA			
9					
10	TOTAL	340 EA			
11					
12					
13					
14					
15					
16					
17					
18					
19					
20					
21					
22					
23					
24					

Sample Estimate Sheet 02260-7

02260 Excavation Support and Protection — Part Two

Sheet Sp-1

Item Soil Nail System For Bank Stabilization													Estimator			
Project ASPE - Sample													Checked		Date Ck'd	
Bid Date 5/99																
Description	Quantity	UM	Unit Hours	Labor Hours	Labor Rate	Total Labor	Unit Price	Material	Unit Price	Equip.	Unit Price	Sub	Unit Price	Total	Unit Price
4" Auger	6,800	LF	.20	1,360	30.00	40,800	6.00	---	---	31,415	4.62	---	---	72,215	10.62
Spoils	27	CY	---	---	---	---	---	---	---	---	---	270.00	10.00	270.00	10
Tendon	19,065	LB	.020	382	25.00	9,555	.50	5,725	.30	1,875	.10	---	---	17,145	0.90
4,000 PSI Grout	24.1	C.Y	1.68	40.5	38.00	1,539	63.86	1,976	82.00	3,014	125.06	---	---	6,529	270.91
Wire Mesh	4,760	S.F.	.03	142.8	25.00	3,575	.75	952	.20	---	---	---	---	4,522	.95
Shotcrete	110	C.Y	1.40	154.0	30.00	4,680	42	8,690	79	2,200	20.00	---	---	15,510	141.00
Tension Plate	340	EA	.17	57.80	25.00	1,445	4.25	1,700	5.00	---	---	---	---	3,145	9.25
Centralizers	1,134	EA	.025	28.35	25.00	709	.63	9,525	8.00	---	---	---	---	10,284	4.02
														Total Cost $129,567 SFCA 4,000 S.F. Cost/Safe $32.40	

Sample Estimate Sheet 02260-8

02310 – GRADING

Introduction

Grading (mass excavation) is one of the first operations considered when beginning site development. It may include site clearing (Section 02230), stockpiling, berming, cutting, filling, shelving, finish grading and other reshaping or modification of existing grades. These methods are mainly utilized for the purpose of creating areas for new construction, promoting better site drainage, wind brakes, noise reduction, preservation of wildlife and other requirements for new construction. Structural excavation for structures and structural footings will not be included in this section.

Type of Measurement Used

There are three basic units of measure, commonly used when estimating grading: square feet, cubic yards and tons. Square feet is used to determine total surface area. Cubic yard is used to measure the volume of materials such as, cut and fill. Trap rock, modified stone, crusher run stone, etc. is calculated in cubic yards and converted to tons. Quantities are taken off as "bank", or "in place" cubic yards. Once the material has been disturbed or excavated, the material will increase in volume (swell), which should be addressed by using a suitable increase factor or multiplier based on the type of material. Conversely, the reverse is true in placing and compacting fill.

Factors Affecting Takeoff and Pricing

There are numerous types of earth moving equipment utilized in the grading process (mass excavation).

John Deere - Contractors and Loggers

Most earthmoving equipment has standard productivity rates available from the manufacturer. Sizing the proper equipment to the scale of the project is very important. Limited space, ability to access the site, equipment/crew production rates, underground utilities, erosion control and weather conditions during construction, are among the items that should be considered when estimating an earthmoving project.

Phasing and sequencing in the process of earthwork is very important in determining efficiency and costs. The locations of suitable borrow and spoil sites should also be determined.

A region may include several types of soil types and composition. Geotechnical consultants are vital in determining this information as well as providing crucial information and determining any other existing conditions, using various tests such as, sieve analysis, compaction tests and boring samples. Generally these reports make recommendations and suggestions to the design team and are included in the bid documents. The owner or design professional during the pre-construction phase of the project generally pays for the cost of these services. Reports generated by these consultants may also include recommendations on the following:

- ✓ Instructions on stripping and stockpiling of topsoil
- ✓ Soils that require removal and replacement
- ✓ Soils that require removal and re-compaction
- ✓ Pier and/or caisson depths and bearing capacity
- ✓ The quality and quantity of rock to be removed
- ✓ The suitability of the existing excavated materials being used for fill and structural backfill
- ✓ Inspection and testing requirements

These reports may also include instructions on using granular material (rock/crushed stone) for backfill. The estimator should be aware of water conditions, which may require dewatering and/or major dewatering systems such as well-points, sumps, etc., as well as the potential for flooding.

Calculating Quantities

A wide variety of estimating software is available to assist in the quantity survey. Most programs provide for different soil types, which require separate quantity calculations and methods of excavation. If site clearing is not required, stripping of the topsoil (a four to six inch depth, unless noted otherwise) is the first process. The excavation process usually is done in layers of the different types of soil, i.e. clay, sand, rock, etc. Each stratum is quantified and priced using pricing developed from the equipment and labor production rates and past historical data. Additionally any borrow material (imported fill) or spoil (excess excavation) must be quantified and priced to reflect the hauling costs associated with bringing in and hauling away materials.

One method of calculating quantities is by the use of a digitizer, which is a computer device. Cross-section and borrow-pit (grid chart) are manual methods of calculating quantities. The manual calculation method is described in detail in publications such as Moving the Earth: The Workbook of Excavation by Herbert L. Nichols, et al, Estimating Earthwork Quantities by Daniel Benn Atcheson, and other texts on excavating.

Pricing

After quantities are determined, the next process is calculation of production rates and selection of equipment types. Equipment manufacturers generally provide average production rates for various types of their equipment. Local conditions and your company's historical data may modify these production rates. This information will be the basis for developing your unit costs.

There are two methods to account for equipment costs, rental or purchase. Even if the equipment is purchased, an internal rental rate may be charged to the job. The rental rate should include all expenses, such as fuel, daily maintenance, replacement costs, and operator's wages as well as labor burden. Equipment should be suited to the application required, using the wrong piece of equipment will lead to inefficient operation, and consequential higher cost. Generally past experience, equipment manufacturers and rental companies can help you determine the best equipment for your application.

Mobilization and de-mobilization includes all labor, materials, and equipment necessary to move on to and off of the site. This may require the cost of special permits for oversized hauling, temporary easements and utility charges. This may also include the construction, and removal, of temporary roads, temporary parking lots, lay down areas and temporary office locations, in order to facilitate new construction. Erosion Controls such as silt fences, geo-textile fabrics, diversion berms, temporary detention areas, or other mandated temporary controls might also be required.

Governmental and environmental agencies require bonds, fees and permits to be paid prior to the start of any earthwork project. Other costs may include wetland plantings, safety precautions to avoid contaminating existing bodies of water, protection of existing conditions, and other environmental protection.

Specific comments referring to hazardous material and unforeseen conditions should be included as clarifications to your estimate, unless costs have been included. Costs for soils tests, inspectors, engineers or consultants may be required by the bid documents and should be included in the estimate. Specific instructions may be included to salvage or protect trees and other landscaping and should be addressed in the estimate.

Access to the site is an important consideration, roads, bridges, tunnels, etc, should be examined and determination made that the weight of construction equipment will not exceed the legal limits. Different longer routes may be required for access if that the equipment will not fit through, over or under obstructions. Temporary shoring or other temporary construction may be required. Consideration should be given to the proximity of railroads, power, telephone, other overhead obstructions, underground obstructions, hiking/biking trails, and water above and below grade. Extremely dry may require watering to control dust. Wet conditions may require cleaning of roads, streets or other access routes as well as the site. Weather conditions, equipment repairs, change in soil conditions, etc. are the indirect costs that must be factored into the unit prices. Other factors could include unfavorable working conditions, such as, steep slopes, difficult access, rock, water control, flooding, etc.

02315 EXCAVATION & FILL – TRENCHING

INTRODUCTION

Trench excavation and backfill is one of the main tasks in any pipeline or utility installation. Conditions may vary from rural cities to urban communities to cross-country areas. Trench excavation can be performed by hand, with a backhoe or with trenchers. Various types of ground will be encountered, such as dirt, rock, sand, or clay. Some of the other factors to consider when excavating include shoring requirements, ground water, soil placement and soil haul off. Each project will have it's own unique factors which must be considered.

TYPES AND METHODS OF MEASUREMENTS

Trench excavation and backfill is broken down into two basic measurements: lineal feet (LF) and cubic yards (CY).

Lineal Feet

The original takeoff of the trench will be measured in lineal feet. To determine the lineal feet, scale the drawings or use the station marks on the plan to come up with the actual lineal feet of trench.

Cubic Yards

The excavated and backfill materials are measured in cubic yards. To calculate the cubic yardage of excavation use the following formula: Length x Width x Depth / 27. To calculate haul-off or backfill, a swell or compaction factor should be applied to the excavated quantity. Typical factors can be as follows: Common earth — 30%; Moist loam, sand clay — 40%; Sand and gravel — 18%. The formula for CY of backfill for common earth would be CY (of excavation) x 1.3 (swell factor) = CY (of backfill).

FACTORS AFFECTING TAKEOFF AND PRICING

It is recommended that a thorough site examination be made prior to preparing the estimate to determine the existence of any varying site conditions and/or obstructions that might have an impact on the methods for accomplishing the work and productivity at which the work can be completed. Several items that must be considered are the existing terrain which will affect the types of equipment selected for the project (i.e. track vs. rubber-tired, etc.), the soil type which will affect the equipment selected as well as determine the need for imported backfill material and disposal of unsuitable excavated material, the existence of rock which may require blasting or other means of removal, ground water conditions which may require pumping or other dewatering methods, and the existence of other utilities either overhead or underground which again could affect equipment selection as well as limit productive. The time of year must be considered which will greatly affect production. In the winter consideration should be given to working with frozen ground. In some areas heavy spring rains can be also detrimental due to the effect of yielding soils and required dewatering activities.

OVERVIEW OF LABOR, EQUIPMENT, MATERIALS, AND QUANTITY

In developing an estimate for trenching it must be determined what type of equipment is to be used for the excavation and backfill. Many factors come into play when selecting equipment to be used. Some considerations are: equipment available, is it owned or rented, quantity of trenching to be accomplished, location, whether or not it is a congested site or open site, type of soils, etc. Equipment selection will determine what the products will be and help determine the crew needed to perform the tasks required.

02315 Excavation & Fill - Trenching

The determination must be made if excavated material is suitable for backfill or if imported material is to be brought in and excavated material hauled away. Also determine if there is a requirement for a bedding material in the trench. It must also be determined, based on depth of excavation, if shoring or layback is required. When calculating quantities for backfill and spoil removal, be sure to consider the volume the pipe, footing, etc. is displacing.

SAMPLE ESTIMATE

The sample estimate is Trench Excavation and backfill for a partial site storm system and is based on the following assumptions.

Line 1.   STA "A" - STA "B"   156' 24 IN RCP   Average 6' depth
Line 2.   STA "B" — STA"C"   119' 24 IN RCP   Average 6' depth
Line 3.   STA "C" - STA "D   200' 30 IN RCP   Average 6' depth

Site prep has been completed.

Layout has been completed.

No obstructions either above of below grade exist.

A 1:2 lay-back will be used.

The excavated material is suitable for backfill.

1-1/2 cy hydraulic backhoe.

Tamp backfill with vibrating plate.

The following quantities will be generated:

Excavation	CUYD
Hard Trim Bottom	SQFT
Backfill	CUYD
Compaction	CUYD
Load/Haul from Site	CUYD

EXCAVATION 763 CUYD

Line 1. 156.0 (length) x 7.0 (width including layback) x 6.0 (depth)/27 = 243 CY
Line 2. 119.0   x 7.0   x 6.0   / 27 = 186 CY
Line 3. 200.0   x 7.5   x 6.0   / 27 = 334 CY

HAND TRIM BOTTOM 1050 SQFT

Line 1. 156.0 (length) x 2.0 (width) = 312 SQFT
Line 2. 119.0   x 2.0   = 238 SQFT
Line 3. 200.0   x 2.5   = 500 SQFT

BACKFILL 905 CUYD

Line 1. 243 CY(excavation) - 18CY (pipe displacement)* 1.3 (compaction factor) =293CY
Line 2. 186 CY   - 14CY   * 1.3   =224CY
Line 3. 334 CY   - 36CY   * 1.3   =388CY

COMPACTION 905 CUYD

LOAD & HAUL FROM SITE 87 CY

Excavated Material 763CY * 1.3 (swell factor) = 992CY - Backfill 905CY = 87CY

See Figure 1. for sample estimate.

CONCLUSION

As with any section of the estimate, the estimate should refer to company historical data to help in evaluation of the final price for the estimate. This information can be invaluable when considering all the variables required for trench excavation and backfill. Also in the final evaluation make sure proper consideration has been given to the special risk conditions associated with trenching: type of soil, time of year, equipment selection, safety (shoring and layback requirements), ground water, traffic restrictions, and above or below grade obstructions.

02315 Excavation & Fill - Trenching — Part Two

Estimate Detail

Company Name: American Society of Professional Estimators
Project: Standard Estimating Practice, Sixth Edition
Address: Wheaton, MD 20902
System/Work Area: Trench Excavation & Backfill — Estimate # 6th Edition
CSI Section: 02315 — Bid Date July 2004

Estimator: Standard Board — Date: January 2003
Pricing: Standard Board — Date: January 2003
Extended: Standard Board — Date: January 2003
Summary: Standard Board — Date: January 2003
Verified: Standard Board — Date: January 2003

Item or Description			Material				Labor		Equipment		Sub		Total
DESCRIPTION	QUANTITY	UNIT	UNIT $	COST	HR/UNIT	HOURS	UNIT $	COST	UNIT $	COST	UNIT $	COST	COST
Trenche Excavation	763	CY		-			1.10	839	1.50	1,145		-	1,984
Hand Trim Bottom of Trench	1,050	SF		-			0.45	473		-		-	473
Trench Backfill	905	CY		-			.98	887	0.65	588		-	1,475
Compaction of Backfill	905	CY		-			2.75	2,489	0.25	226		-	2,715
Load & Haul Surplus	87	CY		-			1.85	161	3.85	335		-	496
Subtotals				-		-		4,848		2,294		-	7,142
Labor Burden on labor %							50%	2,424					2,424
Small Tools %							5%	242					242
Total Direct Cost								7,515		2,294		-	9,809
Overhead 10%													981
Profit 5%													539
Insurance & Bond 2%													227
Sales Tax on Material 6%													-
Total Subcontractor Cost													$11,556
Notes:													

02465-BORED PILES – DRILLED CAISSONS

Introduction

Drilled concrete piers and shafts are poured-in-place reinforced concrete footings. These piers are located under both the exterior and interior grade beams and also under column locations. The pier holes are drilled and excavated, the reinforcing steel is put in place and the concrete is poured into the piers. These piers typically vary in depth and will have a shaft diameter, or "plinth" ranging from 12 IN to 48 IN and larger. The diameter of the under-ream, or bell, of the pier is typically 2 to 3 times larger than the diameter of the shaft. The angle of under-ream is usually 45 degrees; however, a 60 degree under-ream is sometimes specified by the structural engineer.

Types and Methods of Measurement

The diameter of the shaft and the bell or under-ream of concrete drilled piers is measured in inches. The size of the piers is usually noted on drawings by listing the diameter of the shaft in inches over the diameter of the bell in inches (See Figure 02465-1). The depth of the pier is measured in lineal feet. The volume of the piers is measured in cubic feet, which can be converted, to cubic yards. Tables are available and examples are included that list the volume of pier in cubic feet for any given size and depth. (See Figure 02465-2)

ASTM STANDARD REINFORCING BARS

Bar Size Designation	Weight Lbs/lf	Dia. In Inches
#2	0.167	0.250
#3	0.376	0.375
#4	0.68	0.500
#5	1.043	0.625
#6	1.502	0.750
#7	2.044	0.875
#8	2.670	1.000
#9	3.400	1.128
#10	4.303	1.270
#11	5.313	1.410
#12	7.650	1.693
#14	13.600	2.257

Figure 02465-2

The reinforcing steel in the piers is measured in the weight of steel required and the concrete material to be placed in the piers is measured in cubic yards.

Factors Affecting Takeoff and Pricing

- **Soil Conditions**

The most important factor that will affect takeoff and pricing is the existing soil conditions. Most projects have a geotechnical report included in the specifications, which describes in great detail the characteristics of the existing soil. The geotechnical report is compiled by an engineering firm after a site investigation is completed where borings are taken at various locations on the site and analyzed for the presence ground water, rock, limestone, hard or soft clays, sand, etc. The engineer will make recommendations based on their analysis as to how the foundations should be designed. If there is an indication that the soil is unstable or there is a presence of ground water, the piers will need to be cased to prevent the sides of the shaft walls from caving. Casing is a steel liner that is placed in the pier shaft to keep the sides from caving in until the concrete is placed. The casing may remain in place or be removed. The drilling subcontractor may provide a unit price for casing the piers or will include casing in their bids if all piers are to be cased. It will be necessary to increase the diameter of the pier shaft by 2 IN if the piers are going to be cased and removed to account for the additional concrete required as the casing is removed. The presence of rock will also affect the pricing. More time is required to drill each pier, which results in increased labor and equipment costs. In some cases physical inspection will be required, thus requiring the minimum shaft to be 36 IN in diameter.

- **Site Accessibility and Condition**

On a small confined site with limited access, it can be difficult to get a drilling truck to the location where the piers are to be located. Some renovation projects require piers to be placed in an existing building or structure where overhead space is limited, thus, requiring a drilling rig small enough to drill in small confined areas. Many drilling subcontractors do not have these small drilling rigs, which can affect the pricing and scheduling of the project. If a project is done in phases where the drilling subcontractor has to make more than one trip to the site, there are usually charges for each additional mobilization.

Overview of Labor, Material, Equipment and Indirect Costs

A sample estimate (Example 02465-1 and 02465-2) is attached. This estimate is broken down into the following categories: labor, material, subcontract, and equipment.

- **Material** — Determine the depth or bearing elevation of each pier. Usually a detail or note is provided in the structural drawings listing the elevations of the top and bottom of the pier. Charts (Fig. B) are provided that gives the volume in cubic feet for each pier size for any given depth. In many areas local drilling contractors will have chart, tables and/or booklets available for calculating volumes of piers and bells. The volume is converted into cubic yards to determine the amount of concrete required for the piers. Typically there is a pier schedule in the structural drawings that list the number and size or diameter of the vertical steel reinforcing required in each pier size. (Fig. C) The pier schedule also gives the size of stirrups, or reinforcing bars spaced horizontally from top to bottom, which are tied to the vertical bars. The lineal feet of each size of reinforcing steel are converted to the weight of reinforcing steel required for the piers. Fig. D gives the weight per lineal foot of each size or diameter of reinforcing steel. After the piers are drilled, the spoils, or soil removed from the hole must be hauled away or relocated on site. The quantity of dirt or spoils to be removed should be calculated and a factor applied (depending on soil conditions) to convert the soil from "bank" cubic yards to "loose" cubic yards. The material quantities can be totaled and a waste factor should be added to the concrete and reinforcing steel.

- **Labor** — Productivity rate will be applied on a per unit basis to the material quantities. Labor productivity can be obtained from various publications that provide cost data or from historical cost data from previous projects. The labor required to layout the location of the piers is applied per unit; layout and tie the reinforcing steel is per the weight of reinforcing steel; place the concrete and remove the remaining spoil dirt applied per cubic yard.

- **Subcontract** — Foundation drilling contractors will provide subcontract pricing to drill the piers. The subcontractors need to be aware of the existing soil conditions, the depth of the piers and if there is a requirement for casing the piers.

- **Equipment** - Equipment will be required to either remove the spoils from the site or spread out on site. Equipment may be required to place reinforcing steel into the pier. Accessibility limitations may require the use of special equipment.

Special Risk Considerations

The estimator should review the characteristics of the jobsite before finalizing the estimate. Existing conditions can increase the risk of variations between estimated costs and actual costs. The presence of unknown underground structures or utilities should be considered. In addition to the items listed above, phasing of the project needs to be identified. It is very important to review the soils report and any other geotechnical information when preparing the estimate.

Drill Pier Volume Charts
16 IN CHART
Volumes shown are for shaft and bell

Shaft/Bell	\<td colspan="11">DEPTH
	4'	5'	6'	7'	8'	9'	10'	11'	12'	13'	14'
16 x 20	7.00	8.40	9.80	11.20	12.60	14.00	15.40	16.80	18.20	19.60	21.00
16 x 22	7.40	8.80	10.20	11.60	13.00	14.40	15.80	17.20	18.60	20.00	21.40
16 x 24	7.82	9.22	10.62	12.02	13.42	14.82	16.22	17.62	19.02	20.42	21.82
16 x 26	8.28	9.68	11.08	12.48	13.88	15.28	16.67	18.08	19.48	20.88	22.28
16 x 28	8.77	10.17	11.57	12.97	14.37	15.77	17.17	18.57	19.97	21.37	22.77
16 x 30	9.28	10.68	12.08	13.47	14.88	16.28	17.68	19.08	20.48	21.88	23.28
16 x 32	9.83	11.23	12.63	14.03	153.42	16.83	18.23	19.6.	21.03	22.43	23.83
16 x 34	10.41	11.81	13.21	14.61	16.01	17.41	18.81	20.21	21.61	23.01	24.41
16 x 36	11.02	12.42	13.82	15.22	16.62	18.02	19.42	20.82	22.22	23.62	25.02
16 x 38	11.65	13.05	14.45	15.85	17.25	18.65	20.05	21.45	22.85	24.25	25.65
16 x 40	12.32	13.72	15.12	16.52	17.92	19.32	20.72	22.12	23.52	24.92	26.32
16 x 42	13.01	14.41	15.81	17.21	18.61	20.01	21.41	22.81	24.21	25.61	27.01
16 x 44	13.75	15.15	16.55	17.95	19.35	20.75	22.15	23.55	24.95	26.35	27.75
16 x 46	14.51	15.91	17.31	18.71	20.11	21.51	22.91	24.31	25.71	27.11	28.51
16 x 48	15.29	16.69	18.09	19.49	20.89	22.29	23.69	25.09	26.49	27.89	29.29

(16 IN dia. shaft: volume of shaft = 1.41 c.f. per l.f. depth)

18 IN CHART (partial)
Volumes shown are for shaft and bell

Shaft/Bell	DEPTH										
	4'	5'	6'	7'	8'	9'	10'	11'	12'	13'	14'
18 x 36	13.26	15.03	16.80	18.57	20.34	22.11	23.88	25.65	27.42	29.19	30.96
18 x 38	14.03	15.80	17.57	19.34	21.11	22.88	24.65	26.42	28.19	29.96	31.73
18 x 40	14.84	16.61	18.38	20.15	21.92	23.69	25.46	27.23	29.00	30.77	32.54
18 x 42	15.69	17.46	19.23	21.00	22.77	24.54	26.31	28.08	29.85	31.62	33.39
18 x 44	16.56	18.33	20.10	21.87	23.64	25.41	27.18	28.95	30.72	32.49	34.26

(18 IN dia. shaft: volume of shaft = 1.77 c.f. per l.f. depth)

20 IN CHART (partial)
Volumes shown are for shaft and bell

Shaft/Bell	DEPTH										
	4'	5'	6'	7'	8'	9'	10'	11'	12'	13'	14'
20 x 46	19.71	21.89	24.07	26.25	28.43	30.61	32.79	34.97	37.15	39.32	41.51
20 x 48	20.75	22.93	25.11	27.29	29.47	31.65	33.83	36.01	38.19	40.37	42.55
20 x 50	21.82	24.00	26.18	28.36	30.54	32.72	34.90	37.08	39.26	41.44	43.62

(20 IN dia. shaft: volume of shaft = 2.18 c.f. per l.f. depth)

22 IN CHART (partial)
Volumes shown are for shaft and bell

Shaft/Bell	DEPTH										
	4'	5'	6'	7'	8'	9'	10'	11'	12'	13'	14'
22 x 50	24.31	26.95	29.59	32.23	34.87	37.51	40.15	42.79	45.43	48.07	50.71
22 x 52	25.51	28.15	30.79	33.43	36.07	38.71	41.35	43.99	46.63	49.27	51.91
22 x 54	26.74	29.38	32.02	34.66	37.30	39.94	42.58	45.22	47.86	50.50	53.14

(22 IN dia. shaft: volume of shaft = 2.64 c.f. per l.f. depth)

24 IN CHART (partial)
Volumes shown are for shaft and bell

Shaft/Bell	DEPTH									
	5'	6'	7'	8'	9'	10'	11'	12'	13'	14'
24 x 58	35.46	38.60	41.74	44.88	48.02	51.16	54.30	57.44	60.58	63.72
24 x 60	36.91	40.05	43.19	46.33	49.47	52.61	55.75	58.89	62.03	65.17
24 x 62	38.42	41.56	44.70	47.84	50.98	54.12	57.26	60.40	63.54	66.68

(24 IN dia. shaft: volume of shaft = 3.14 c.f. per l.f. depth)

Figure 02465-2

EXAMPLE 02465-1

Quantity Survey

Company Name Sample Estimate – ASPE Drill Piers
Project WXY – New Bldg, single phase, easy access
Address _____
System/Work Area Drill Piers   Est # _____
CSI Section _____ Bid Date _____

Estimator _____ Date _____
Pricing _____ Date _____
Extended _____ Date _____
Summary _____ Date _____
Verified _____ Date _____

Size & Re-steel Size Shaft/Bell/Depth	Quan.	Unit	CONCRETE Volume	Total Vol. (c.f.)	# Bars	Size	RE-STEEL Weight (per l.f.)	Length (l.f.)	Total Wt. (lbs)
16/36 8'-0"	4	ea	16.62	66.48	5	5	1.043	10.00	209
5 # 5's, #3 Ties @ 12" c.c.					10	3	0.376	4.2	63
18/40 8'-0"	8	ea	21.92	175.36	5	6	1.502	10.00	601
5 # 6's, #3 TIES @ 12" C.C									
20/48 8'-0"	6	ea	29.47	176.82	6	6	1.502	10.00	541
6 # 6's, #3 Ties @ 12" c.c.									
22/52 8'-0"	3	ea	36.07	108.21	7	6	1.502	10.00	541
7 # 6's, #3 Ties @ 12" c.c.									
24/60 8'-0"	2	ea	46.33	92.66	7	6	1.502	10.00	210
7 # 6's, #3 Ties @ 12" c.c.									
SUBTOTALS				619.53					2,310.66
SUMMARY									
Concrete	620	cf							
	/27	cf/cy							
	23	cy							
	x1.1	waste factor							
	25	**Total c.y.**							
Re-steel	2,311	lb.							
	/2000	lb/ton							
	1.16	ton							
	x 1.1	waste factor							
	1.27	**Total tons**							
Excess soil/spoils	620								
	/27								
	23								
	x 1.4	swell factor							
	32	**Total c.y.**							

02465 Bored Piles – Drilled Caissons — Part Two

Estimate Detail

Example 02465-2

Company Name **American Society of Professional Estimators**
Project **XYZ – New Bldg, single phase, easy access**
Address ___
System/Work Area **Drill Piers**
CSI Section ___
Estimate # ___
Bid Date **5/25/01**

Estimator ___ Date: ___
Pricing ___ Date: ___
Extended ___ Date: ___
Summary ___ Date: ___
Verified ___ Date: ___

Item or Description			Material		Labor			Equipment		Subcont.	
DESCRIPTION	QUANTITY	UNIT	UNIT $	Total $	HR/UNIT	HOURS	UNIT $	Total $	UNIT $	Total $	UNIT $	Total $
Layout piers	23	ea					15.00	345				345
Remove spoils/excess	42	cy					5.00	210	10.00	420		630
Drilling - Subcontract	184	vft									5.00	920	920
Pier concrete	33	cy	55.00	1,815			30.00	990				2,805
Reinforcing steel	1.3	ton	500.00	650			250.00	325				975
TOTALS				$2,465				$1,870		$420		$920	$5,675

NOTE:
Total Cost of $5,675.00 = $247.00/pier
Total cost is before Labor, Burden, Taxes, OH&P, etc.

212 — Standard Estimating Practice

02465 - Bored Piles -- Auger Cast Grout Piles

Introduction

Auger cast-in-place piles are a type of drilled pier used when conventional footing designs such as grade beams continuous footings and pad footings are unable to meet the bearing capacity with the existing soil type. They are also used to stabilize soil masses in order to minimize any disturbance to adjacent structures. The method used to place auger cast-in-place piles starts with a crane, concrete pump, and a specially designed hollow stemmed auger. The auger is drilled into the ground to the specified pile depth. Grout is then injected through the shaft of the hollow stemmed auger as the auger is being withdrawn. During this process the grout exerts a positive upward pressure on the earth-filled auger and the surrounding wall of the shaft to prevent the shaft from collapsing and earth in the auger from contaminating the grout. Auger cast-in-place piles are often chosen in lieu of driven pile to void the vibration or hammering noises associated with driving pile.[1] Auger cast-in-place piles are commonly referred to as auger cast pile, intruded mortar piles, augerpress pile, augured pressured grouted piles, auger pile foundations, continuous flight auger piles, grouted bored piles, auger grout-injected piles, drilled or augured uncased piles, and uncased cast-in-place concrete piles.[2]

Types of Measurement

The auger cast-in-place piles are best quantified by adding the total lineal feet of pile to be drilled. Concrete material for the auger cast-in-place piles is calculated by the cubic yard. Because of the process used to place auger cast-in-place piles it is important to add the appropriate waste factor to the concrete material. The top 18 IN of the auger cast is typically formed and is the same diameter as the pile. Each is the measurement used for the formwork at the top of the pier.

Factors Affecting Takeoff and Pricing

- A typical auger cast-in-place crew can drill and place between 500 and 800 lineal feet of pile in a day. The variance in productivity depends on how many times the crane must be moved, site conditions, and the configuration of the pile. Most auger cast-in-place crews travel, so on projects where less than a full day of drilling is required, the project will charged for a full day of drilling to cover the equipment, per diem, and labor cost of a crew. These costs will still be incurred whether 200 or 800 lineal feet of pile are drilled in a day.

- The building codes and soil types in different areas of the country will determine the minimum distance between adjacent auger cast-in-place piles. Typically, the minimum distance allowed between adjacent auger cast-in-place pile is between 8 FT and 12 FT if cast on the same day but can be as small as 4 FT. In some locations it is acceptable to drill two piles that are less than the minimum distance if a previously drilled pile is directly between the two piles being drilled.

- The configuration of the pile can affect the daily crew productivity. Sample configurations are shown in Figure A. Two auger cast-in-place clusters, with 5 piles each, can take 3 or 5 days to complete depending on the configuration. The "previously adjacent pile minimum distance rule" is interpreted differently depending on local requirements. Often the rule does not apply if a previously placed pile is between two piles that are less than the adjacent minimum distance rule. Some requirements will not allow you to place two piles within a distance less than the minimum distance rule even if there is a pile that was place previously between two piles to be drilled. If you are able to drill two piles with a previously drilled pile between them, a pile cluster of four in a square configuration will take 4 days (See Example 2). A 5-pile cluster in a square configuration with a center pile (See Example 3) will take 3 days. Drilling the center pile in Example 3 the first day allows you to drill 2 piles in the same cluster the following day only if the previously placed pile is directly between the 2 piles being drilled.

- It is important that the concrete in each pile sets up before an adjacent pile can be drilled. This time can be from 2 hours to 2 days depending on local requirements, soil types, and project specifications and can be a factor in the amount of pile drilled each day.

- The diameter of the pile will affect the productivity rate of drilling and grouting. With each incremental increase in the diameter of the pile, the drilling productivity rates per lineal foot will decrease. The rate at which the drilling productivity decreases from the increased pile diameter will depend on the soil types. The rate at which grout is pumped through the auger shaft usually remains constant. With an increase in diameter, the amount of concrete to fill the shaft per lineal foot also increases, which reduces the amount of shaft length that can be grouted per hour. Although not common, some projects will require multiple pile diameters. Each pile diameter will have different productivity rates. In addition to different productivity rates, additional time must be added to the cost of the project to change augers from one size to another.

- The depth of the piles influences the size of the crane needed to drill each pier. With increased depth, the use of a larger crane will be required. With a larger crane, the cost per lineal foot of drilling will increase due to the higher crane costs.

- When figuring concrete material for each auger cast-in-place pile, a waste factor must be added to the total volume of each pile. The process of injecting the grout creates a positive lateral pressure. This pressure mixes the grout into the earth filled auger. The concrete, which mixes with the earth, must be removed to meet the design criteria for the pile. The concrete amount that typically mixes with the soil and is used to prime the pump equals approximately 35% for the waste factor Specific soil and site conditions must be reviewed to develop the proper waste factor. Removal of the spoils is generally excluded by the pile contractor and is left to the general contractor or site excavation contractor to be completed.

Drilling and Grouting Pile Process

When estimating most foundation systems, typically the excavation and forming are separated from the concrete placement. Auger cast-in-place piles are unique in that the grout is placed during the removal of the earth as the auger is being withdrawn from the shaft. For this reason, the estimated productivity for drilling and placing the concrete is calculated as one process. The depth of each pile shown on the structural drawings and listed in the pile schedule is based on borings taken at the site. The actual depth of the pile may vary from the pile depth listed on the pile schedule because of the rate of refusal. The rate of refusal determines the final depth of the pile. An auger cast- in-place contractor's bid should reflect a reduction for pile depths less than and an increase for pile depths greater than the lengths stated on the schedule. Placing auger cast-in-place pile is accomplished by injecting grout through the shaft of the hollow stemmed auger after it has reached the required depth or meets the required rate of refusal criteria. Once the auger has reached the required depth, the plug in the auger stem is removed by grout pressure.

The auger will rotate as grout is being pumped into the pile. The injected grout creates a positive pressure on the surrounding walls of the shaft and an upward pressure on the earth filled auger. The positive pressure on the walls prevents the sidewalls from collapsing and causes the concrete to mix with the earth in the auger. The grout and earth mixture is commonly known as spoils. The spoils will have to be completely removed from the pile to ensure the concrete mix is not weakened from soil contamination. The spoils are typically excluded in the cost of the auger cast-in-place pile. By creating a positive pressure, which mixes the grout into the earth filled auger; a large percentage of the grout will be wasted in the soils. If the wasted grout is not considered during the grout material takeoff, the material estimated for the grout will be significantly under estimated. The following examples will demonstrate the analysis for auger cast-in-place pile and will show how some factors may influence the productivity and cost of the pile.

Overview of Labor, Material and Equipment

The following three sample estimates will be analyzed with each individual pile going to a depth of 45 FT - 0 IN, with a diameter of 16 IN, and the minimum allowable distance between adjacent drilled piles at 12 FT with a time blank of one day and an average drill rate of 60 LF/HR. For this estimate, it will be acceptable to drill two piles with less than the minimum allowable distance of 12 FT between adjacent drilled piles only if an existing pile is directly between the 2 piles being drilled. The drill rate includes drilling, grouting, sleeve installation, reinforcing steel installation; dipping and relocation of equipment between pile clusters. The three sample estimates will have a different number of piles in each configuration. Each estimate will have six pile clusters of a single configuration.

- Labor—The estimates will show how the configuration of the pile can influence the labor productivity of each pile cluster. For each example we will assume a possible labor output rate of 720 LF per crew day. Each member of the crew will be paid for 8 hours whether the crew drills for 4 or 8 hours. Because the crew will be paid for 8 hours regardless of the number of actual hours worked it will be necessary to calculate the cost per crew day.

Labor – Auger Cast-in-Place Pile Crew (Superintendent, Crane Operator, Concrete Pump Operator, Driller and 3 Laborers)			
Description	Hourly Rates	Quantity	Total
Superintendent	$35.00	1	$35.00
Crane Operator	$28.00	1	$28.00
Concrete Pump Operator	$20.00	1	$20.00
Driller	$25.00	1	$25.00
Laborers	$18.00	3	$54.00
Hourly Crew Rate			$162.00
Daily Crew Rate	**$162.00**	**8**	**$1,296.00**

To show how the factors previously mentioned can influence cost, the following sample projects will be analyzed.

- Estimate 1: Pile cluster of 3 in a triangle pattern (Figure 1)

- Estimate 2: Pile cluster of 4 in a square pattern (Figure 2)

- Estimate 3: Pile cluster of 5 in a square pattern with an individual pile centered in the square. (Figure 3)

The first step is to analyze each cluster and determine the order at which each pile will be drilled. The cluster configuration strongly influences the number of piles that can be drilled in a single day. The number next to each pile in Figure 1–3 represents the order in which each pile will be drilled. The following tables will illustrate the number of piles per pile configuration that can be drilled per day.

Example 1 – Labor Productivity – Pile Cluster of 3				
Day	Quantity of Pile	Unit	Length	Total Length Drilled Per Day
Day 1	6	Each	60	360
Day 2	6	Each	60	360
Day 3	6	Each	60	360
Total Project Length of Pile Drilled				1080

The total length of pile drilled in example 1 was 1080 LF and the project duration was 3 days, thus, the average drilled per day is 360 LF.

Example 2 – Labor Productivity – Pile Cluster of 4				
Day	Quantity of Pile	Unit	Length	Total Length Drilled Per Day
Day 1	6	Each	60	360
Day 2	6	Each	60	360
Day 3	6	Each	60	360
Day 4	6	Each	60	360
Total Project Length of Pile Drilled				1440

The total length of pile drilled in Example 2 was 1440 LF and the project duration was 4 days, thus, the average drilled per day is 360 LF.

Example 3 – Labor Productivity – Pile Cluster of 5				
Day	Quantity of Pile	Unit	Length	Total Length Drilled Per Day
Day 1	6	Each	60	360
Day 2	6	Each	60	720
Day 3	6	Each	60	720
Total Project Length of Pile Drilled				1800

The total length of pile drilled in Example 3 was 1800 LF and the project duration was 3 days, thus, the average drilled per day is 600 LF.

Using the previous productivity rates and crew day cost, costs per example are:

Labor – Auger Cast-in-Place Pile					
Description	Crew Days	Cost per Crew Day	Total Labor Cost	Total LF	Cost per LF
Example	3	$1,296.00	$3,888.00	1080	$3.60
Example	4	$1,296.00	$5,184.00	1440	$3.60
Example	3	$1,296.00	$3,888.00	1800	$2.16

- **Grout Material**—The grout consists of portland cement, sand, water and admixtures. The grout mix shall be mixed to produce a grout capable of maintaining the solids in suspension, fill voids in adjacent soils and be pumped without difficulty.[3] The grout material is calculated by finding the volume of the shaft.

Volume of a cylinder: p × r² × cylinder length

Volume of shaft per lineal foot: 3.24 × .667 × 1 = 2.09 CU FT/LF

Since concrete is ordered by the cubic yard, the grout volume needs to be converted from cubic feet to cubic yards.

Cubic feet to cubic yards: Cubic feet , 27 = Cubic Yards

Cubic yards of grout per lineal foot for a 16 IN diameter shaft:

2.09 CF/LF , 27 CF/CY = .078 CY/LF

Material – Grout Cost					
Description	LF of Pile	CY Per LF + 35% Waste	Total CY	Cost Per Cy	Total Cost	Cost Per LF
Example 1	1080	.078 x 1.35	114	$80	$9,120	$8.44
Example 2	1440	.078 x 1.35	152	$80	$12,160	$8.44
Example 3	1800	.078 x 1.35	190	$80	$15,200	$8.44

Standard Estimating Practice

- **Equipment**—The equipment costs are calculated as a daily rate regardless of the number of hours per day the equipment is used. The equipment used to drill and grout the pile is a crane, concrete pump, and a set of piling equipment. The equipment used for each project will vary depending on the depth of the pile, diameter of the pile and side condition.

The daily equipment costs are calculated as follows:

Equipment – Auger Cast-in-Place Pile			
Description	Daily Rate	Quantity	Total
45 Ton Crane	$1,050	1	$1,050
40 CY/HR Concrete Pump	$1,000	1	$1,000
Piling Equipment	$200	1	$200
		Daily Equipment Rate	$2,250

Using the previous equipment rates and equipment cost per day, the equipment cost per example calculates as follows:

Equipment – Auger Cast-in-Place Pile					
Description	Days	Cost Per Day	Total Equipment Cost	Total LF	Cost Per LF
Example 1	3	$2,250	$6,750	1080	$6.25
Example 2	4	$2,250	$9,000	1440	$6.25
Example 3	5	$2,250	$6,750	1800	$3.75

Most Auger Cast-In-Place Pile contractors will work over a large area of the country and often internationally. The cost to mobilize the equipment and crew will vary depending on the distance needed to travel.

Conclusion

The previous information shows the importance of understanding all the conditions and requirements from one project to another. Even though the material costs remain consistent per lineal foot of pile, the estimator should have full understanding of the project before applying labor productivity rates and equipment cost to the project. It is not always true that a project with more lineal footage has a lower cost per lineal foot of pile. Example 2 had more lineal feet of pile that Example 1, but the cost per lineal foot remains the same. It would be poor judgment by an estimator to apply labor and equipment cost per lineal foot of pile without considering local codes and regulations, site condition, pile depth, pile diameter, number of pie and the pile cluster configuration.

When utilizing historical cost to compare projects, it becomes very important to fully understand and compare the projects based on local codes and regulation, site conditions, pile depth, number of pile and pile cluster configuration rater than relying completely on the cost per lineal feet of pile. These factors may have a large influence of the daily productivity rate, which in turn affects the lineal foot costs.

Glossary

Auger Refusal is defined as the rate of auger penetration of less than 1 LF per minute of drilling.

Auger Spoil is a mixture of concrete and subsurface material that is produced during the concrete placement.

The formwork is a pile sleeve made of 28 gauge galvanized steel that is 18 IN long and the same diameter of the pile.

Reference: Auger Cast-In-Place Manual, Deep Foundation Institute

[1] Auger Cast-In-Place Piles Manual—Page 4

[2] Auger Cast-In-Place Piles Manual—Page 5

[3] Auger Cast-In Place Piles Manual—Page 4

02530 SANITARY SEWERAGE

Introduction

Eight-inch PVC sewer pipe is a common material that is specified for municipal sewer main systems, which act as a collector pipe for individual house or building connections that tie into the main. Although the material itself remains the same the conditions in which it is installed can vary greatly, which provides the problem to the estimator. If the sewer main were to be installed in the same soil, depth, moisture, work area, temperature, etc., the estimator would have little trouble to establish an average production rate that his crews could meet. However, the reality is every job is different and each job has its own parameters that must be accounted for to derive an accurate cost for the work.

TYPE OF MEASUREMENT USED

There are several portions to the items that must be measured to capture the quantities associated with the proposed pipe work. The first part of the process is to measure the amount of 8 IN PVC required for the job, in units of linear feet (LF). The next phase is to count the amount of wyes that will be used for the building connections. The wyes can also be taken off under the task of building or house connections, the estimator should choose a system he is comfortable with and stick to it to avoid omissions.

A critical measurement is to obtain the quantity of soil that will have to be excavated to install the pipe. The first step is to look at the profiles to find an average depth for the run of pipe to be estimated. Once the depth is determined than the average end area of the trench can be calculated based on the decision to slope, bench or shore the excavation, if required. If the trench was 10' deep in type "C" soil there are a couple of options that are suitable, based on OHSA (Occupational Health and Safety Act) requirements for this type of trench. Two such options are to slope the trench walls or provide shoring of the vertical trench wall. The decision of which method to use will depend on each companies resources and preferences. However, if a trench shield was figured for the project a simple formula of the average depth (feet) * width of the trench box (feet) would equal the average end area, which multiplied by the linear feet of pipe would result in the bank cubic feet of soil to be excavated. Soil is most often measured in cubic yards, so the estimator would divide the cubic feet by 27 to calculate the number of bank cubic yards.

Next measurement will be the amount of bedding required around the pipe. A typical installation of 8 IN PVC pipe will call for 6 IN of stone underneath the pipe and to terminate stone at 6 IN over the top of the pipe. That specification would require 1.67 FT of stone * the width of trench * length of trench / 27 to equal the cubic yards of stone required for bedding, with one adjustment deducting the volume of the pipe. In most small pipe installations it is acceptable to neglect the volume of the pipe, but once the diameter of the pipe becomes too large it can begin to make a significant difference in the overall price of the job, if the estimator does not deduct the equal amount of stone.

The backfill bank volume can be calculated by subtracting the bedding volume from the excavated volume and the spoil bank cubic yards will be equal to the bedding volume.

All of these items hold true until the factors of soil swell and shrinkage are applied to the bank or measured quantities, which gives the estimator adjusted quantities that better represent the actual work to be done. The effects of soil shrink and swell will be illustrated in the sample takeoff.

FACTORS AFFECTING TAKEOFF AND PRICING

The soil type will also dictate the rate at which the soil swells during excavation or shrinks during compaction. Certain soil types will expand and contract significantly more than others, requiring different amounts of work to handle the same bank volumes of material. Experience or a soils lab can help the estimator make reasonable predictions about the rate of swell and shrinkage.

Rain becomes a problem when an area has a very wet spring and the typical water table for the area becomes higher than normal. Installing pipe well into the grounds natural water table can become quite difficult and expensive. Beyond the water table issue the rain mostly hinders proper backfill, when trying to obtain a certain compaction density with the wet native material that was excavated from the trench.

Any extreme in the weather can and will cause trouble with production rates. Workers become very slow in extreme heat and often fail to reach realistic production rates, while equipment can become the problem in very cold weather. An engine that won't start or frozen tracks on an excavator can leave an entire crew without the means of performing their work.

Obstructions

Obstructions can be any item that is in the path of the proposed sewer line installation that will have to be excavated around to get the pipe in its proper place. These obstructions are often existing utility lines that cross the path of the new line. The example in this paper deals with a new site that will not encounter any obstructions, but pipe is installed every day around other utility lines. Tracking the time it takes for your crew to properly locate and excavate around an obstruction is important to determine how much time will be taken away from the crews pipe production in a day. An estimator will usually adjust the anticipated rate of production for a day to compensate for these obstructions.

OVERVIEW OF LABOR, EQUIPMENT AND MATERIALS

Labor & Equipment

The labor and equipment to install an 8 IN PVC sewer main is best described by looking at the individual pieces that make up the installation crew. Once the measurement phase is complete the estimator will have a good understanding of the work to be performed and the type of equipment that would be required to install the pipe. A large part of knowing what crew will work best is experience and past performance of the company's crews with similar projects. If the estimator is new to this type of work he can consult equipment performance guidelines that can be found in books such as the Caterpillar Performance handbook, that provide the equipment's abilities with relation to different types of soils.

The pricing of the labor will have to be done based on the type of project. If the project is a union job than the unions current wage rates must be used vs. an open shop project were the current market rates of non-union labor must be determined to provide a competitive bid, but allow enough money to complete the work. The labor rate decisions must be based on the type of work being bid. Prevailing wage, union and non-union projects all have different circumstances that effect the overall labor rate.

In this example the estimator wants to try and find the most cost effective excavator to perform the trenching and backfill work for this small project. Once the excavator is chosen than the remaining crew can be built based upon the production rate of the main excavator performing the trenching work. A typical crew for a small job would consist of a minimum of five men and three pieces of equipment, but it is important for the estimator not to limit himself to one type of crew. There are many configurations of different crews that can yield better results for certain situations. Installing pipe in a busy roadway will require many more man and equipment than installing pipe across an empty field and if this is not recognized by the estimator he may either over price the project or lose a lot of money on the job.

Material

The materials used on the project are easily obtained through local suppliers of pipe, aggregates and select fill if required. The estimator should call for frequent pricing on the PVC materials, especially during times of increased petroleum prices, because PVC is a petroleum-based product. In the past year prices have risen approximately $1.00 per linear foot on 8 IN PVC pipe, which on a large project could turn into a costly mistake if the price was not adjusted accordingly.

SAMPLE PRODUCTION PROCESS

To best illustrate the entire process a sample production process will be performed with a spreadsheet application that I built for our company's production evaluation. The spreadsheet was based on information that I obtained from multiply sources. Experience and the following list are the major contributing sources I used to develop this spreadsheet:

1. Caterpillar Performance Handbook Edition 29
2. Construction Estimating Institute Workshop Handbook
3. Pipe & Excavation Contracting by Dave Roberts

Sample Spreadsheet:

The following example will demonstrate a step-by-step process to estimate a reasonable production rate that can be obtained by a pipe crew using a Cat 320 BL excavator to perform the trenching & bedding tasks. The sample production process will follow the spreadsheet format that can be found on pages 19, 20 and 21.

Section 1 - Project information is for data purposes only, to help the estimator identify the section of trench being analyzed and the materials required.

Section 2 — Trench information are the items that the estimator must takeoff from the plans and input into the spreadsheet. The end area of the trench is taken from the beginning of the trench and the end of the trench and then averaged to get an approximate cross section of the proposed trench. The cross section will vary based upon how the estimator intends on performing a legal and safe excavation. He can slope or bench the trench walls back or he can use a trench shield, which has been chosen for this example. A skilled crew that has worked with trench boxes can usually obtain better production rates using the trench box then sloping the walls back once the trench becomes deep enough, especially if restoration needs to be performed. The end area for this trench looks like the following:

End Area No. 1: 9 FT depth * 5 FT width = 45sf

End Area No. 2: 10 FT depth * 5 FT width = 50sf

No 1 (45sf) + No 2 (50sf) = 100sf/2 locations = **Average End Area of 47.5sf**

The swell and shrinkage (compaction) factors depend on the soil type and can be determined by a soil engineer or from past experience.

Section 3 — Trench and bedding calculations are derived from the previous information provided by the estimator. The results give the estimator a good "feel" for the amount of work that must take place and will aid in his decision in selecting an excavator or backhoe for a particular project.

Section 4 — Equipment selection and work skills is were the estimators experience really comes into play. The estimator will select the machine that will perform the excavation for the pipe work and what size bucket capacity will be used. If the estimator is new to this type of work it would be wise to try several different machines to check the production out comes. For this example a Cat 320 excavator will be used that shows a 25 second cycle time based on soil type and depth. The cycle time was obtained by referring to the Caterpillar Performance Hand Book, which provides performance levels for their equipment. The cycle times are based on the best professional operators under ideal conditions, so allowances must be made to adjust to real field conditions[1]. Work will usually only be efficient for approximately 50 minutes out of 1 hour and the average operator through the course of a day will only be about 75% as efficient as the operators used to develop these cycle times. These are only guidelines and each estimator should check his crews to see what they are actually capable off performing in the field.

Section 5 — Trenching productions can then be calculated to see how long the trenching should take to perform based on the current equipment selection.

Section 6 — Pipe installation cycle time is the section where the estimator must complete the picture to develop production rates of pipe that can be installed in a specific time. The pipe length is important because it creates the length of the set, which is the complete placement of one length or joint of pipe. The following items 6.2 through 6.10 provide the remaining information to complete the cycle time of the set. The times that are shown here are based upon past experience and actual times that were recorded in the field.

Section 7 — Allows time for setting structures or manholes into place, so that the pipe laying can continue. The time for the structure is only based on the crew time used to set it in place and does not include any other work that may need to be performed to complete the structure. It is entirely up to the estimator if he wants to account for this time now or make an allowance in the structures to cover the cost. The important part is to pick a method and stick with it, to eliminate omissions.

Section 8 — Summary of production gives the exact amount of time that it should take to complete the project. However, this section allows the estimator to adjust the time to better suit his company's schedule or requirements. Instead of having the crew scheduled for 1.95 days the estimator can make a judgment to extend the estimated time to 2 days to cover the cost of the crew if they can not move onto other work.

Section 9 — Spoil removal is not used in our example, but if the project being bid was a long a busy roadway that had no room for spoil, the estimator could plan how many trucks he would need to keep up with the removal of spoil. The time elements are based upon the distance the truck must travel and at what rate the spoil is going to be generated.

---

[1] Caterpillar Performance Handbook Edition 29 by Caterpillar

Section 10 — Costing the work is only to provide the estimator with a reasonable figure that can be used as a check in his final bid price. The crew cost is generated from sections 11, 12, & 13 that use common rates for labor and equipment that build a potential installation crew for the proposed job. The crew cost is then added with the remaining components to develop an overall cost that can be divided by the total length of pipe to achieve the unit cost for the pipe.

Section 14 — The note section is for the estimator to address any potential problems or to comment on his choices so another member of his estimating team can pick up where he left off if need be.

DEVELOPING CREWS AND PRICING

Although other machine choices could have been mode, the Cat 320 BL excavator will be used to construct the crew for the project. The crew for the example will be made up of the resources used in sections 11 & 12 in the sample production worksheet. The Loader choice of a Cat 938 was based on the amount of backfill that will have to be handled and pipe bedding that will have to be supplied to the excavator. The weight of the pipe length itself is light enough for a single worker to handle by himself and requires no other equipment or time of the excavator to set the pipe. The 8 FT * 16 FT trench box allows for 8 FT of wall to be shored with 2 FT of wall exposed at the bottom, which is within legal limits of 3 feet. Pipe lasers are the standard in providing grade and alignment for laying pipe. The amount of time that is saved from not having to check grades for string lines set a long side the trench line will quickly pay for the cost of the laser.

The labor chosen for the crew is about the minimum that would be used on this type of work. There are two operators, two laborers and a working foreman. The operators are essential in maintaining a good production rates, the excavator operator will be the major factor that controls the rate of progress. Almost as important is the pipe layer that must guide the operator to the final grade with minor hand trimming by his helper. The top man/foreman will help provide a constant flow of information between the operators and laborers, acting as a conductor to lead the crew to a productive workday.

As with the equipment different combinations of labor can be utilized to best suit the individual companies resources. However, for this example will use a typical crew structure. Certain companies will provide another laborer to handle the job of top man, so the foreman can have more time to plan the day's work and anticipate any potential problems before the crew reaches that point. A union agreement may also require that additional labor is used in the crew's composition. An oilier may be required to work with the excavator operator, if the machine is over a certain size or an additional operator may be required to run the walk behind trench compactor. In general the estimator needs to be familiar with the requirements of the crew to properly estimate the labor cost to install the pipe.

The pricing portion of the equipment will reflect local rental rates or internal ownership cost of the equipment. It is a cost that must be accounted for based on the company's methods of obtaining equipment. If the company satisfies all of its equipment needs by renting, than the estimator will have to apply the appropriate rental rate for each machine, but if the company owns the equipment the estimator must establish the ownership costs for each machine. The cost may be obtained from the accounting department if they track the individual machine cost or can be estimated with various guides that provide average ownership costs for equipment.

The labor pricing may come from different sources. Public projects will typically require the contractor to use the most current prevailing wage rates set by the state or government. If a job is union the contractor will have to conform to the wage rates that have been negotiated by the union vs. an open shop project that will have to pay the market value for non-union labor. The labor stipulations should be clearly understood before a final price is established.

APPROACH TO APPLYING INDIRECT COST, OVERHEAD AND PROFIT

Indirect costs are tasks that must take place for the work to be completed, but have no specific line item in the proposal to the owner. Many of these items can be grouped under general conditions if the bid documents allow for such an item. If a general conditions bid item is not provided than the estimator must decide where he can incorporate these items into the bid. There are numerous ways to apply the indirect cost to the bid. The sum of the indirect cost can be spread across each bid item in different relations. The estimator may choose to base the spread on percentages of the total price, total labor, total subcontracts, total materials or any combination of these items. If a project is estimated with large mobilization cost for the beginning of the job, the estimator may choose to spread the cost only with the first few bid items that would be performed first on the job, so the contractor could recover his money in a timely fashion.

Overhead for a utility contractor is usually on the high side in comparison to a general contractor that subcontracts the majority of its work. There seems to be various methods of applying overhead rates to a bid. Many estimators apply a certain percentage to the bid cost to cover their overhead, based on past overhead cost or projected budgets. Although this method is common an analysis of the projects potential time to complete should be reviewed, to see if the overhead cost of the same time period can be covered by the percentage applied to the bid.

Current market conditions will be one of the major items that control how much potential profit can be added to a bid and still be competitive. Being able to obtain a high profit may be out weighed by the necessity of a firm to acquire new work in lean times. In contrast strong economic times or a high-risk job may warrant the estimator to add additional profit to a bid in an effort to capitalize on the market condition.

SPECIAL RISK CONSIDERATIONS

Underground construction of utilities is filled with risk that the estimator must try to compensate for or provide conditions to deal with the risks. Rock excavation and dewatering are two items that need to have special conditions in a bid to handle these items. Rock excavation is often a separate line item with an associated unit price that is paid by the cubic yard for removal, but the estimator still needs to evaluate what type and how much rock may be encountered. The type and quantity of rock will determine the method and the price that can be used for the rock removal.

Exact water conditions are very hard to determine without employing the use of test pits, which can still be misleading depending their locations. Ground water can surprise even the most seasoned veterans and must be carefully evaluated to see if dewatering operations will be required to install the pipe. The process may be as simple as a small trash pump and some additional gravel or as complicated as well point system and deep wells to lower the water table below the pipe's proposed grade. The amount of water that can be handled by a simple pump and stone will only be a small cost and may not even hurt the production of the crew, but water that requires a well point system can be very expensive to install and dramatically cut down on production.

Part Two 02530 Sanitary Sewerage

ASPE Estimating
Pipe Production Worksheet

Estimator :
Project Name:
Project Location:

6.0 Pipe Installation Cycle Time

6.1 Length of Joint	13	LF	6.11 Excavate Set	20.33	Min.	
6.2 Place Bedding	2	Minutes	6.12 Place Bedding	2	Min.	
6.3 Place Pipe	2	Minutes	6.13 Place Pipe	2	Min.	
6.4 Top Bedding	2	Minutes	6.14 Top Bed/B'fill	2	Min.	
6.5 Re-set Machine:	0.5	Minutes	6.15 Re-set Machine	0.5	Min.	
6.6 Re-set Shoring:	1	Minutes	6.16 Re-set Shoring	1	Min.	
6.7 Laser Set Ups	1	Each	6.17 Lazer / Set	0.65	Min.	
6.8 Each set up time	20	Minutes	6.18 Other Time	0	Min.	
6.9 Other Time	0	Minutes	Total Cycle Time / Set	28.48	Min.	
6.10 Hrs. per day	8	Hours	Production per Hr.	27.39	LF	

7.0 Allowing for Time at Structure Locations

7.1 No. of Structures	2	Each	7.5 Structure Hrs.	1	Hrs.
7.2 Time / Structure	30	Minutes	7.6 Conflict Hrs.	0	Hrs.
7.3 No. of Conflicts	0	Each	7.7 Pipe Hrs.	14.60	Hrs.
7.4 Time / Conflicts	20	Minutes	7.8 Adj. Total Hrs.	15.60	Hrs.

8.0 Summary of Production

8.1 Adj. Total Hrs.	15.60	Hrs.	
8.2 Total Days	1.95	Days =	205.07 Pre adjusted LF / Day
8.3 Adj. Days	2	Adj. Days	
8.4 Daily Prod:	200	Adj. Daily Production Rate	
8.5 Hourly Prod:	25	Adj. Hourly Production Rate	

9.0 Spoil Removal

9.1	3.64	Hrs. to fill One 15 CY Tandem
9.2	0	Hrs. Round Trip To Haul Area
9.3	0.00	Number of Trucks Required to Keep up w/Spoil Output
9.4	0	Adj. Number of Trucks

10.0 Cost Section

	Rate	Quan	Unit	Total Cost
10.1 Crew cost	$263.00	16	Hr.	$4,208.00
10.2 8" PVC Pipe	$3.25	400	LF	$1,300 w/fittings & Acc.
10.3 ¾" Stone	$10.50	173.68	Ton	$1,823.64
10.4 Rented Trucks	$55.00	0	Hr.	$0.00
10.5 Other Cost	$0.00	0	Hr.	$0.00
10.6 Dewatering	$0.00	0	LS	$0.00
		Total Cost		$7,331.64
		Total Cost per LF		$18.33 8" PVC Pipe

Note: These are only direct cost and do not include overhead or profit

Standard Estimating Practice

02530 Sanitary Sewerage

ASPE Estimating

Estimator:
Project Name:
Project Location:

Pipe Crew Worksheet

11.0 Labor Selection for Crew: Prevailing Wage

No.	Description	Quan.	Hr. Rate	Total $
11.1	Foreman	0	$45.00	$0.00
11.2	Excavator Operator	1	$42.50	$42.50
11.3	Loader Operator	1	$40.00	440.00
11.4	Pipe Layer	1	$25.00	$25.00
11.5	Topman/Foreman	1	$22.50	$22.50
11.6	Helper	1	$19.00	$19.00
11.7	Other	0	$0.00	$0.00
			Total Hourly Labor Rate:	**$149.00**

12.0 Equipment Selection for Crew

No.	Description	Quan.	Hr. Rate	Total $
11.1	Cat 320 Excavator	1	$41.50	$41.50
11.2	Cat 928 Loader	1	$32.00	$32.00
11.3	Ramax Compactor	1	$12.50	$12.50
11.4	8' * 16' Trench Shield	1	$15.00	$15.00
11.5	Pipe Lazer	1	$8.00	$8.00
11.6	3" Trash Pump	0	$10.00	$0.00
11.7	Misc. Tools	1	$5.00	$5.00
11.8	Other	1	$0.00	$0.00
			Total Hourly Equipment Rate:	**$114.00**

13.0 Total Crew Cost

13.1 Total Hourly Labor and Equipment Cost   **$263.00**

Note: These are direct costs only and do not include overhead and profit

14.0 Notes

GLOSSARY

SF – Square foot used to measure area.

SY – Square yard used to measure area.

CF – Cubic foot used to measure volume.

CY – Cubic yard used to measure volume.

LCY – Loose cubic yard of soil

CCY – Compacted cubic yard of soil

PVC – Lightweight, but strong plastic pipe used for sewer and water installations.

Swell – The amount that excavated soil expands in volume when removed from its bank state.

Shrinkage – The amount that loose soil will decrease in volume when compacted to a desired percentage.

Dewatering – methods used to remove water from the excavated trenches to install the pipe.

REFERENCES

1. Caterpillar Performance Handbook Edition 29 by Caterpillar
2. Pipe & Excavation Contracting by Dave Roberts
3. Earthmoving Equipment Production Rates and Costs by Daniel Atcheson
4. Construction Estimating Institute Workshop Materials.

02530 Sanitary Sewerage — Part Two

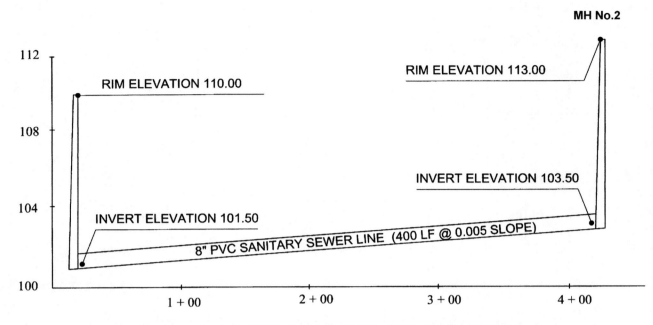

8" PVC SANITARY SEWER PROFILE
Scale: Horizontal: 1" = 60'
Vertical: 1" = 4'

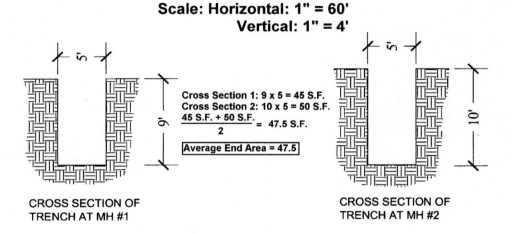

Cross Section 1: 9 x 5 = 45 S.F.
Cross Section 2: 10 x 5 = 50 S.F.
$$\frac{45 \text{ S.F.} + 50 \text{ S.F.}}{2} = 47.5 \text{ S.F.}$$
Average End Area = 47.5

CROSS SECTION OF TRENCH AT MH #1

CROSS SECTION OF TRENCH AT MH #2

Part Two 02530 Sanitary Sewerage

ASPE Estimating

Sanitary Sewer Take off Sheet

Item No.	Description	L	W	D	Quantity	Units
1	8" PVC Pipe				400	LF
2	Bedding Material	400	4	1.67	99	CY
3	Trench cross section no. 1		5	9	45	SF
4	Trench cross section no. 2		5	10	50	SF
	Average end area section:				47.5	SF
5	Trench excavation (400' *47.5)/27'/CY				704	CY
6	Trench Backfill				605	CY
	704 CY of excavation – 99 CY of bedding					
7	Adjusted Trench Excavation (704CY * 20% Swell)				845	LCY
8	Adjusted Trench Backfill				806	LCY
	LCY for backfill = 605 * ((1.20)/(1.0-(.95-.85))					
9	Spoil Adjusted Excavation- Adjusted Backfill				39	LCY

Standard Estimating Practice

02530 Sanitary Sewerage — Part Two

ASPE Estimating
Pipe Production Worksheet

Estimator :
Project Name:
Project Location:

1.0 Project Information:

1.1	Trench Section	MH1 – MH2	1.5	Rev. Date:	05/15/00
1.2	Pipe Type	8" PVC Pipe	1.6	Bid Date:	05/01/00
1.3	Bedding Type	¾" Stone	1.7	Soil Type	Clay

2.0 Trench Information

2.1	Average Depth:	400	feet	2.7	Org. Soil Density	85%	percent
2.3	Avg. End Area 1	9.5	feet	2.8	Comp. Rate %	95%	percent
2.4	Avg. End Area 2	45	sf	2.9	Swell Rate %	20%	feet
2.5	Avg. End Area	50	sf	2.10	Bedding Depth:	1.67	feet
2.6	Pipe Diameter	47.50	sf	2.11	Bedding width:	4.00	feet
		0.67	feet	2.12	Bedding Area	6.68	sf

3.0 Trench & Bedding Calculations

3.1	703.70	Bank Cubic Yards of Excavation
3.2	844.44	Loose Cubic Yards of Excavation
3.3	583.85	Bank Cubic Yards of Backfill
3.4	778.46	Compacted Cubic Yards of Backfill
3.5	65.98	Loose Cubic Yards of Spoil
3.6	0.16	Loose Cubic Yards of Spoil Per. Foot
3.7	90.94	Linear feet to produce one 15 CY Tandem
3.8	98.96	Bank Cubic Yards of Bedding
3.9	115.79	Loose Cubic Yards of Bedding
3.10	173.68	Tons of Bedding required

4.0 Equipment Selection & Work Skills

4.1	Excavator Type:	Cat 320 BL		4.5	Min. of work / Hr.	50	Minutes
4.2	Bucket Capacity	1	CY	4.6	Avg. operator skill	75	Percent
4.3	Spill Reduction:	10%	Percent	4.7	Load Trucks	0	Second
4.4	Cycle Time:	25	Seconds	4.8	Min. Reach	6.65	VLF

5.0 Trenching Production & Time Calculations

5.1	0.90	LCY Bucket Capasity
5.2	0.42	Minutes per Cycle
5.3	144.00	Cycles per Hr.
5.4	144.00	Adj. Cycles per Hr. if Loading Trucks
5.5	120.00	Adj. Cycles for min. of Work per Hr.
5.6	90.00	Adj. Cycles for Operator Skill
5.7	81.00	Loose Cubic Yards per Hr.
5.8	2.11	Loose Cubic Yards per Linear Foot of Trench
5.9	38.37	Linear Foot of Trench Per Hr. to be Excavated
5.10	0.64	Linear Foot of Trench per Minute to be excavated
5.11	10.43	Total Hrs to Excavated
5.12	625.51	Total Minutes to Excavate

Page 1

Part Two 02530 Sanitary Sewerage

ASPE 8" PVC Sewer Pipe   05/29/2000 12:35
DIRECT COST REPORT

Activity Detail	Desc Pcs	Quantity	Unit Unit Cost	Labor	Perm Material	Constr Matl/Exp	Equipment	Subcontract	Total

BID ITEM = 1
Description = 8" PVC Sanitary Sewer Pipe   Unit=   Takeoff Quan: 400.000   Engr Quan: 400.000

1.1 sanitary sewer: 10' deep Quan:		400.00	LF Hrs/Shft: 8.00 Cal: 40		WC: NJ0001				
02532 ok	Sewer Main Pipe: no haul	16.00 CH			Prod: 200.0000 US				
2AG0-3/4	¾ in @ 106%	149.00	TON		10.150		1,603		1,603
2DP3	8"PVC @ 106%	405.60	LF		3.000		1,290		1,290
8E CAT 320	Cat 320 Excavator	1.00	16.00 HR		38.250			612	612
8P B1	Trench Box 8' * 16'	1.00	16.00 HR		12.000			192	192
8S&L2	Pipe Laser	1.00	16.00		10.500			168	168
8WL CAT 938	Cat 938 Wheel Loader	1.00	16.00 HR		36.250			580	580
L2	Laborer, semiskilled	1.00	16.00 MH	19.70		527			527
L5	Laborer, grade & Pipe	1.00	16.00 MH	21.150		554			554
L9	Laborer, foreman & Layout	1.00	16.00 MH	21.700		564			564
O3	Operator, excavator	1.00	16.00 MH	29.120		753			753
O6	Operator, Loader	1.00	16.00 MH	23.990		660			660
$7,502.98	0.2000 MH/LF		80.00 MH	[4.626]		3,058	2,893	1,552	7,503
25.0000	Units/Hr 200.0000 Un/Shift		5.0000 Unit/MH			7.65	7.23	3.88	18.76

➤ Item Totals: 1   -8" PVC Sanitary Sewer Pipe

$7,502.98	0.2000 MH?		80.00 MH	[4.626]		3,058	2,893	1,552	7,503
18.757			400			7.65	7.23	3.88	18.76

$7,502.98	***Report Totals***		80.00 MH			3,058	2,893	1,552	7,503

-----Report Notes:-----
The estimate was prepared with TAKEOFF Quantities.
This report shows TAKEOFF Quantities in the detail.

Bid Date: 06/01/00 Owner: ASPE Engineering Firm: XYZ Engineering
                             Estimator-In-Charge:
*on units of MH indicate average labor unit cost was used rather than base rate.
[ ] in the Unit Cost Column = Labor Unit Cost Without Labor Burdens
In equipment detail, rent % and operating % not = 100% are represented as XXX%YYY=Rent% and YYY=Operating%
-----Calendar Codes-----

Calendars are found in crew and labor codes and have the format XXXdY where XXX = The Calendar and Y = The Starting Day of the Week with Day 1 = Monday, etc.

40	40 Hour Week / 8 hr/day
45	45 Hour Week / 9 hr/day
45T	40 hr work / 5hr travel
50	50 Hour Week / 10 hr/day
50T	40 hour work / 20 hr travel

Standard Estimating Practice

| ASPE | ASPE 8" PVC Sewer Pipe | | | | | | | Page 1 05/29/2000 13:18 | |
| Default User | | | **ESTIMATE SUMMARY (BID PRICES)** | | | | | | |
Bid # Quantity	Client Bid # Unit	Total Cost	Total Cost Unit Price	Markup	Balanced Bid Total	Balanced Bid Unit Price		Bid Price	Bid Total
1 400.00		7,502.98	8" PVC Sanitary Sewer Pipe 18.76	20% 1,501	9,004	22.51	F	22.510	9,004.00
	TOTALS:	7,502.98		1,501	9,004				9,004.00

Code between Balanced Bid & Bid Price: U=Unbalanced, F=Frozen, C=Closing Bid tem (item to absorb unbalancing differences).

| Part Two | 02530 Sanitary Sewerage |

Page 2

ASPE   ASPE 8" PVC Sewer Pipe   05/29/2000   13:18
Default User   **ESTIMATE SUMMARY (BID PRICES)**

| Bid # | Client Bid # | Total | Total Cost | | Balanced | Balanced Bid | Bid | Bid |
Quantity	Unit	Cost	Unit Price	Markup	Bid Total	Unit Price	Price	Total

    BID COSTS ————▶ 7,503
    ACTUAL MARKUP ————▶ 1,501
    TOTAL BID ————▶ 9,004

Spread Indirects on : TOTAL COST   Spread Markup on: TOTAL COST   Spread Add-ons & Bond on: TOTAL COST

Default Markup%
    Labor: 20.00
    Burden: 20.00
    Perm Matl: 20.00
    Const Matl: 20.00
    Sub: 20.00
    Eqp Op Exp: 20.00
    Co. Equip: 20.00

-----ESTIMATE NOTES:-----
Bid Date: 06/01/2000   Owner: ASPE   Engineering Firm: XYZ Engineering
     Estimator-In_Charge   1100022   Desired Bid (if specified) = 0

02750 RIGID PAVEMENT

Introduction

Rigid Pavement covers cement concrete pavement, including reinforcing, joints, finishing, and curing. Related sections are 03200 Reinforcing Steel and 03300 Cast-in-Place Concrete.

Types and Methods of Measurements

Concrete pavement is typically measured by the square yard (SY) to calculate labor and equipment costs. Concrete is purchased by the cubic yard (CY). Calculate the quantity and add a 5 to 10% waste factor (as determined by historical costs). Measure the lineal feet of expansion joints to determine the amount of joint filler and number of tie-bars required.

Reinforcing can be either steel bar which is measured by the pound (LB), or wire mesh, which is measured by the square foot (SF).

Concrete finishing is measured by the square foot (SF).

Specific Factors Affecting Estimate

Work performed prior to the placement of concrete pavement starts with the earthwork, or cut and fill, to bring the roadway or pavement area to subgrade. Unsuitable materials will then be replaced with compactable fill; then the subgrade will be compacted to the density required in the specifications. Curbs and/or gutters, if any, are placed next. Once the stone subbase is placed to the thickness required by the design, paving operations can begin.

The key factors in determining daily production are the pavement thickness in inches, and the width of the pavement area between longitudinal joints. For example, a 10-foot paving width will have a lower labor unit cost than an 8-foot paving width. A typical lane width for a highway is 12 feet, although paving widths up to 24 feet are not uncommon. The lower the thickness in inches, the more area can be paved in a day. The method of construction is also critical. A slip-form paving machine is typically used for highways and large concrete aprons, and will have a much larger daily production than conventionally formed pavement with edge forms. See the attached picture for a GOMACO paver in action. Visible in the picture is the string line set up to ensure that the paver holds a straight line. You can also see that forms are not required, since the concrete mix will be stiff enough to stay in place. Another factor that will affect the daily production is the type of reinforcement. Wire mesh is faster to install on a square footage basis than bar reinforcement. Straight rectangular runs will have a higher production than an irregular shape. Small quantities that require less than a day's work for a crew will also have a higher unit price. As with all concrete work, additional cost is required for winter curing. Inclement weather can effect production.

Typically, 3,000 PSI concrete is required. Additional costs will be incurred for higher strength concrete and additives such as super-plasticizers or accelerators.

Organization of the Estimate

Takeoff the quantity by square feet, which is usually the length times the width, and convert to square yards by dividing the quantity of square feet by nine. Irregular areas can add to the difficulty. A computer takeoff program with a digitizer can be useful. Calculate the quantity of cubic yards by taking the area in square feet times the depth in feet and divide by 27. Use this quantity to determine the amount of concrete to be purchased. Measure the length and spacing of joints to calculate the amount of joint filler. Calculate the number of tie bars and dowels by dividing the spacing into the length of the joint.

Overview of Labor, Material, and Equipment

Labor

The labor crew mix includes laborers, rebar installers, finishers, and operators. A typical composite crew consists of a labor foreman, 4 laborers, 2 equipment operators, 1 rebar installer, and 3 cement finishers. The cost of this crew will depend on the labor rates. Small areas where a paving machine is not practical will have a different crew mix. Any requirement for state or federal prevailing wages is another factor to consider. At an average of $50 per hour, this crew costs $550 per hour. A crew can expect to place 2,000 to 3,000 square yards per day, depending on conditions.

Materials

Concrete, typically 3000 PSI, is purchased by the cubic yard. Additives such as super-plasticizers, ice, accelerators, and pigments are extra. Bar reinforcement is purchased by the pound, and mesh reinforcement is purchased by the square foot. Other materials include curing materials, joint filler, rip strips for control joints, and tie bars. Edge forms are necessary if the pavement is conventionally formed.

Equipment

Equipment includes a concrete paving machine and a grader for final grading of stone subbase. A concrete saw is used for cutting control joints. Conventionally placed pavement will require hand screeds and finishing tools. Depending on the size of the pour, method of placement, and accessibility, concrete pumps may be used.

Pricing

Per the attached example, consider a straight 12 FT × 1500 FT section of 8 IN thick pavement, 18,000 SF, or 2,000 SY. This can be considered a typical day's production for the crew. The example is for a wire mesh reinforced pavement, with expansion joints 40 feet apart. The first calculation is for the labor costs. The labor rates in the example include the base rate, fringe benefits, and the contractor's taxes and insurance.

The concrete quantity for purchase includes a 10% waste factor. Wire mesh is purchased by the square foot or square (100 SF). Curing materials, although purchased by the gallon, is presented here as a square foot cost. Tie bars for longitudinal joints will be spaced four feet on-center, and dowels for transverse joints will be spaced 12 IN on-center. Joint filler is purchased by the lineal foot.

Equipment consists of a grader for fine grading the stone subbase, a slip-form paving machine for placing concrete, and a concrete saw for cutting contraction joints.

Miscellaneous Pertinent Information

Other items related to pavement construction are adjacent shoulders, curbs, and drainage.

Conclusion

It is important to consider all the productivity and cost factors when estimating the cost of concrete pavement. As shown above, the size of the project and the amount of production expected on a daily basis is critical.

02770 CONCRETE CURBS AND GUTTERS

Introduction

The Main CSI Division is 2, Site Construction, and the main subsection is 02700 Bases, Pavements, and Appurtenances.

The specific section is 02770, Curbs and Gutters

Related specification sections are 03200 Reinforcing and 03300 Cast in place concrete.

This section covers cast in place concrete curbs and integral curbs and gutters.

This section can also cover bituminous, precast, and stone curbs.

Types and Methods of Measurements:

Curbs are typically measured by the linear foot (LF) to calculate labor costs.

Concrete is purchased by the cubic yard (CY). Calculate the quantity and add a waste factor, typically 5 to 10%. Determine the spacing of expansion joints to determine the amount of joint filler.

Curbs are typically not reinforced, however, in some cases one or two longitudinal bars may be required. If this is the case, measure the length and convert to pounds.

Specific Factors Affecting Estimate:

Work performed prior to the placement of curb includes excavation to the bottom of the curb plus subbase underneath. Unsuitable soils are removed and replaced if necessary; then the subgrade is compacted to the required density. If required, stone subbase is then placed to the required thickness, and compacted.

The cross-section of the curb or curb and gutter is key in determining labor productivity, since it will determine the quantity of formwork and concrete. Curbs are typically 6 IN or 8 IN thick at the top, with a 1 IN batter down to the level of adjacent pavement. The depth is typically 18 IN or 20 IN. Integral curb and gutter has a shallower depth, usually 12 IN to 14 IN, but can be up to 32 IN wide. Another key factor is the amount of straight and radius curb. Forming radius curb is much more labor intensive than straight curb. Reusable steel or wood forms can be used for straight curb, but for radius curb, handmade wood forms must be used. Concrete curb is also sometimes poured with a slip-form machine. If reinforcement is required, it will add to the labor and material costs. Another factor is the spacing of expansion joints. Depressed curb at driveways will add to labor costs. Small quantities requiring less than a day's work for the crew will have a higher unit price.

Typically, 3,000 PSI concrete is required. Additional costs will be incurred for higher strength concrete and additives such as super-plasticizers or accelerators.

Organization of the Estimate/Quantity Survey and Calculation Quantities:

Take-off the quantity by the linear foot. Separate the amount of radius and straight in your estimate, particularly for determining forming costs. Determine the amount of concrete required by multiplying the cross-sectional area times the length and converting to cubic yards. Joint filler will be installed at each expansion joint, typically 20 to 30 feet apart for straight curb, less for radius curb.

Overview of Labor, Material, and Equipment: Labor

The labor crew mix depends on the method of construction selected. Concrete curb or curb and gutter is usually formed with wood or metal, so a composite crew can consist of a foreman, 3 carpenters, 1 cement finisher, and 1 laborer. Union labor, obviously, costs more than non-union. Any requirement for state or federal prevailing wages is another factor to consider. A crew can expect to install 200 – 500 LF of curb per day, depending on conditions.

Concrete curb can also be machine formed, so the crew would be a labor foreman, 3 laborers, 1 cement finisher, and 1 operator. A crew can expect to install 1,000 LF of curb per day, depending on conditions.

Materials

Concrete, typically 3000 PSI, is purchased by the cubic yard. Additives such as super-plasticizers, accelerators, and pigments are extra. Bar reinforcement is purchased by the pound. Forms are made of wood and can be reused several times. Metal forms are also available for straight runs. Other materials include curing material and joint filler.

Equipment

Concrete curb conventionally formed does not require equipment. The other option is to use a curb forming machine.

Pricing or Estimate:

As an example consider 400 LF of 6 IN thick concrete curb with a depth of 20 IN, 250 LF of which is straight and 150 LF is radius. See the attached sketch. This is a typical roadway curb that is actually 6 IN thick at the top tapering to 7 IN thick at the top of pavement. Assume that the curb will be conventionally formed. The first calculation is for the labor costs. The labor rates in the example include the base rate, fringe benefits, and the contractor's taxes and insurance. Figure four crew hours for the straight curb, and four crew hours for the radius curb. The concrete quantity for purchase includes a 10% waste factor. Curing materials, although purchased by the gallon, is presented here as a linear foot cost. Joint filler is purchased by the square foot. Forms are presented as a linear foot cost.

The result of the estimate is that straight curb costs $7.73 per LF, and radius curb costs $14.29 per LF. The unit cost for the full 500 LF is $9.70 per LF.

Miscellaneous Pertinent Information

After the curb has cured and the forms stripped, the curb is backfilled to subgrade on both sides. Then pavement or topsoil can be placed up to the curb.

Conclusion

It is important to consider all the productivity and cost factors when estimating the cost of concrete curb. As shown above, the size of the project and the amount of production expected on a daily basis is critical.

Part Two   02770 Concrete Curb and Gutters

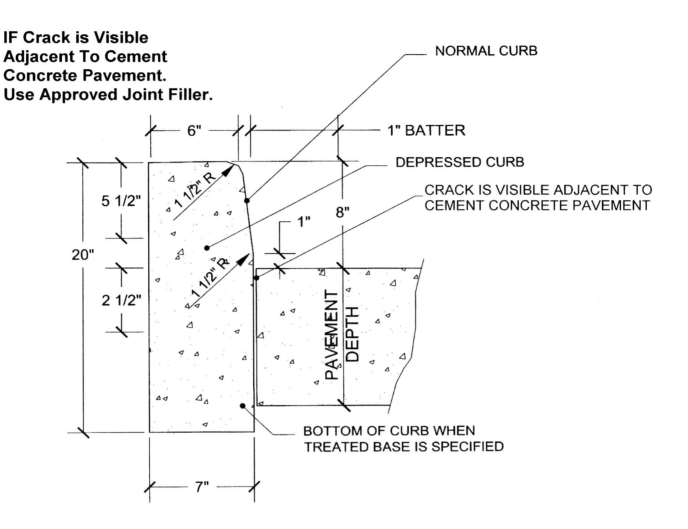

P.C.C. CURB
TYPE 2

ASPE
Concrete Curb & Gutter Sample Estimate

Item	Qty	Unit	Unit $	Total$
Quantity				
Straight – 350 LF				
Radius – 150 LF				
Straight curb				
Crew:				
Labor foreman	4	hrs	$45.00	$180
Laborers	4	hrs	$41.00	$164
Carpenters	12	hrs	$49.00	$588
Cement finisher	4	hrs	$50.00	$200
Material:				
Concrete 3000 psi	10	cy	$70.00	$700.00
Curing material	250	lf	$0.25	$63
Joint Material ½"	10	sf	$2.00	$20
Wood forms	250	lf	$1.75	$438
Radius curb				
Labor foreman	4	hrs	$45.00	$180
Laborers	4	hrs	$41.00	$164
Carpenters	12	hrs	$49.00	$588
Cement finisher	4	hrs	$50.00	$200
Material:				
Concrete 3000 psi	6	cy	$70.00	$420
Curing material	150	lf	$0.25	$38
Joint Material ½"	6	sf	$2.00	$12
Wood forms	150	lf	$1.75	$263
Total Trade Cost	15.0%			**$4,216**
Overhead & Profit				**$632**
Total Cost				**$4,832**
LF cost straight				**$7.73**
LF cost radius				**$14.29**
LF cost total				**$9.70**

02775 SIDEWALKS

Introduction/Description:

The Main CSI Division is 2, Site Construction, and the main subsection is 02700 Bases, Pavements, and Appurtenances.

The specific section is 02775, Sidewalks

Related specification sections are 02720 Unbound Base Courses and Ballasts, 03200 Reinforcing, and 03300 Cast in place concrete.

This section covers Portland cement concrete sidewalks, including reinforcing, joints, finishing, and curing.

Types and Methods of Measurements:

Concrete sidewalks are typically measured by the square foot (SF) to calculate labor and equipment costs.

Concrete is purchased by the cubic yard (CY). Calculate the quantity and add a 5 to 10% waste factor. Measure the linear feet of expansion joints to determine the amount of joint filler.

Reinforcing, if required, is typically wire mesh, which is measured by the square foot (SF).

Concrete finishing is measured by the square foot (SF).

Specific Factors Affecting Estimate:

Work performed prior to the placement of concrete sidewalk starts with the earthwork, or cut and fill, to bring the pavement area to subgrade. Unsuitable materials will then be replaced with compactable fill; then the subgrade will be compacted to the density required in the specifications. Once the stone subbase is placed to the thickness required by the design, reinforcing and concrete can be placed.

The key factors in determining daily production are the concrete thickness in inches, and the width of the sidewalk. Sidewalks are typically four to six inches thick. Usually, for a given thickness of concrete, the wider the walkway is, the higher the daily production will be. The most common sidewalk width is four feet, but widths up to eight feet or higher are not uncommon. Another factor that will affect the daily production is whether wire mesh reinforcement is required. Irregular shapes will also affect production. Small quantities that require less than a day's work for a crew will also have a higher unit price. Also, the type of finish will affect the pricing of the estimate (i.e.: colored concrete, stamped concrete, scored, etc.)

Typically, 3,000 PSI concrete is required. Additional costs will be incurred for higher strength concrete and additives such as super-plasticizers, accelerators, or pigments.

Organization of the Estimate/Quantity Survey and Calculation Quantities:

Take-off the quantity by square feet, which is usually the length times the width. Irregular areas can add to the difficulty. A computer take-off program with a digitizer can be useful. Calculate the quantity of cubic yards by taking the area in square feet times the depth in feet and divide by 27. Use this quantity to determine the amount of concrete to be purchased. Measure the length and spacing of joints to calculate the amount of joint filler.

Overview of Labor, Material, and Equipment:

Labor

The labor crew mix includes laborers, carpenters, and finishers. A rebar installer is required to set the wire mesh. The carpenter will set the edge forms. A typical composite crew consists of a labor foreman, 2 laborers, 2 finishers, 1 rebar installer, and 1 carpenter. The cost of this crew will depend on the labor rates. Union labor, obviously, costs more than non-union. Any requirement for state or federal prevailing wages is another factor to consider. At an average of $50 per hour, this crew costs $350 per hour. A crew can expect to place 800 to 1500 square feet per day, depending on conditions.

Materials

Concrete, typically 3000 PSI, is purchased by the cubic yard. Additives such as super-plasticizers, ice, accelerators, and pigments are extra. Mesh reinforcement is purchased by the square foot. Other materials include curing materials, joint filler, and edge forms.

Equipment

Typically, none is required. A small grader and roller can be used to prep the subbase.

Pricing

Consider a straight 300-foot section of four-inch thick sidewalk, four feet wide, or 1200 SF. This can be considered a typical day's production for the crew. The example is for a wire mesh reinforced sidewalk, with expansion joints 20 feet apart. The first calculation is for the labor costs. The labor rates in the example include the base rate, fringe benefits, and the contractor's taxes and insurance.

The concrete quantity for purchase includes a 10% waste factor. Wire mesh is purchased by the square foot or square (100 SF). Curing materials, although purchased by the gallon, is presented here as a square foot cost. Joint filler and edge forms are purchased by the linear foot.

Miscellaneous Pertinent Information

Other items related to sidewalk construction are adjacent pavements and curbs, and drainage.

Conclusion

It is important to consider all the productivity and cost factors when estimating the cost of concrete sidewalk. As shown above, the size of the project and the amount of production expected on a daily basis is critical.

Part Two | 02775 Sidewalks

ASPE

Cement Concrete Sidewalk Sample Estimate

Item	Qty	Unit	Unit $	Total $
Quantity = 1,200 sf, 4" thick				
Crew:				
Labor foreman	8	hrs	$45.00	$360
Laborers	16	hrs	$41.00	$656
Carpenters	8	hrs	$57.00	$456
Rebar installer	8	hrs	$63.00	$504
Cement finisher	16	hrs	$50.00	$800
Material:				
Concrete 3000 psi	14	cy	$70.00	$980
Wire mesh	1000	sf	$035	$350
Curing material	1000	sf	$0.05	$50
Joint Material ½"	60	lf	$2.00	$120
Edge forms	660	lf	$1.00	
				$660
Total Trade Cost	15.0%			**$4,936**
Overhead & Profit				**$740**
Total Cost				**$5,676**
SF Cost				**$4.73**

Standard Estimating Practice

03050 – BASIC CONCRETE MATERIALS AND METHODS

Hot & Cold Weather Procedures

This section addresses the cost effects of placing concrete under hot & cold weather conditions, also know as hot weather concreting and cold weather concreting. Heat and humidity are the factors that must be controlled in order to provide the proper concrete mix and final product in hot weather while freezing is the major factor in control in cold weather. In this section we will examine the problems associated with hot and cold weather concrete placing and the related costs associated with preventing problems. For detailed information on takeoff and other specifics related to concrete you should refer to other sections of this manual. For a more in depth discussion of the particulars of cold weather concreting you should consult the American Concrete Institute - Manual of Concrete Practice - Part 2.

The costs of this work should be treated as a separate operation and all costs associated with it listed separately and summarized as a specific line item in your concrete estimate.

Among the additional costs that must be included will be the additional cost for adding heating or cooling to the ready mix materials. Adjustments to the concrete design for the specific condition should be reviewed with the engineer, inspector and testing authority.

As soon as the determination has been made that hot or cold weather will be a factor, preparations for this work should be anticipated. Consideration should be given to starting the pour earlier or delaying the pour into the later, warmer times of the day. This should be reviewed with the ready mix supplier to see what, if any, additional costs will be encountered. The concrete placing crew may also require additional cost in order to place and finish the concrete. Additionally costs for lighting, safety, and overtime should be addressed. Prior to placing concrete the forms, reinforcing steel, subgrade, and any other surfaces that will be in contact with the concrete should be cooled (with water and/or a fog or mist spray) or pre-heated. Temporary wind screens may also be required to help block the wind and contain the heat or slow the effect of evaporation. After the concrete is placed, continued heating and protection should occur along with blankets, straw coverings, or other appropriate measures to insure the prevention of freezing of the concrete and to alleviate any cold weather related problems of the concrete or, continued coiling and curing should occur along with shading or other appropriate measures to alleviate the rapid drying of the concrete.

The various methods of heating will depend upon the geographic areas and the temperature extremes that will be encountered. In some areas LP gas will be used and in other areas where natural gas is available it may be cheaper to use that as the fuel of choice. In some cases oil, steam or other various methods may be used for heating. Enclosures may be as simple as a polyethylene wrap or as complicated as a pre-manufactured panel system. In other cases a blanket wrap will be sufficient.

Caution should be used in the heating process to insure that proper venting has occurred. If not, the potential for damage to the concrete from carbon dioxide could cause softness in the surface of the finished concrete. All heating should be to the total environment and the total concrete. If heating is not done carefully, flash heating could occur and the mix would not be uniform.

Other methods that help in the cold weather concrete process include using various types of concrete, such as Type III cement (High Early), additional cement, or, the addition of accelerators. As previously noted, this should all be decided with the concurrence of the engineer.

Obviously the magnitude and scope of additional costs will vary from project to project. If for example the project is a single story slab on grade structure, the costs might be minimal. However, if the project is a multistory building with slabs on metal deck the costs will be considerably more. In the latter case the exposed underside of the metal deck is a major factor and needs to be taken into account with regard to enclosing and heating this underside the night before placing the concrete or shading the deck to help keep cool. This is not as complicated as it seems. In most cases the geographic areas in which this is a consideration, it has been dealt with over the years and local experience and practices will be very helpful in assisting you in developing pricing for this portion of the work.

Specific costs that need to be addressed include but are not limited to:

- Increased material (ready mix) cost
- Costs for premium time
- Costs for heating materials, forms and the enclosures
- Costs for water, fogging and misting
- Costs for additives to the concrete
- Costs for heating of all surfaces to be in contact with the concrete to be placed
- Costs for windscreens, enclosures and other enclosing materials
- Costs for heating fuel, equipment and devices
- Costs for curing material, equipment and application
- Any additional costs directly attributable to the cost of cooling concrete and its environment

03100 – CONCRETE FORMS AND ACCESSORIES

Introduction

Cast-in-place concrete presents many benefits as a low cost choice for a building's structural system. The versatility, aesthetic appearance, durability, fire resistance, time of construction, maintenance cost and cost of initial investments make cast-in-place concrete a desirable system. The forming system is usually the single, biggest cost component of a concrete frame.

The concrete formwork digest published by the concrete reinforcing steel institute (CRSI) defines formwork as "a temporary structure used to contain fresh concrete to form it to the required shape and dimensions, and to support it until it is able to support itself. Formwork includes the surface in contact with the concrete and all necessary supports." This section will attempt to give the reader an overview of factors and consideration that impact formwork costs.

Analyzing the Job

In analyzing a concrete formwork job there are three basic principles of constructability: repetition, dimensional standards and dimensional consistency.

Changes in layout directly impact the labor cost associated with erecting concrete formwork. In the field, delays are experienced as production is slowed so that plans can be interpreted and measurements verified. Equipment changes or modifications may be required to accommodate the changes in layout. This decrease in productivity caused by a non-repetitive design increases the total cost of the structure.

Repetition in design facilitates the optimum flow of production. Look for building layouts that are repeated from bay to bay and floor to floor. When building layouts are repeated, equipment changes can be recycled easily and a production line flow of work is obtained. Design repetition increases jobsite labor productivity and decreases the total cost of concrete forming.

The construction industry has standardized member sizes, correspondingly, standard size forms are readily available from manufactured formwork suppliers. The fabrication of custom built forms will increase the total cost of formwork as this expense will be fully charged to the project. However, standardized forms can be utilized on many projects and the associated cost pro-rated by usage.

Standard nominal lumber dimensions are also important to cost control. Structural members should reflect dimensional lumber sizes. These dimensional standards eliminate costly carpentry requiring sawing and piecing together. Consistency of dimensional standards reduces the formwork cost by controlling waste and increasing productivity.

In general, when analyzing concrete formwork, consistency and simplicity yield savings, while complexity increases cost. Check for dimensional consistency by analyzing the depth of horizontal construction, column dimensions from floor to floor and story heights. The majority of formwork cost is usually associated with the horizontal elements with less cost impact produced by vertical support and lateral restraint except in high-rise construction. Simplicity and design consistency will bring the formwork in at lower cost.

Consideration should be given to the type of usage of the building being analyzed (i.e., parking garage, laboratory, office building, etc,). Each building type requires comparison with similar building types. Productivity levels will vary significantly for each building type, even if construction materials are the same. The quality of workmanship and concrete finishes can be determined by analyzing the building type. For example, areas with structural systems exposed to public view usually require a higher quality workmanship and smoother finishes at the slab soffit (with the exception of parking garages). The cost impact is that the higher quality of workmanship required yields lower productivity and smoother finishes indicate the need for high quality form facing materials. Both of these items equal added expense to be charged to the project. However, exposed concrete may be more cost effective than an applied finish material.

The project schedule is critical when analyzing the job and its formwork requirements. Important items to consider when studying the project schedule are pour sequence, crane availability, specification requirements on stripping time, pour rates, concrete strength, etc. Overtime requirements caused by fast cycle projects as well as weather conditions will impact forming costs.

The amount of equipment required depends on the make-up of the project and the construction timetable. A basic concept is to furnish a minimum amount of equipment consistent with maximum field performance. Reuse of equipment reduces formwork costs.

Plans and Specifications

The plans and specifications play an important role when the estimator is analyzing the formwork cost on a project. The specifications outline technical requirements for the job while the plans provide the architectural features and the structural intent of the project. CSI Master Format Section 03100 is currently set aside for concrete formwork. This section addresses the formwork material requirements and erection procedures that apply to the project. Frequently, ACI 347, CRSI and U.S. product standard PS-1 are referenced for standards of comparison. This section of the specifications as well as Sections 03200 and 03300 address form type, concrete type, finish requirements, architectural concrete, pour limitations, erection procedures, stripping times, reshore requirements, camber and special considerations. Each item must be studied for its impact on the formwork material and erection costs.

The architectural plans define the building function. As discussed earlier, this is important in the estimating process. Building function will give some indications as to quality of workmanship desired as well as concrete finishes. The architectural floor plans will provide finish schedules, elevator and stair conditions, openings in the floor and column and wall layout. A mental walk through the building can be accomplished. As this is done, note below grade conditions, access to basement areas and crawl-spaces, opening in the walls, depressions in the slab on grade, elevator pits, docks, wells, etc. The building elevations will help the estimator visualize the exterior conditions. The estimator must be cognizant of exposed or critical dimensions, window and skin treatments, minimum tolerances and accent areas spelled out in the building elevations. The floor-to-floor height, high work, overhangs, balconies and slab edge conditions are revealed on the building sections and overall details.

The structural plans outline the logical sequence of construction through the plans, elevations, building sections and details. The structural system configuration, bay sizes and layout are usually shown on the structural floor plans. The estimator should be aware of expansion joints, pour joints or pour strips, rebar configuration and penetrations. The overall constructability of the project is expressed in the contract documents and is important to the estimator in determining the cost of concrete formwork.

Horizontal Forming

The horizontal forming of a project represents a major concrete formwork cost component. Once again, the level of dimensional standards and consistency as well as repetition will impact formwork costs. In general, any soffit offset or irregularity may cause a stop-and-start disruption of labor, requiring additional cutting and waste of materials. Formwork costs are increased proportionately. Dimensional consistency of drop panels in both plan and section reduces formwork costs particularly if the dimensions conform to nominal lumber sizes.

Consistent spacing between joists and consistent soffit elevations with the depth of beams equal to the depth of the joists indicate the feasibility of using a cost effective forming system. Additional cost forming features to look for when analyzing concrete formwork are uniform soffit elevations, consistent beam depths designed to accommodate nominal lumber sizes. Variations to these items will increase formwork material and labor cost to the degree of variation.

A careful study must be made of the total horizontal area and quantity of equipment to be furnished on each project. This analysis should include determination of pour sequencing, construction schedule, pour cycles and cure times. Changes in beam sizes, openings, etc. may require furnishing more "working equipment" throughout the project than the actual footprint of the project would indicate. Once this detailed study is completed the required quantities are computed. Generally, furnish the minimum amount of equipment to meet the construction cycle and optimum productivity.

One of the major problems in horizontal forming is to obtain a durable, long-life material for supporting the concrete slab that will give a good concrete finish after many uses. Typically, plyform is used for forming exposed concrete slabs.

The life of the plyform decking has a great influence on costs: the cost of plyform is one of the largest single material cost factors in this type of work when concrete finish is important. The concrete finish specifications must be carefully examined by the estimator in order to determine what form facing material will provide the specified finish.

The cost of forming beams presents a special problem to the estimator. Variations in beam forming costs can be wide due to size inconsistency and degree of difficulty. Some indications of the degree of difficulty are many different beam depths, variable span lengths, high number of beam intersections, beam intersections with varying beam depth, low slab to beam contact area ratio and beam column intersection conditions. In addition, consideration for the placement of inserts, chamfers, dovetail slots, and feature strips, forming of beam pockets for post-tensioning tendons and construction adjacent to the beam are important in determining beam forming cost. A beam forming system can then be chosen as well as the amount of equipment to be furnished.

Vertical Forming

Vertical structural costs in concrete buildings are typically less than the horizontal. Only in high-rise construction does the vertical component for gravity and lateral forces exceed the cost of the floor framing system.

Forming costs of the vertical components - walls, columns, cores - are also highly sensitive to design simplicity and repetition. Elaborate designs can increase labor and material cost significantly.

Formwork costs for vertical components increase when there are changes in height and variations in the size of members from floor-to-floor as well as layout. Conditions at the first lift of vertical construction, footing steps, reinforcing steel configurations and splices further impact vertical forming costs.

Walls can be classified for estimating purposes into at least four areas:

1. Foundation walls - designed primarily to resist the earth or for below ground enclosures. It is important to consider excavation methods, grade conditions, and temporary earth retention.

2. Shear walls - shear walls are used in mid to high-rise construction to resist lateral forces. They are often categorized as exterior shear walls which have at least one face with no adjacent construction or interior shear walls with construction adjacent to both wall faces.

3. Core walls - core walls usually surround elevator or stair openings and can often require more labor on a square foot basis than the rest of the floor construction.

4. Site Walls - these miscellaneous walls normally are outside of the building lines and include retaining walls, planter walls and architectural feature walls. They are often poured after the structure is complete.

The material required to furnish is dependent upon the type of wall forming system to be used on the project. For handset systems consider the number of days necessary to complete the work, the amount of wall concrete that can be poured in an eight hour day either by crane or pumping, strip times and wall height. For each pour requirement, the minimum amount of equipment furnished should be one additional face beyond the pour area for continuity of production. For gang forming wall systems determine the formwork material required by considering the quantity of exterior faces and material movement of interior faces. The cost of panel set-up and dismantling are to be included in the formwork cost.

The various types of walls and forming system utilized will affect wall forming labor costs. Points of consideration when determining wall formwork labor costs are the time of year and the effects of weather, jobsite conditions, wall height, finishing requirements openings, intersections, pilasters, beam pockets, corners and blockouts. Labor costs increase as the degree of difficulty increases.

The fewer changes in column size on a floor and from floor-to-floor equate to lower formwork costs. This holds true for changes in column heights as well. While changes in column orientation don't impact the cost of column formwork, it disrupts the formwork at the horizontal intersection. Special attention should be given to the column face as well as rebar splices and supports. Uniform and symmetrical column patterns facilitate the use of gang or flying form slab systems.

The material requirements for column forming are similar to those of wall forming. Generally, the amount of equipment furnished is dependent upon the type of forming system used as well as material handling equipment available. Handset systems are attractive as a forming system when there is little repetition and many penetrations. Gang systems are used under highly repetitive and multiple use forming situations. Consideration of the project schedule and pour cycles is essential in determining the quantity of column forms to furnish.

There are some important factors when determining column form labor. Columns formed on grade level can be more difficult if footings are below grade. Depending on the column size and height, crane handling may be the only way to place and strip the forms. The column height will affect productivity with greater heights decreasing productivity. Hooked rebars limit column form height and do not allow for setting the column form over the rebar cage while two level column reinforcement may require temporary supports that interfere with the formwork. Inserts cause problems in setting and stripping. In addition locating and maintaining the position of the insert introduces a risk level that is difficult to control. Finally, movement and storage of column forms will impact costs particularly if columns must be stored on the ground between pours.

Formwork is the single largest cost component of a building's concrete structural frame. The estimator's careful analysis of the factors that affect forming costs and early input with the design team can result in an efficient and economical cast-in-place concrete structure. The estimator need only visualize the forms, visualize the field labor required to form various structural members and be aware of the direct proportion between complexity and cost.

The methods for estimating/pricing formwork as a general topic have been presented many times. What has not been fully addressed are the so called "red flags" which an estimator must investigate amidst the multitude of details already needing attention. It is helpful then, to understand the idiosyncrasies which are inherent in the major building types, such as hotels, offices, or residential condominiums. This will enable the estimator to be aware of potential issues of concern. For example, if a cast-in-place structure is intended for use as a mixed-use, high-rise building in a congested location, certain inherent characteristics associated with this type of structure should stand out and require further investigation, should come to mind.

Also, within the areas designated for retail and office space, the floor-to-floor height may be higher and require larger clear floor space areas. The living units may have lower floor-to-floor heights, with less emphasis on large open areas. This could affect the slab shoring system used as it may not be transferable from the retail to residential levels.

It is evident that to be armed with a list of these rules would make estimating schematics or budgets easier as well as providing the estimator with items for consideration which will contribute to a more accurate formwork estimate.

Considerations

The following is a selection of categories which will help the estimator when approaching the analysis of a formwork estimate. By selecting the appropriate building type, along with other categories which pertain to the project under consideration, the relevant so called "red flags" will be available for deliberation. The building types which will be reviewed in this section are office buildings, multistory hotels, and parking structures.

Major Building Groups

The majority of commercial structures commonly designed as cast in place concrete frames can be found to fall under one of the following categories. Once the building's intended function and design is understood, it becomes easier to prevent the neglect of key issues of concern during the preparation of a formwork estimate.

1. Office Buildings - These are usually designed with a higher lobby or ground floor, usually fourteen to seventeen feet in height. The typical office levels are around twelve feet in height, to accommodate the suspended ceilings and overhead duct systems.

    a) Ductile frame - This design utilizes relatively close column spacing with a deep perimeter beam, usually three to five feet deep and interior drop caps, four to six inches deep at each of the interior columns.

        i) Considerations: An important concern is the increased amount of reinforcing steel needed to create the moment resisting frame in these deep beams creating a highly dense rebar cage. Therefore, it is common for the deep exterior beamside to be left off until just prior to pouring the concrete in order to access and install the rebar steel. The installation of the exterior beamside then becomes a highly inefficient operation. In addition, for economy as well as aesthetics, the exterior face of the beam is often exposed architecturally as part of the building's facade or skin. (See Architectural Concrete and Building Configuration)

2. Multistory Hotels - This type of structure usually has a low-rise area for the restaurant, lounge and ballrooms, which is usually structural steel construction. There is generally a high lobby level and a mezzanine floor between the ground floor and the first guest level to accommodate a partial floor of meeting rooms. Typical guest floors are commonly around eight feet in height.

    i) Considerations: The ground level usually requires extensive underslab piping to accommodate the kitchens and service areas and due to the high possibility of an accelerated schedule, it is likely that the slab or grade will be placed later in order to proceed with the completion of the frame, causing the first high level to be formed off of a rough grade. In addition, service areas are normally depressed in many areas for drainage and equipment which adds to the problem at grade with uneven shoring heights. In addition, service areas are normally depressed in many areas for drainage and equipment which adds to the problem at grade with uneven shoring heights. In hotels it is not uncommon for the vinyl wall coverings or paint to be applied directly to the cast-in-place walls. It is also common for the finished ceiling treatment to be applied directly to the concrete soffits. Thus, hotel room wall and ceiling surfaces require an extremely smooth finish, devoid of blemishes, bulges and areas in excess of specified tolerances. (See Architectural Concrete and Multistory Construction)

3. Parking Structures - These are generally designed as open structures with the emphasis on its appearance to the public. This may entail the use of precast panels, perimeter planters and/or architectural concrete features.

    a) Flat Plate - This design system is typically an eight to ten inch thick slab with drop caps at the columns. Shear walls are usually found along the ramps and at the exterior sides of the building.

    b) Long Span or Beam and Slab System - This scheme carries the loads through the use of long span beams, often around sixty feet in length spaced in the range of eighteen to twenty three feet apart, resting on columns or shear walls. When there is a great deal of repetitiveness, a steel beam forming system can be implemented. The floor-to-floor height is often ten feet or more in order to accommodate the generally used three-foot-deep beam system.

i) Considerations: The most confusing aspect of these structures is the layout on two-dimensional prints of the ramping system and the floor-to-ramp-to-floor connections. It is helpful to draw a cross section of each side of the ramp in order to get a full view and count of the connections if not provided by the architect. Due to their open design, the slabs are usually designed to drain the rainwater runoff to centrally located drains. As a result, there can be extensive warping of the slabs, which is often carried through to the soffit system. This may affect the layout of long stringers or the use of steel beam-forming systems.

ii) A consideration which is unique to the long span beam garages and the use of the steel forms, is the movement and cycling of these forms. A great deal of thought and experience is needed to establish the most advantageous layout of the pours. Sequencing the material through ramps and formed areas can create a multitude of problems if not dealt with carefully.

Project Characteristics

1. Confined or congested sites usually implies that the site is restricted by existing buildings and/or property lines. This is primarily the case on construction in inner city locations.

    i) Considerations: The ability to cycle forms as well as the type of system planned for use may be affected if the access for the hoisting equipment. The clearance between adjacent structures, property lines and the new construction may require the use of one sided, blind or shotcrete wall forming.

2. Multistory Construction, buildings in excess of +/- twelve stories are susceptible to factors which are easily overlooked. Despite the appeal of the repetitively designed floor plan, it is often a trade off between the economy of the repetitiveness and the potentially costly problems which arise directly as a result of the structure's height.

    i) Considerations: The perimeter edge and formwork design conditions become extremely important as the building progresses upward. It is important to investigate the possibility of ganging up the edge formwork to minimize the number of loose pieces. This is crucial for the stripping operation and, if not given enough forethought, the cost for stripping can increase as the project progresses. If it is a conventionally formed area, each piece will have to be stripped by hand and a method of catching any runaway forms must be developed in order to avoid the risk of losing a piece over the edge. This is especially crucial in inner city high rises.

    ii) In addition, it is important to verify the floor-to-floor height consistency throughout the structure. This is important since there is the tendency to alter the floor-to-floor height to accommodate a non-typical requirement such as a rooftop restaurant or luxury penthouses.

3. Building configuration - In concrete formwork, building configuration is of extreme importance. Not only does it affect the design and economical use of the forming system, it may affect the field labor as well. The two areas of primary importance are the layout and the reuse of materials. Complex building floor plans would require extensive labor and a greater amount of single use plywood. Even a structure composed of simple rectangles may also require additional planning in the layout of the plywood, joists, and shoring.

Accessories

The form area (SF) for architectural concrete must be accumulated separately. Architectural concrete is that which is to be exposed to view in finished areas of the project, and is usually subject to special form and finish requirements, which may add substantially to the cost of both the form material and the labor to install it.

Form liner is a uniquely shaped mould fastened to the regular form to impart a special texture or finish to the wall. It must be accumulated, for reasons of cost, separately from other wall form quantities.

Rustication (See Figure 03100-1) is a strip fastened to a form to provide an architectural indentation in the wall. It is surveyed by LF and totaled separately.

Figure 03100-1

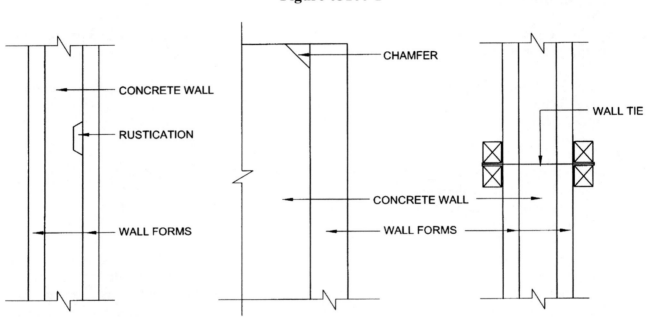

Chamfer is a strip sometimes fastened to forms where the concrete would otherwise form a right angle. It is used to keep concrete edges from being easily chipped. The quantity is measured in LF.

Stripping forms is the removal of the formwork when the concrete has cured sufficiently. It is the same quantity as the formwork put in place.

After forms have been stripped, they must be cleaned and oiled prior to their next use. Quantity is the same as strip forms.

Form ties are devices used both to hold the two sides of a form together and to space them the correct distance apart, maintaining the proper wall width. The number of ties required depends upon both the design of the wall form as well as the type of tie. Quantity is each or piece.

Form ties are generally considered to be steel with breakback and patching requirements to forestall liquid passing through the structure. Steel form ties, without adequate protection, are subject to rusting and staining of finish surfaces. Fiberglass form ties are also available that may not have the breakback/patching requirements and rusting potential.

Estimators should become aware of various available form tie systems and determine from test data and product literature which system is more effective from a material/ labor cost comparison standpoint and finished product appearance requirements.

Wall form accessories include such items as nails, form hardware, and form oil. Some estimators survey and price these items separately, while some include their cost in the unit prices used for material.

As with forms and concrete, the estimator must know what finish is required for the wall segment being surveyed. Finishes for architectural concrete must be segregated.

Accessory requirements for angles, curb angles, dovetail anchor slots, expansion joint fillers, perforated drain pipe, reglets, wood grounds, nailing strips, and waterstops shall be separately defined and quantified.

Angle iron ledgers, floor and roof decks, channel door frames, pit frames and covers, ladder rungs, and miscellaneous fabrications may be furnished and/or installed under Division 05500 - Metal Fabrications. Refer to that division for clarification.

Dampproofing, insulation, and sheet metal may be furnished and/or installed under Divisions 07100 - Dampproofing and Waterproofing, 07200 - Thermail Protection, and 07600 - Flashing and Sheet Metal. Refer to those divisions for clarification.

Grouting and column base grouting shall be measured in each and/or in cubic feet.

Vapor barriers, curing compounds, and bond breakers shall be identified and measured in square feet.

Waterstop - waterstop is a rubber, vinyl, or metal material made into strips and placed in joints as described as follows. Waterstops come in several widths and numerous configurations. (See Figure 03100-2)

Figure 03100-2

Ribbed with center bulb - vinyl ribbed waterstops are used in construction joints where movement between members is anticipated. They are generally used above grade but in fact are a universal design and can be used in any type of joint above or below grade.

Dumbbell - vinyl dumbbell waterstops are used in construction joints where movement between members is not anticipated. They are usually employed below grade and in horizontal applications.

Flat Ribbed - Vinyl flat ribbed waterstops are employed in construction joints where movement between members is not anticipated. They are generally used below grade in footings, walls, slabs, or for breaks in concrete due to limitations in working methods.

Split Ribbed - Vinyl split ribbed waterstops are used in either construction or expansion joints where elimination of split form work is feasible. The split leg of the waterstop is spread open and nailed to the bulkhead. When the form work is removed following completion of first pour, the splitleg is joined together and the second pour completed.

Waterstop should be surveyed by size, type, and configuration. It's quantity is measured in LF. For example, 6 IN center bulb PVC waterstop.

Other Concerns

1. Architectural concrete commonly involves one or more of the following: forming material to be used must be of a higher quality than that specified for standard concrete formwork; surfaces must be void of air pockets, bleeding, rock pockets or other surface blemishes; tie hole pattern as well as face plywood may be required to conform to a specific alignment or layout; construction joint and cold joint locations may be restricted and require chamfering or other joint defining treatment; and tolerances are usually more stringent.

    i) Considerations: It is worthwhile to investigate the availability and practicality of the higher grade of form facing material specified as well as any options which can achieve the same or better end product.

    ii) It is also important to remember that in order to achieve the smooth surface, super plasticizers may be used as an admixture to the ready-mix and may require "beefing up" of the formwork to accommodate the full liquid head. Also, in an effort to eliminate the air pockets, there is a tendency to over vibrate when placing concrete, which may damage the form facing material and require replacement or remedial work to the concrete surface after forms have been removed.

    iii) Though the layout of tie holes and plywood sheets in accordance with a preset scheme is not usually a great burden, it is still prudent to investigate their impact. The placing of plywood forms is particularly difficult on footings or sloped surfaces. The primary impact of a regimented joint layout is the unusually high amount of remedial work necessary after forms are removed to rectify misaligned panel joints which were misformed or moved during pouring. Extensive use of reveals and joint chamfers should be considered in light of the time consuming nature of joint layout and installation.

iv) The American Concrete Institute (ACI) publishes a document entitled, "Recommended Practice for Concrete Formwork, (ACI 347-78) (reaffirmed 1984)," which provides a standard for tolerances for cast-in-place concrete. Typically, this Standard is found as a reference in the bid documents; however, it is not uncommon, especially in a structure where there is a great deal of exposed concrete or architecturally finished concrete, the tolerances may have been made more stringent. In order to achieve these more exacting requirements, the labor costs must increase to account for the extra work which will be necessary to achieve these tolerances. It is important to note that it is also possible that the tolerances specified may be virtually impossible to achieve in view of the tolerances allowed in the fabrication of the materials which may be used. The manufacturers of plywood and lumber allow tolerances which may affect the final requirements.

2. Cambering is a typical requirement in conventional construction. This provides additional compensation for deflections. Cambering usually entails the forming of an upside-down bowl-like warpage in each bay and often can be achieved simply by jacking up the shores after all of the forms are in place.

i) Considerations: The amount of cambering required is important as it may affect the layout of the forming system and in severe cases, the cutting of the forms may even be necessary. Occasionally, stringent design conditions, there may be reference to reshoring. This refers to the method of stripping one bay at a time and reshoring immediately or leaving a designated number of shores in place without disturbing the soffit forms. Often, for conventionally reinforced slabs, either of these options may be used in order to strip and release the majority of slab forms after seven days of curing.

Waste Allowances should be determined from historical data including the number of uses of the forms.

Takeoff Procedures

Formwork requirements for continuous wall footing, grade beams, wall footings with integral stem walls, footings, slabs on grade, blockouts, bulkheads, and other foundation structures shall be quantified in square feet.

Round/square column and pier formwork shall be measured in square feet or lineal feet as applicable.

Scaffolding for support of or access to form work shall be included in the formwork estimate and defined in detail.

Chamfer, chases, grooves, molding, expansion joint material less than one foot wide, and articulation material shall be measured in lineal feet.

Walls

Most formwork quantities are determined using two of the three dimensions used for calculating concrete volume. Concrete uses L, W, and H. Formwork uses L and H. In a wall segment measuring $100 \times 1.0 \times 10$, only the length, 100, and the height, 10, are used.

If the wall must be formed on both sides, the formula for square feet of forms required is length, 100, times height, times 2 sides = 2000 square feet. SF = L × H × 2.

This quantity is called wall form area, square feet of contact area, or simply wall forms. If the ends of the wall must be formed, this quantity must be added: 2(ends) × 1.0 × 10.0 = 20 SF.

If there are openings in the wall, as opposed to the practice with concrete volume, they are usually not deducted from form quantity. Forms for openings, unless they are very large, are continued over the opening to minimize cutting form materials. Additional forms must be placed around the perimeter of the opening, however, to prevent concrete from entering the opening. These additional forms are called blockouts or boxouts and the formula for calculating their quantity is [2(L) + 2(H)]W, where L is the length of the opening, H is the height of the opening, and W is the width of the wall. A 3.0 × 1.0 × 7.0 door opening would require 20 SF of boxout. Boxouts can also be surveyed in lineal feet, where the formula for the perimeter of the opening, 2L + 2H, is used. (See Figure 03100-3A)

Quantities for ledge forms must also be included as required. Ledges are usually formed by adding a box to the inside of a wall form, thus causing the same contact area to be formed twice. The formula used is L × H, where L is the length of the ledge segment and H is its height. These quantities are accumulated under the heading ledge forms. Some estimators also add the width of the ledge to the formula, which is certainly acceptable, (W + H)L. (See Figure 03100-3B)

Figure 03100-3

Similarly, haunch or corbel forms must be included, if present. In this instance, the primary forms are interrupted at the haunch, and the additional footage, though not formed twice, as in a ledge form, serves to offset the cost of interrupting the primary forms. The quantity equals length times height and is accumulated under "haunch forms." Thus, the previous 100 × 10 wall with a 2 ft high haunch will have added to it's quantity as "haunch forms" 100 × 2.0, or 200 SF of haunch forms. Wall forms and haunch forms should be totaled separately. (See Figure 03100-4A)

Pilaster forms are calculated by using the formula 2WH, where W is the pilaster projection width and H is the projection height. Quantities are accumulated under "pilaster forms." Some estimators will also add the length of the pilaster, L, to the equation to offset the added cost caused by interrupting the wall forms by the pilaster. (See Figure 03100-4B)

Battered wall form area is calculated by the formula 2LH where H is the height of the sloped side. This dimension may be calculated using the formula for the hypotenuse of a right triangle, or it may be scaled. In the event the wall is battered on only one side, the average height of the wall must be determined and used as the height for calculation.

Figure 03100-4

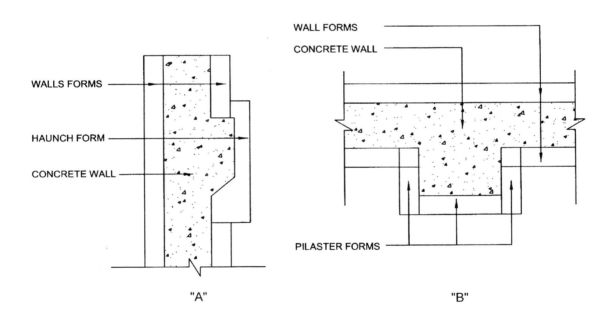

Circle and curved wall forms are also calculated using the 2LH formula for area. L, however, must be calculated as above in the determination of concrete quantities. The diameter of the inside surface of a curved wall will, of course, be less than the diameter of the outside surface. The average diameter of the two wall surfaces is the distance halfway between them. Thus the diameter is taken to the midpoint of the wall, as with the calculation of concrete quantities.

Bulkheads are used at the ends of any wall segment being poured at one time. The length of these individual pours varies from job to job. Generally not more than 100 LF of a wall should be poured without a bulkhead. Bulkhead area would be the same as end wall forms previously discussed, but usually accumulated separately as they require drilling for rebar and keys.

One-sided forms, where required, are calculated by the formula LH and accumulated under "one sided wall forms."

Columns

For column forms, the length dimension is the perimeter of the column. Multiplied by the height, it is accumulated in SF and listed on the pricing sheet as "column forms." Case 1, section A-A, will read 8.0 × 9.0; case 2, 6.67 × 9.0; and case 3, 3.14 × 2.0 × 9.0 (See Figure 03110-5). The lower portion of the column with the capital is surveyed similarly. The upper portion's average circumference is determined, then multiplied by one-half the slant height of the cap (which usually is scaled), or ½ H [(D1+ D2)(3.14)], (H = the slant height and D1 and D2 = the diameters of the bottom and top of the capital respectively). These formulas the estimator should memorize, so that given a line of survey, he may be able to apply both the concrete volume and the form area formulas without writing two separate lines.

Figure 03110-5

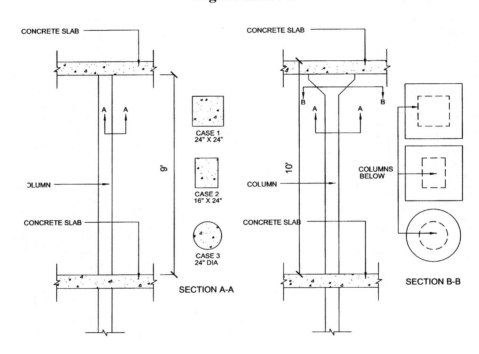

Beams

Formwork for concrete beams consists of beam bottoms and beam sides (See Figure 03100-6). Beam bottom area is the contact area of the width of the beam times its length. Beam side area is the contact area of the depth of the beam below the slab times its length. The beam bottom area of girder G1 is 128 LF times 1.5 FT. The beam side area is typically 2 × 128 × 2.33; 2.33 being the depth of the beam below the slab. This is the case with interior beams. The exterior beams, however, are a somewhat special case. As can be seen in section B-B, the inside beam side conforms to the typical case, while the outside beam side extends to the top of the slab. The depth of the inside side is 2.33 FT, while that of the outside side is 3.0 FT. Adding the two, the total beam side depth for G1 is 5.33 FT. This is the beam side depth factor, or "form factor". As with concrete, the formwork dimensions must be adjusted for corners and intersections.

The estimator will have noticed that the lengths used above for beam bottoms effectively run the beam bottoms through the columns, when in fact the beam bottoms stop at the column, or, at least, are interrupted by it. There are several possible beam column intersections, some of which are in Figures 03100-7 and 03100-8. While the estimator must recognize the individual cases as he encounters them, the practice of running the forms through the columns is an acceptable practice, as it gives a slightly larger quantity, thereby providing additional money to apply to the interruption caused by the column. Where two beams intersect at a column, one of their lengths must be adjusted as noted above.

Figure 03100-6

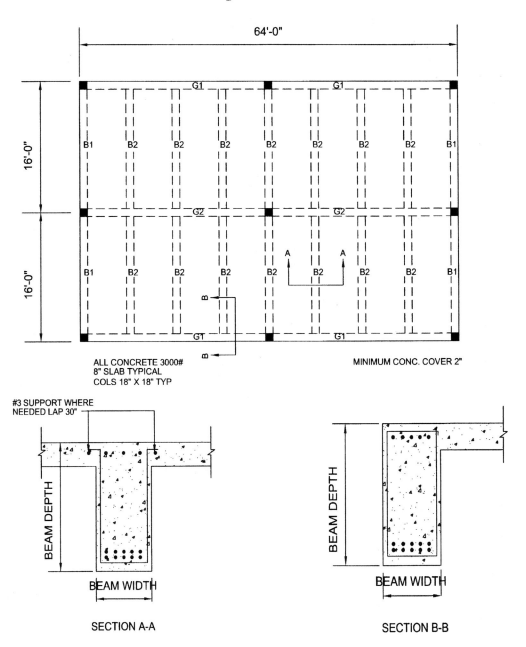

Figure 03100-7

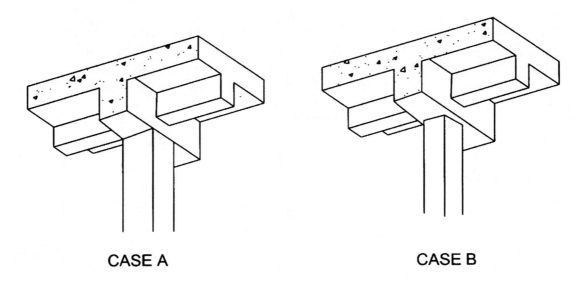

CASE ACASE B

Figure 03100-8

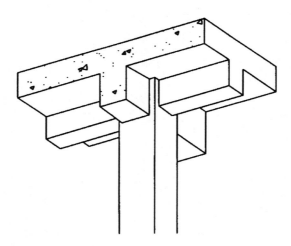

In Case A, the column placement height would stop at the bottom of the lower beam, the deepest beam would be surveyed through the column and the shallow beams to the edge of the deep beam. Case B would be surveyed the same way. Case C, however, presents a situation in which the estimator may wish to take the column height to the bottom of the floor, then stop both beams at the column.

A note to indicate the approximate shoring height of this particular set of beams should be included. This should be done for later calculating the cost of shoring. Beam quantities should be aggregated in height increments of 5–6 FT (i.e., Beams 0–6 LF shoring height, beams 6–12 LF shoring height, etc.), as this is the usual height of a shoring tower. The girders and beams have here been surveyed in their smallest length increments, although larger ones are acceptable.

Beam width is as given in the schedule (Figure 03200-16), and beam depth is that portion below the slab. A beam side form factor, to remind the estimator that the beam sides in these two cases are not simply twice the height of the beam, should be developed. Where no form factor is indicated, the estimator knows that the standard condition prevails.

When extending these takeoff lines, the estimator, as he did with columns, must remember the appropriate formula for the item under consideration. The formula for concrete volume, $V = L \times W \times D$, the formula for beam bottoms, $SFCA = L \times W$, and the formula for beam sides, $SFCA = L \times 2H$ must all be used when calculating their respective quantities. Since no special finish requirements are noted, the estimator simply uses the sum of quantities for beam bottom and beam sides and posts them under "P & P" (point and patch).

Slabs

Concrete slabs are the structural elements which carry the occupant load of a structure, the space for which the structure is being built. The simplest of these, usually, is the slab on grade (SOG). It rests directly on the ground, and is the simplest to survey and price. (See Figure 03100-9)

Figure 03100-9

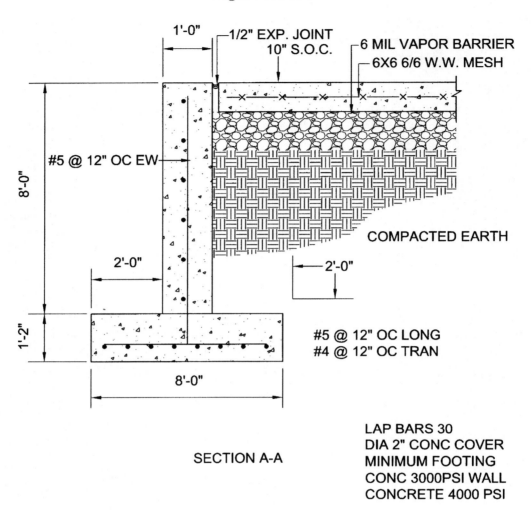

SECTION A-A

LAP BARS 30
DIA 2" CONC COVER
MINIMUM FOOTING
CONC 3000PSI WALL
CONCRETE 4000 PSI

The slab on grade in Figure 9 is 10 IN thick and reinforced with 6 × 6 wire mesh. This particular mesh is also designated 6 × 6 W2.9 × W2.9 under the most recent change of nomenclature. The slab is also underlain with 6 mil vapor barrier and 6 IN stone.

The estimator will survey SOG concrete, forms, screeds, edge forms, finish, reinforcing, and miscellaneous items associated with the slab.

With this information, the estimator has been able to quantify the slab on grade. Concrete volume is L × W × H, as before; finish and fine grade are L × W; mesh and vapor barrier are L × W × 1.1, stone is L × W × D, the miscellaneous item "expansion joint" is added as LF.

The suspended slab, however, is not quite so simple. A great deal depends upon the type of slab being surveyed.

The first type to be considered is the flat slab.

Any particular slab need not have a spandrel beam and it need not have "drop panels", the square portions which project below the bottom of the slab. Drop panels are also called "drop heads", and less frequently, "capitals." In such cases, the quantity survey items for these members would not be required as part of the flat slab.

Flat slab bottom, or "soffit", forms are necessary for the bottom surface of the area, the outside beam side providing the perimeter stop for the liquid concrete. The beam forms will be surveyed as shown previously, then their bottom quantity deducted from the gross slab soffit quantity. The horizontal surface of the drop heads should also be calculated and identified separately, for later pricing adjustment. In addition, the edge of the drop panel must be identified as a separate quantity and cost item.

Slab concrete is, again, calculated by L × W × D, and grouped by strength and type.

Slab finish is calculated using L × W and grouped by type of finish as specified; trowel, broom, etc.

03200 – REINFORCING STEEL

The cost of reinforcing steel is affected by the grade (tensile strength) of the material, whether the bars are deformed (have ridges or indentation on their surface) or smooth, the extent to which they must be fabricated (shaped to a specified length and bent into designed shapes), and whether they have a plain, galvanized or epoxy coated.

Survey the reinforcing shown in Figure 03200-1 for cases A and B, assuming a footing segment length of 200.0 LF. Laps are specified as 3.0 FT, stock length rebar is 40.0 FT, and minimum coverage is two inches. Accumulate bars by grade, size, and weight.

Figure 03200-1

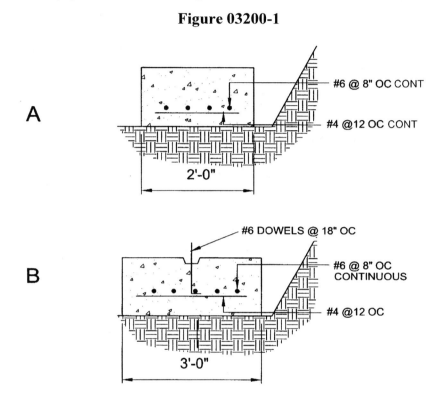

Example A: Longitudinal Bars: 4 #5 × 199.67 LF. (200 FT - 2 IN cover at each end) add 20 laps × 3.0 FT. (See Figure 03200-2)

Figure 03200-2

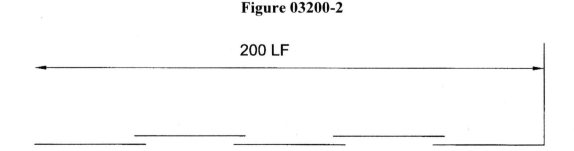

Transverse Bars: 199.67 / 1.33 + 150 spaces which = 151 bars in summary — 151 #4 × 1.67 LF.

Example B: Longitudinal Bars: 5 #6 × 199.67 LF (32 IN / 8 IN = 4 spaces which = 5 bars. Length same as example A.) add 25 laps × 3.0 LF

Note that there are six bars shown, but only five required. The number required should always be calculated, rather than relying on a graphic representation.

Transverse Bars: Length – 2.67 LF

No. required – 199.67 / 1.0 = 200 spaces which = 201 bars, in summary – 201 #4 × 2.67 LF

Dowels: Length – 3.5 LF

No. required – 199.67 / 1.50 = 133 spaces which = 134 bars, in summary – 134 #6 × 3.5 LF

Reinforcing steel is surveyed in lineal feet, which is then converted into pounds, hundred weight (CWT), or tons using the following chart:

BAR SIZE	WEIGHT LB/LF	DIAMETER (IN)	CROSS SECTION (IN)	PERIMETER (IN)
#3	.376	.375	.11	1.178
#4	.668	.500	.20	1.571
#5	1.043	.625	.31	1.963
#6	1.502	.750	.44	2.356
#7	2.044	.875	.60	2.749
#8	2.670	1.000	.79	3.142
#9	3.400	1.128	1.00	3.544
#10	4.303	1.270	1.27	3.990
#11	5.313	1.410	1.56	4.430
#14	7.650	1.693	2.25	5.320
#18	13.600	2.257	4.00	7.090

Footing reinforcing, both size and spacing, is usually given in a footing section as in Figure 03200-3.

Figure 03200-3

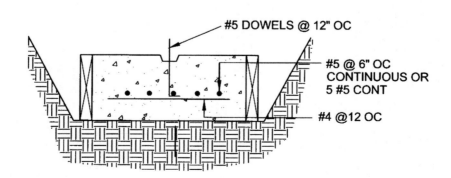

Reinforcing Steel Placement

The designation #5 @ 6 continuous refers to the longitudinal steel and means space #5 rebar at 6 IN on center across the footing width and run it continuously for the length of the footing. 5 #5 cont means in this footing, no matter what its width, run 5 #5 continuously for the footing length. #4 @ 12 IN means place a #4 bar at 12 IN on center across the longitudinal steel. This bar is called the transverse steel and one of its primary purposes is to keep the longitudinal steel in the correct position as the concrete is being placed. #5 dowel @ 12 IN means place a #5 rebar bent @ a 90 degree angle @ 12 IN on center down the length of the footing, tying it to the longitudinal steel. "DWL" is the abbreviation for dowel. Meaning a bar which will penetrate the next construction system (in this case probably a concrete wall) and provide rebar continuity by overlapping and being tied to the reinforcing in the next system.

For reinforcing steel to function properly, it must have proper concrete coverage. This coverage varies for different design conditions and is either called out in the design documents or is specified by reference to a recognized rebar standard, usually the American Concrete Institutes' standard practice for detailing reinforced concrete structures. ACI SP -66. In the case of footing steel, minimum coverage might be 3 IN from the sides of the footing and 2 IN from the ground. This information is important, since it determines the length of the rebar.

To determine how many longitudinal bars will be required in a 30 IN wide footing, look at Figure 03200-4.

Figure 03200-4

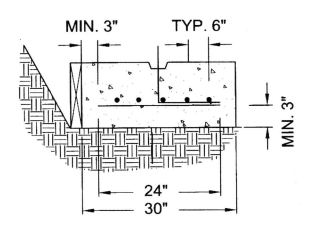

With a coverage of 3 IN required from the side of the footing, a space of 24 IN remains between the two outside bars. Since the bars are spaced at 6 IN OC, there will be 4 spaces requiring 3 bars for the intermediate area, or 5 bars altogether. The transverse bars can be determined to be 24 IN long, since they go from the leftmost bar to the bar at the extreme right.

Where footing length dictates rebar shorter than stock length(s), the bar must be cut to the appropriate length. For example, a footing segment of 34 FT 6 IN with the coverage as given above would require a longitudinal bar length of 34 FT 0 IN. If footing length dictates a length longer than stock length, the bar must be lapped to maintain rebar continuity. (See Figure 03200-5)

Figure 03200-5

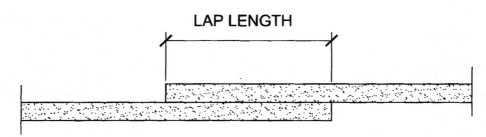

These laps must be taken into consideration when surveying rebar quantities. Required lap lengths are usually specified. If none are specified, a lap length of 40 times the diameter of the bar under consideration is usually adequate.

Figure 03200-6

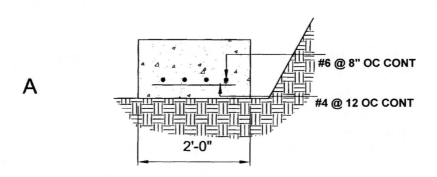

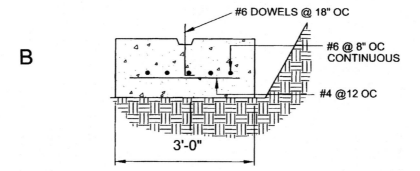

Survey the reinforcing shown in Figure 03200-6 for cases A and B, assuming a footing segment length of 200 LF. Laps are specified as 3LF, if stock length rebar is 40 LF, minimum coverage is 2 IN. Accumulate bars by grade, size, and weight.

Figure 03200-7

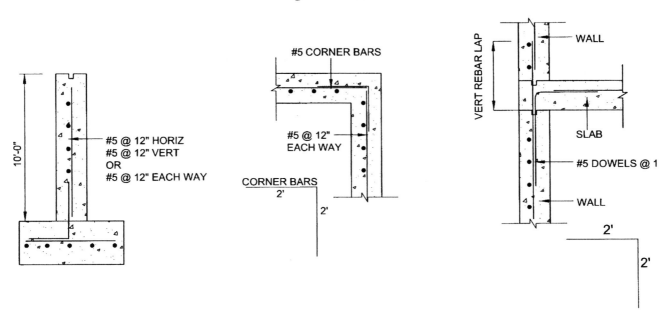

CORNERS

(See Figure 03200-7) This example points out corner bar configuration. Many estimators, however, to save time, take the outside perimeter of the footing as the length dimension, effectively doubling the corner quantities. The error introduced, it is felt, is quite small when compared to the uncertainties of pricing, and the time saved is a beneficial trade-off.

Reinforcing steel, when placed in a footing, must be anchored so that it will not be displaced when placing the concrete. To hold the rebar the correct distance from the ground, manufactured chairs or concrete bricks may be used. To keep the rebar from moving sideways or floating, rebar templates may be needed. These may be made from many materials, usually from existing scrap to minimize material cost, and are usually surveyed as "rebar templates in lineal feet, with the quantity being equal to the length of the footing. (See Figure 03200-8)

Figure 03200-8

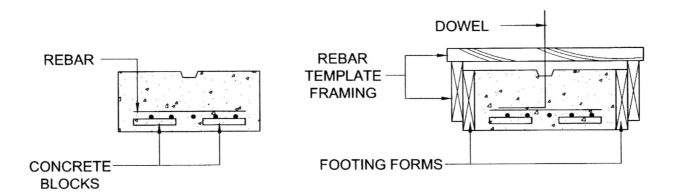

03200 Reinforcing Steel

Wall reinforcing steel is surveyed in pounds and summarized in tons or hundredweight (CWT).

For example, given a 100 LF wall segment with section as above:

Recall that dowel steel and continuous footing steel have been previously surveyed with footings.

To calculate the weight of the vertical bars, their number must be determined, multiplied by their length and their weight per pound. In the example, there are 101 vertical bars, each bar is approximately 9.83 LF in length and weighs 1.043 LBS/LF. The total weight of the vertical steel is 1036 LBS. If the horizontal steel is assumed to come in 40 LF lengths, the 99.67 LF, (assuming 2 IN concrete coverage) will require two laps of 30 diameters as noted in the specifications (30 × 5/8 IN = 1.58 LF EA), and the total length of the horizontal steel will be 99.67 + 1.58 + 1.58 = 102.9 LF. The number of horizontal bars is 11, and the weight is 1.043 #/LF. The total horizontal weight is 1181 LBS.

To recap:

Vertical steel	1036 LBS
Horizontal steel	1181 LBS
Total	2217 LBS divided by 2000 = 1.11 tons.

There are several typical conditions for which rebar must also be added.

The number of corner bars equals the number of horizontal bars (in the above example, 11), their length is 4.0 LF, and the weight of the bar is again 1.043#/LF. The total weight of the corner bars is 46 LBS. Corners should be surveyed in the same fashion as corners in footings.

The height of the vertical bar must be increased to add the projection into the wall above. Slab dowels must be calculated by number required, their length, and their weight per foot.

Slab dowels and corner bars should be segregated from straight horizontal and vertical rebar quantities, posted on the quantity survey sheet under "heavy" bending.

The estimator will have noted that in the corner plan given, surveying both wall segments will result in the counting of the corner vertical bar twice. This bar must be deducted from one of the wall segments for the count of vertical bars to be correct. In the interest of takeoff speed and the small quantity variance, this is often not done. The estimator must determine company policy with respect to this practice. In this text, the bar will not be counted twice.

Consider also the haunch: If the haunch runs the full length of the wall segment, the horizontal bars will be of equal length to the horizontal wall reinforcing, if this is not the case, they must be calculated only for the length of the haunch. The length of the irregularly shaped haunch tie may be calculated as follows:

The number of ties is calculated as above for the number of vertical wall bars, or as specified. The weights of horizontal haunch steel and haunch ties may be calculated as above, with the ties being segregated.

Figure 03200-9

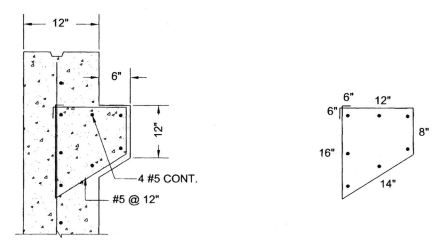

Rebar for columns is shown below:

Figure 03200-10

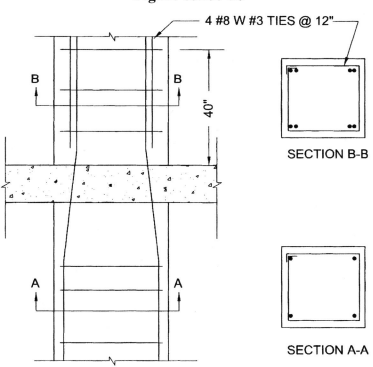

The purpose of the ties is to hold the vertical bar in place until the concrete is placed. To survey the vertical bar, the estimator must know, as with all rebar, the grade of steel, its size, its length and the number of bars with this identical set of characteristics. In the example above the size of the bar is #8, the assumed grade is 60, and there are four bars per column. The length of the bar is the floor to floor height plus 40 IN, the required lap. Assuming the floor to floor height is 10.0 FT, the vertical bar is thus 13.33 LF. To calculate the length of the tie, the dimensions of the column and the required concrete coverage must be determined. Assuming the coverage to be 2 IN and the column to be 16 IN square, the tie would be 4+12+12+12+12+4 = 56 IN = 4.67 LF. The 4 IN hook on each end of the tie is determined by referring to a standard hook detail chart. Since the ties are 12 IN on center, there will be 10 ties per floor, assuming a 10.0 floor to floor height and a slab thickness of 12. IN To recap:

Per column 4 # 8 × 13.33 LF vert

10 #3 × 4.67 LF ties

Given the data in Figures 03200-9, 03200-10, and 03200-11, survey the columns.

Figure 03200-11

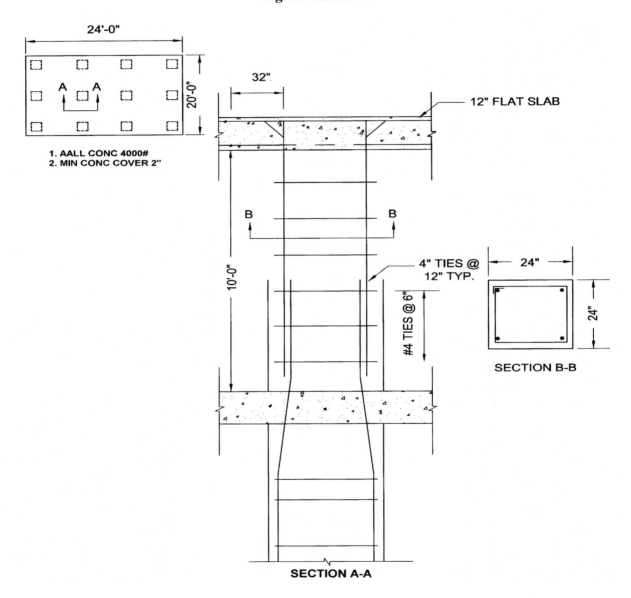

There are 12 columns, each 2.0 × 2.0 × 9.0. There are 4 #10 vert bars whose length is 9.83 + 2.67 = 12.5 LF. There are 11 #3 ties × 7.42 LF.

Surveying reinforcing steel for beams offers a somewhat more complicated task. The principle reason for this is the diagrammatic nature of the elevations and sections of rebar in beams, the necessity of using tables to communicate different beam conditions, and the industry standards promulgated by CRSI and ACI, all of which may not be shown on a particular project, but which are included by reference.

Consider girder G1. The "A" bar, which is shown in the table (See Figure 03210-16) to be 5 #9, penetrates 6 IN into the face of each column, which gives a dimension of L1 + 1.0 FT. L1 is 29.75 ft, and the length of the "A" bar is 30.75 ft. Since the columns are 1.5″ square, the "B" bars, 5 #8, are L1 + 6 IN - 27 IN, or 28.0 ft.

Figure 03200-12

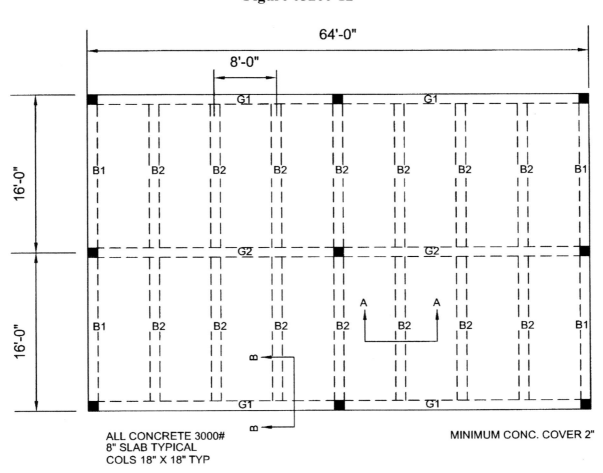

The top bars are 5 #9, a hooked bar at the outside end and a straight bar over the center column. The hooked bar is 25% of L1 + the length of the hook, taken to be a standard hook since it is not dimensioned. Twenty five percent of L1 is 7.44 FT and the standard 180 degree hook for a #9 is 1.25 FT, for a total bar length of 8.69 FT.

Figure 03200-15

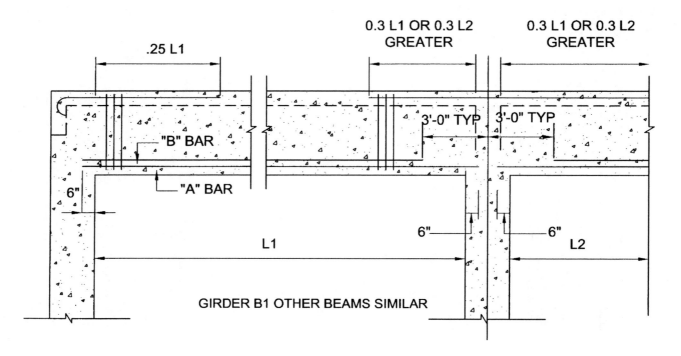

Figure 03200-16

Mark	Size		Bottom		Top	Stirrups	
	W	D	"A" Bars	"B" Bars		#-Size	Spacing F
G1	18	36	5 #9	5 #8	5 #9 @ non-continuous end 5 #9 @ columns	18 #4	1 @ 3, 3 @ 6, 5 @ 12
G2	18	36	6 #9	6 #8	6 #9 @ non-continuous end 6 #9 @ columns	18 #4	1 @ 3, 3 @ 6, 5 @ 12
B1	12	22	3 #7	3 #7	3 #7 each end	14 #3	1 @ 3, 3 @ 6, 3 @ 12
B2	12	22	4 #7	4 #7	4 #7 each end	14 #4	1 @ 3, 3 @ 6, 3 @ 12

Beam and Girder Schedule

The center top bar, to the column centerline, is 30% of L1 and L2, whichever is greater, plus one-half the width of the intersecting beam. In the case of girder G1, both L1 and L2 are 29.75 IN. 30% of the length is 8.93 FT. The beam framing into the girder is beam B2 at 12 IN wide. Adding half of this, or 6 IN, to 8.93 gives a dimension of 9.43 FT to the centerline. Some estimators take the bar to its total length, but because of differing beam combinations, this is not recommended.

Note in section A-A (Figure 03200-13) the designation #3 support bar where needed. This bar is used to stabilize the stirrups in the case where the stirrups extend beyond the top beam bars. A quick calculation in this case shows that the stirrups go only to 55 IN inside the face of the support, while the hook bar goes over 7.0 FT and the center bar over 9.0 FT. Therefore, no stirrup support bar is needed. This condition must be checked each time a beam is taken off, however. The estimator must bear in mind that all beam or girder designations which have different spans must be taken on separately, since the lengths of the bars depend upon not only the span of the member in question, but also perhaps the span of the adjoining member.

In Figure 03200-17 is a flat slab design with column shear heads. Reinforcing steel is placed in two directions in "column bands" and "middle bands." The designation shown for the column and middle bands appear on structural drawings only.

Looking at the east-west middle band between column lines B and C and beginning at column line 1, following is the east-west middle strip rebar (Rebar is symmetrical about bay 2 1/2):

E-W		N-S
1 to 1-1/4	10 - #6 Top	5 - #6 Bottom
1-1/4 to 1-3/4	10 - #6 Bottom	10 - #6 Bottom
1-3/4 to 2-1/4	8 - #6 Top	11 - #6 Bottom
2-1/4 to 2-3/4	8 - #6 Bottom	10 - #6 Bottom
2-3/4 to 3-1/4	8 - #6 Top	11 - #6 Bottom
3-1/4 to 3-3/4	10 - #6 Bottom	10 - #6 Bottom
3-3/4 to 4	10 - #6 Top	5 - #6 Bottom

Figure 03200-17

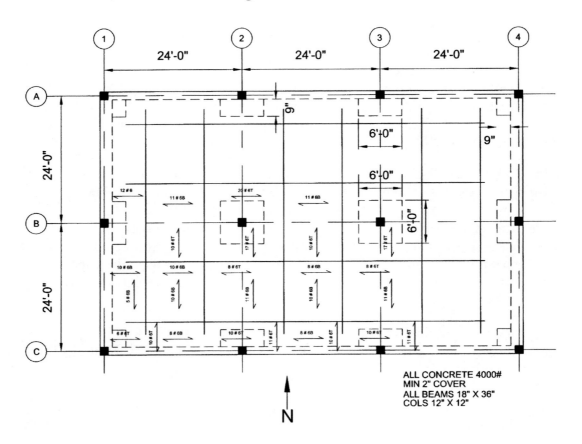

Figure 03200-18

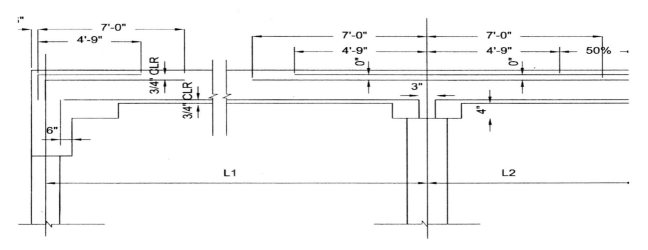

TYPICAL COLUMN STRIP - FLAT SLAB

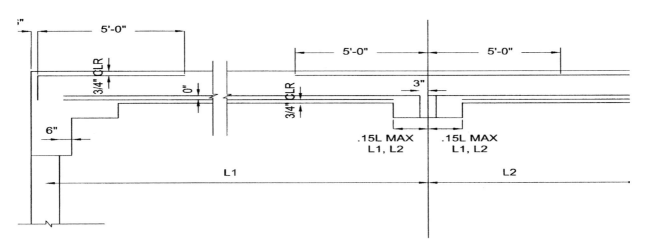

TYPICAL MIDDLE STRIP - FLAT SLAB

Refer to the diagram of the typical middle strip (See Figure 03200-18). Top bars are shown extending 5.0 FT into the interior from the outside column line, and centering on the interior column line with a length of 10.0 FT. Since the specifications call for 2 IN cover, the outer bars will be 5.0 + 4 IN + 12 IN, or 6.33 FT. The 8 IN dimension comes from the standard hook chart. Fifty percent of the bottom bars extend from a point 6 IN into the spandrel beam to 3 IN from the next column line. With L1 being 16.0 FT., these bars are therefore 15.25 FT. long (16.0 - 6"-3"). The other fifty percent also begin at a point 6 IN into the spandrel, but extend only to a point which is 15% of the length of the column to column distance, L1, from the next column line. L1 in this instance is 16.0 FT., so 15% of L1 is 2.4 FT. The length of these bars, then, is 14.1 FT (16.0 - 6"- 2.4').

The estimator should carefully note where L1 and L2 occur on the particular plan being surveyed. This is because the bar lengths will vary in accordance with the particular column to column spacing. In the middle strip just discussed, the bottom bars for the next bay are a function of L2, not of L1, and this will often result in a different bar length for the next bottom bars. Even though the column spacing in the present example is the same, the bottom bar lengths will not be the same. Fifty percent of the bottom bars in L2 will be 16.0 - 6 IN, or 15.5 FT., and fifty percent will be 16.0 - 15% L2 on both ends, or 16,0 - 2.4 - 2.4, or 11.2 FT. It is important to recognize where L1 and L2 occur, also, so that the appropriate lengths may be determined. L1 is a non-continuous bay, while L2 is a continuous bay. Thus, in the east-west direction, the bay from column line 1-2 is an L1, that from column 2-3 is an L2, and that from 3-4 is an L1. In the north-south direction, both bays are L1, since both are non-continuous.

03300 – CAST-IN-PLACE CONCRETE

Takeoff Procedures

Concrete shall be separately defined and quantified by mix design types and location. Requirements for integral admixtures and color shall be stated in cubic yards.

Concrete transportation, placing, finishing, and curing shall be separately identified including labor requirements.

Aggregate base courses for slabs on grade shall be defined in cubic yards including factors for compaction. Conversion to tons may subsequently be made for pricing purposes.

Rough screeding, wood float finish, steel trowel finish, hardeners, surface applied colors, curing, broom finish, special finish, and exposed aggregate finish shall be identified and quantified.

Screeds, construction joints, control joints, pre-formed control joints, nosings, saw cuts, special edgings, keyways, and expansion joint material less than one foot wide shall be described and quantified.

Concrete requirements for continuous wall footings, grade beams, wall footings with integral stem walls, square footings, slabs on grade, structural mats on grade, ganged formed structural walls and curved walls, single waler system concrete walls and curved walls, round/square columns and piers, one and two way elevated slabs/beams, one and two way pan-joist elevated slabs, flat plates/slabs, domed covers, stairways and pile caps/beams shall be individually quantified.

Concrete placing equipment shall be defined and quantified.

Hot and cold weather protection for all concrete shall be defined and quantified.

Formwork, accessories, and reinforcing steel shall be defined and quantified as indicated in the appropriate sections of this manual.

Waste Allowances

Concrete waste (or overage) may result from uneven ground surface, forms which are slightly out of alignment, or a desire to have enough material on hand to complete the scheduled pour since the expense of a crew waiting for the last few yards to arrive far outweighs the cost of those yards. In addition, there will always be some material left in the concrete truck which cannot be utilized. An appropriate waste factor should be added according to jobsite conditions and historical data. Footing trenches in rock may require an unusually high waste factor. Waste may also be added to the other items surveyed.

Concrete Footings

Concrete footings are used to transfer the vertical loads of walls and columns to the earth. The concentrated load supported by the wall or column is spread through the footing to a greater area of earth, thus lessening the load per unit of bearing area, and using the bearing capacity of the soil effectively.

Footings may be continuous, as under a wall, or they may be pad footings, also called pads, as under a column. (See Figure 03310-1)

Figure 03300-1

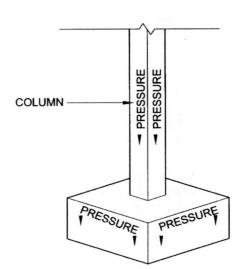

Concrete volume in cubic feet = length × width × depth (in feet) = cubic feet/27 = cubic yards.

Finish = length × width in square feet.

Given a continuous footing 100.0 (L) × 3.0 (W) × 1.5 (D) calculate:

Concrete volume = (100.0 × 3 × 1.5) = 450 CF/27 = 16.67 CY

Form quantity = 2 (100.0 + 3) × 1.5 = 309 SF

Footing finish = 100 × 3 = 300 SF

Continuous footings should be segregated from pad footings on the quantity survey sheets.

In some weather conditions, after footings are placed, they must be protected and cured. In cold weather, protection may involve covering the concrete with insulated tarps so the concrete will not freeze. In hot weather, protection may include spraying with a curing compound to keep it from drying out too quickly. These items are usually surveyed as "protect and cure footings," and measured in square feet, and is the same quantity as finishing footings.

Concrete Walls

Concrete walls are used primarily as structural elements. They support other structural elements such as floors, resist lateral pressures from the earth or wind, and add stiffness to a building frame.

Concrete walls have the three primary dimensions of length, width, and height. From this basic common characteristic, however, individual examples of concrete walls may have as many different names as different characteristics.

Figure 03300-2

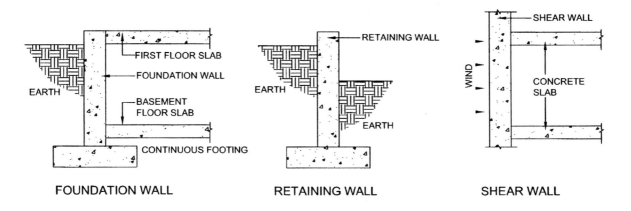

Figure 03300-3

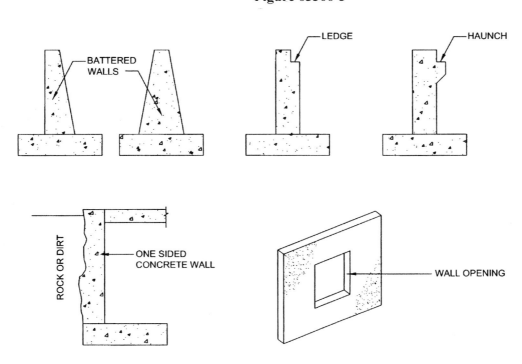

While the shape of a concrete wall is most often rectangular, it may also be circular (as, for example, in a tank or fountain), curved for artistic effect, or irregularly shaped. It may be structural (not seen in the finished structure but covered by another material) or architectural (exposed to view in the finished structure). It may be the same width throughout its height or its width may change. The latter is called a battered wall. It may be battered on one side or on both sides. Walls may have ledges, which are used for supporting an architectural facing material such as brick or stone. Such supports can also be achieved using a wall with a haunch or corbel. Some walls may even be one-sided walls where one surface of the wall is placed directly against earth or rock. Walls may have openings which permit installation of doors, windows, louvers, and similar items, and they may have pilasters, offsets in the wall used to give the wall added strength.

Figure 03300-4

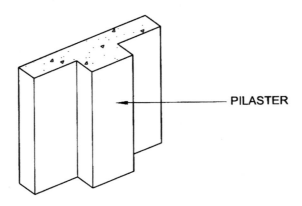

In order to prepare a complete quantity survey for concrete walls, the estimator must be able to:

- ✓ Categorize the wall.
- ✓ Recognize the shape.
- ✓ Recognize any special features.
- ✓ Determine any special finishes.
- ✓ Determine the strength of the concrete required for the wall.

Having recognized the basic characteristics of the wall, the estimator may then begin to accumulate quantity survey data, being careful to group similar wall types. He determines the basic dimensions of the wall segment in question, its length, width, and height. From these dimensions, the estimator can then generate the required quantities for concrete and finishes.

Concrete Data - The formula for concrete quantity is V=LWH, where L is the length of the wall segment in feet, W is the width of the segment in feet, and H is the height of the segment in feet. The result will be expressed in CF, which must then be converted into CY.

For example, a wall which is 100 LF long, 1.0 ft wide, and 10.0 ft high has a volume of 100 × 1.0 × 10.0 = 1000 CF. If there are openings of significant size in the wall, they should be deducted. A 3.0 × 1.0 × 7.0 opening results in 21 CF being deducted from the wall volume. Similarly, if there are ledges in the wall, they should be deducted. A ledge of 100 × .33 × 2.0 results in a deduction of 67 CF. A haunch or corbel must be added to the volume. A haunch which is 100 × .5 × 1.33 = 67 CF addition. Pilasters must be added. A pilaster which is 1.33 × .67 × 10 = 9 CF addition.

To correctly determine the volume of a battered wall, the average width of the wall must be determined. Each segment should be surveyed when any of these conditions change.

For circular walls, the length may be determined using the formula L = PI × D, where D is the diameter of the circle at the centerline of the wall. For example, the length with a diameter of 15.0 FT = 3.14 × 15 = 48 LF.

For curved walls, the length of the segment must be determined by multiplying the preceding formula by the percentage of the full circle the segment represents. For example, a curved segment representing one-fourth of a full circle - 3.14 × D × .25.

For irregularly shaped walls, an approximation of one of the basic formulas must be used. For example, with a curved segment which is also battered, the average width of the segment can be found, then used to assist in finding the length of the segment.

Figure 03300-5

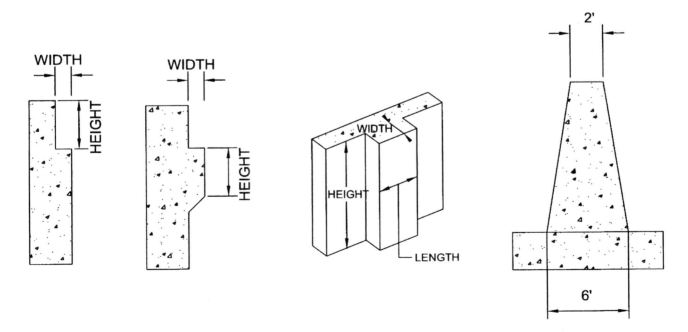

Walls which do not receive a special finish may require pointing and patching. Pointing and patching is the process of running a cement paste into small holes and honeycomb left in the surface of the concrete due to air pockets in the plastic concrete, to improper vibration when placing the concrete, and tie hole voids. Pointing and patching, as well as the other finishes discussed later, is calculated using the same formula as for formwork, LH. This quantity must, of course, be doubled if both sides of a wall are pointed and patched.

If other finishes are required, their area must be deducted from pointing and patching. Here are some finishes which might be used:

Sacking is the process of running cement paste on the wall with a coarse cloth fabric, such as burlap. This gives the concrete a textured appearance.

Rubbing is the process of rubbing freshly placed concrete, usually the day after the concrete is poured, with an emery stone. Rubbing may require small amounts of paste, but it primarily works by bringing some of the concrete's paste to the surface, where it can be given a smooth dense finish with almost no pin holes.

Sandblasting involves directing a sharp particle stream, using air pressure, at the cured wall. The force of the particles wears away the outer skin layer of the concrete and exposes the aggregate beneath. It may be classified as light (wearing away a small amount of surface), medium (wearing away more), or heavy (exposing large aggregate).

Retarding the concrete involves adding a chemical to the concrete which causes the concrete's rate of cure to be slowed. When the forms are removed, usually the next day, the surface grout can be washed away by spraying it with high pressure water.

Bushhammering the concrete surface involves chiseling the surface paste away with a small electric or air-driven cutting tool.

Form liners may also be used for special finishes. (See 03100 Concrete Forms and Accessories)

All of these finishes are surveyed in SF, and should be surveyed concurrently with forms and concrete.

Columns

Concrete columns are almost exclusively used to support a suspended structural element such as a floor or beam. Unlike a wall, whose length typically exceeds its width, a column's length and width are typically close in dimension. A column may be various shapes and depending upon the structural requirements of the building and its esthetics. It may not change in dimension throughout its height, or it may contain a capital at its top. (See Figure 03100-5)

Column concrete volume is calculated, as with walls and footings, using the dimensions LWH. Length and width, in this case, are the column's size, measured in feet. The height is the net height of the column from the top of the surface below to the bottom of the surface which the column supports. For example, assume the floor to floor height in the examples above is 10 FT, with a slab thickness of 1 FT. The net column height for the column at section A-A (Figure 03100-5) is 9.0 FT. The dimensions recorded for case 1 then are 2.0 × 2.0 × 9.0 for LWH. For case 2, 1.33 × 2.0 × 9.0, and for Case 3, 1.0 × 1.0 × 3.14 × 9.0.

For a column with a capital, the lower portion of the column must be surveyed, then the upper, or capital portion. Assume the lower portion of the column has a height of 7.0 FT, while the upper portion is 2.0 FT high. The formula for the lower portion is as above with the height dimension being 7.0 FT. The upper portion is calculated using the formula for a frustum of a cone, $V = 0.2618\ H\ (D^2 + Dd + d^2)$ plus the vertical top where V = volume and H = height and D = diameter of large base and d = diameter of small base, the answer is expressed in cubic feet.

Beams

Refer to the floor plan and Figures 03300-6 through 03300-8.

Figure 03300-6

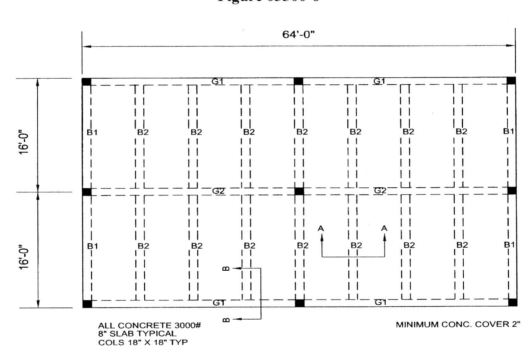

03300 Cast-in-Place Concrete

Figure 03300-7

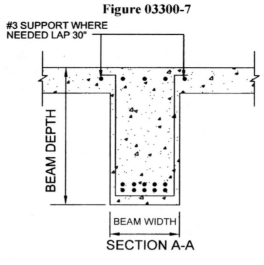

SECTION A-A

Figure 03300-8

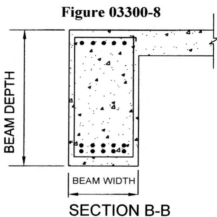

SECTION B-B

Figure 03300-9

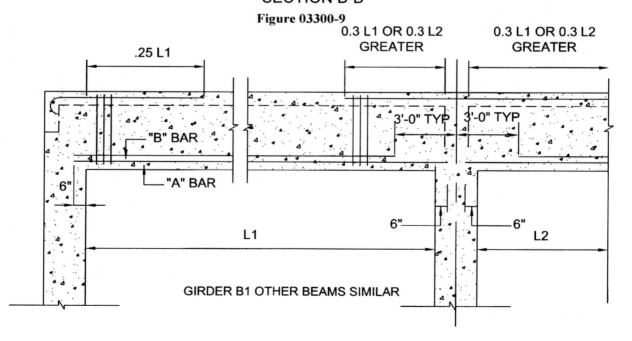

GIRDER B1 OTHER BEAMS SIMILAR

Figure 03300-10

Mark	Size		Bottom		Top	Stirrups	
	W	D	"A" Bars	"B" Bars		#-Size	Spacing F
G1	18	36	5 #9	5 #8	5 #9 @non-continuous end 5 #9 @ columns	18 #4	1 @ 3, 3 @ 6, 5 @ 12
G2	18	36	6 #9	6 #8	6 #9 @non-continuous end 6 #9 @ columns	18 #4	1 @ 3, 3 @ 6, 5 @ 12
B1	12	22	3 #7	3 #7	3 #7 each end	14 #3	1 @ 3, 3 @ 6, 3 @ 12
B2	12	22	4 #7	4 #7	4 #7 each end	14 #3	1 @ 3, 3 @ 6, 3 @ 12

Beam an Girder Schedule

The plan view shows a flat slab 64.0 FT × 32.0 FT, supported by a series of drop beams and girders which in turn are supported by nine 18 IN × 18 IN columns. The beam and girder schedule gives the size of the beams, marked B1 and B2, as 12 IN wide and 22 IN deep, while the girders are 18 IN wide and 36 IN deep. Note from section A-A (Figure 03300-9) that the depth of the beam is measured from the top of the slab, a fact which can affect the estimator's quantity survey substantially.

The difference between a "beam" and a "girder" is, for the estimator, one largely of semantics. A girder is a beam which is typically longer from support to support, wider, deeper and subject to heavier loads than the "typical" beam. For quantity survey purposes, they are treated the same.

There are several different types of slabs with which beams may be used (a flat slab is shown in the figure. Some beams do not occur with slabs and are therefore "isolated" beams. While surveyed using the same procedures given here, different types of beams must be quantified separately. Beams which project below their associated slabs are called "drop" beams, and those which do not are called "flush" beams. The properties of each are the same, except the flush beam does not have beam sides.

A beam's depth is measured from the top of slab elevation. Therefore, the top portion of the beam, as in Section A-A (Figure 03300-7), is in the same plane as the slab. When surveying the concrete, formwork, and finishes, only the portion of the beam projecting below the slab is considered.. The beams are 22 IN deep and the girders 36 IN, and given the 8 IN deep slab, their depth below the slab is 14 IN and 28 IN, respectively.

For surveying the concrete volume of the girder G1, the width dimension is 1.5 FT, the depth dimension is 2.33 FT, and the length dimension is any convenient choice. For example, it could be 128.0 FT, 2 × 64 FT, or 4 × 32 FT, as the estimator wishes. A similar condition holds true for the other girders and beams.

Note the length is surveyed from outside dimension to outside dimension for girder G1. Length dimensions for the other beams must be adjusted so that corners and intersections are not surveyed twice.

Slabs

Concrete slabs are the structural elements which carry the occupant load of a structure, the space for which the structure is being built. The simplest of these, usually, is the slab on grade. It rests directly on the ground, and is the simplest to survey and price.

Figure 03300-11

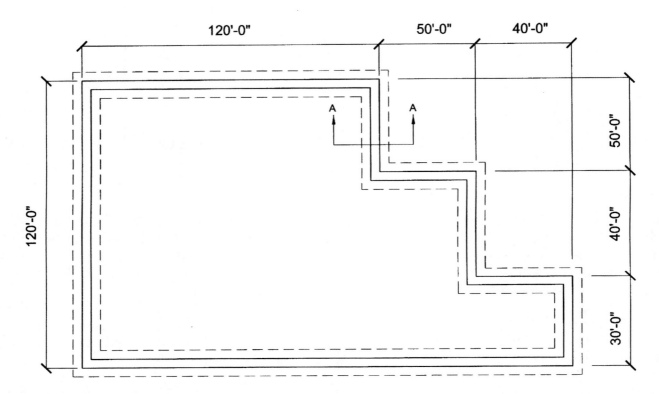

The slab on grade here is 10 IN thick, is reinforced with 6 × 6-6/6 wire mesh. This particular mesh is also designated 6 × 6 W2.9 × W2.9 under the most recent change of nomenclature. The slab is also underlain with 6 mil vapor barrier and 6 IN stone sub base.

The estimator will survey SOG concrete, forms, finish, reinforcing, and miscellaneous items associated with the slab.

Figure 03300-12

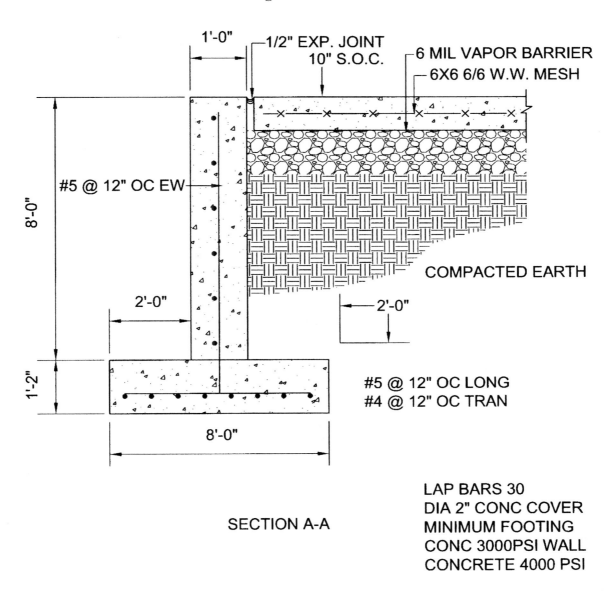

SECTION A-A

LAP BARS 30
DIA 2" CONC COVER
MINIMUM FOOTING
CONC 3000PSI WALL
CONCRETE 4000 PSI

With this information, the estimator has been able to quantify the slab on grade. Concrete volume is LWH; finish and fine grade are LW; mesh and vapor barrier are LW plus lap and waste allowances = SF, stone is LWD × compaction factor/27 = CY, the item "expansion joint" is measured in LF.

The suspended slab, however, is not quite so simple. A great deal depends upon the type of slab being surveyed.

The first type to be considered is the flat slab. See the illustration in Figure 03300-13.

Any particular slab need not have spandrel beam, as is shown here, and it need not have "drop panels." The square portions which project below the bottom of the slab. Drop panels are also called "drop heads," and less frequently, "capitals." In such cases, the quantity survey items for these members would not be required as part of the flat slab.

Figure 03300-13

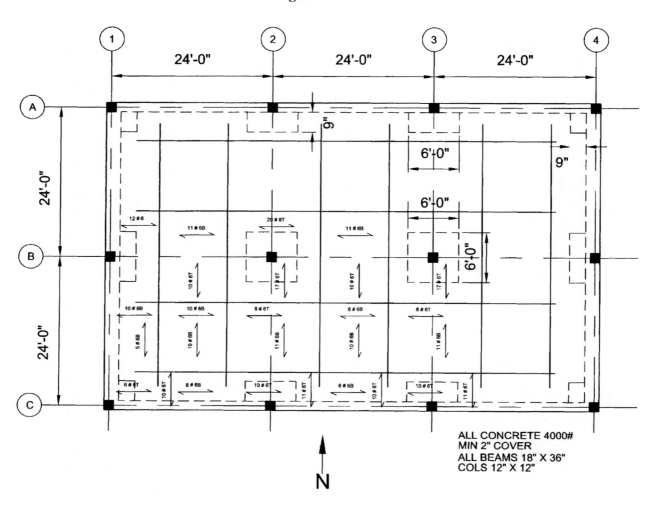

Slab concrete is, again, calculated by LWD, and accumulated by strength and type.

Slab finish is calculated using LW and is accumulated by type of finish as specified; trowel, broom, etc.

03410 - PLANT – PRECAST STRUCTURAL CONCRETE

Introduction

In the 1960's building costs began increasing at a faster rate than most industrial products. One of the main reasons for this increase was the cost associated with labor. Shop construction of building components help to offset these labor costs. With this in mind precast concrete construction developed rapidly in an attempt to minimize on–site labor costs by construction materials in less labor intensive environments. Precast construction is used in all major types of structures: industrial buildings, residential and office buildings, bridges, etc. Precast plank is typically used for roof or floor systems. Reinforced cast-in-place concrete toppings are frequently used in conjunction with the precast to provide continuity and monolithic surface. This topic is covered in more detail in other sections of this manual. Other types of structural precast concrete are similarly installed including single tees, double tees, and beams and columns. These types of precast construction are most prevalent in parking structures. Manufacturing the precast will not be included in this discussion.

Types of Measurements

Precast concrete is typically measured in pieces, weight, and square feet of surface for solid and hollow core planks, single tees and double tees. Columns and beams generally measured in pieces and weight. It is necessary to know the weight in order to determine the size of crane required to set the precast.

Specific Factors Affecting the Estimate

A site with material storage areas will allow for efficient material handling operations. A "tight" site with limited or no storage areas will require multiple handling operations or "just in time" deliveries. If the bottom of the plank is exposed and being utilized as the ceiling for the area below, special texturing of the bottom plank surface or caulking of the underside joints may be required. Specifications must be checked to determine if the precast plank supplier is to be approved by the Precast Concrete Institute (PCI). This may limit the choices in material suppliers. Since the number of plants is limited, it is essential that the estimator inquire about delivery times. A material premium may be required to expedite production. The overall project schedule must be considered. If the precast plank system occurs on various levels the estimator must determine if multiple mobilizations will be required.

In merit shop areas the erection crew is made up of qualified craftsmen. In strongly union markets, the crew is usually a crew composed of carpenters, ironworkers, and operating engineers. A laborer may also be required in some areas.

Organization of the Estimate

Time must be spent during the estimating phase planning the sequencing and time of construction. The cost to install precast floor or roof systems includes labor, material, equipment, overhead, contractor fees and specified markups. Material costs are obtained by requesting quotations from material suppliers. Equipment costs must be considered for all aspects of the work.

Overview of Labor, Material and Equipment

The direct cost to install a precast plank floor or roof system includes labor, material, and equipment. To price labor, the estimator must determine the labor cost associated with each worker and apply this cost to the number of work hours required to complete each unit of work. The labor rate includes the worker's salary and benefits with appropriate adjustments to determine the appropriate wage rate for each worker. The work hours per unit are determined by experience of the estimator or by historical data collected from completed projects. Equipment costs must be considered for all aspects of the work. Equipment can be rented or owned. Contractor owned equipment, depreciation, repairs; maintenance, insurance, storage and mobilization must be considered. Rented equipment costs are obtained from equipment suppliers. The lifting phase of floor or roof planks is the erection and placement of the planks into place. The capacity of the crane the erector uses can impact the cost of the plank system. The crane selection should be determined by referencing a crane-lifting chart, which is available from crane service companies or manufacturers. Repositioning the crane and crew will also affect the time to set the panels. To determine the crane production time divide the total number of pieces by the number of pieces estimated to be set each day. It is important to obtain site information to determine if obstructions, i.e. railroad tracks, power lines, etc. will limit the locations of the crane for erection of the precast.

Labor costs associated with precast plank are usually calculated using a crew composed of a foreman, numerous skilled workers and a crane operator. The estimator must determine the size of the crew and the cost of the crew. The quantity of precast is then divided by the productivity rate for setting multiplied by the crew rate to determine the labor costs. Material costs are typically solicited from precast manufacturers. These suppliers will calculate the quantity of precast and multiply this quantity by the unit cost of the each piece. The unit cost will account for shipping of the precast to the project site. To determine equipment cost the estimator must determine the size of crane needed. The height of lift and reach need to be considered when choosing a crane. The productivity rate for setting the precast must be multiplied by the unit cost of the crane to determine the crane cost. Mobilization of the crane must be included.

Once the precast planks have been set the joints are grouted to tie the planks together and create a uniform surface. The material cost is obtained from a concrete supplier and the grout is pumped to the floor or roof area and screeded across the joint. The pump rental time is determined by dividing the total square feet of plank area by the anticipated grout production per day.

Estimate

The following sample project addresses the above mentioned labor, material and equipment costs as they pertain to a precast plank floor or roof system. A general contractor is constructing a two story building and has request a quotation to install the precast plank floor system as shown on the Figure 03410-1 and sections 1 through 3. The scope of work includes supply and installation for the precast plank and reinforced concrete topping. The general contractor and other subcontractors will provide the masonry bearing walls, structural steel and other wall and finish materials.

Precast Concrete:

The length and width of the floor is 100 feet and 50 feet requiring 5,000 square feet of plank.

The quantity of plank, 5,000 SF, is divided by the productivity rate, 700 square feet per hour or 7.14 hours (use 8 hours) to set the planks, and multiplied by the crew rate of $278.45 per hour, resulting in a total labor cost of $ 2,228.00.

The area of plank, 5,000 SF, is multiplied by the unit cost of the plank; $5.00 per square foot, resulting in a total material cost of $25,000.00.

The daily rental rate of the crane, $1563.00 per day, is multiplied by the number of days required, which results in an equipment cost of $1563.00.

Grout PrecastPlank Joints:

The volume of grout is calculated by multiplying the length, 100 feet, by the width, 50 feet, and divided by 800 square feet (assume 1 cubic yard per 800 SF of 4 FT plank). The grout required is 6.25 cubic yards of grout.

Assume the crew cost $95.00 per hour. The crew can place 15,000 square feet per day. At this rate it will take 3 hours to pour the grout (use 4 hrs.) Therefore, the labor cost equals 4 times $95.00 per hour or $380.00

The cost of the grout is $75.00 per cubic yard. A 5% waste factor is anticipated. The concrete material cost is $75.00 per cubic yard multiplied by 6.25 cubic yards and 1.05 or $492.00.

The grout will be pumped to the 2nd floor. The rental rate for the pump is $85.00 per hour. This rate multiplied by the previously determined 4 hours results in an equipment cost of $340.00.

The overhead cost is 10% of the total labor, material and equipment cost. The markup is 10% of the total labor, material and equipment cost.

Example 03410-1

SAMPLE ESTIMATE SHEET FOR PRECAST FLOOR OR ROOF PLANKS

A General Contractor is constructing a 2-story building and has approached JLK Construction to install the precast plank floor system as shown on the attached sketches SK-1 through SK-4. JLK Construction's scope of work includes supply and installation for the precast plank and reinforced concrete topping. The General Contractor and other sub-contractors will provide the masonry bearing walls, structural steel and other wall and finishing materials.

Item or Description				Material			Labor				Equipment			Sub		Total
DESCRIPTION		Quantity	Unit	Unit $	Cost	Hr/Unit	Hours	Rate/Hr	Cost		Unit $	Cost	Unit $	Cost		Cost
1	FORMWORK	400	SF	0.75	300	0.0313	12.52	21.00	263		0.0875	35	-	-		598
2	CONCRETE REINFORCING	70	LBS	1.00	70	0.0057	0.40	21.00	8		0.00	-	-	-		78
3	CONCRETE PLACEMENT INCL 5% WASTE	28.3	CY	60.00	2,300	2.9220	111.9	24.75	2,772		34.24	1,312	-	-		6,384
4	PRECAST CONCRETE	5,000	SF	5.00	25,000	0.0144	72.00	30.94	2,228		0.31	1,563	-	-		28,791
					-				-			-		-		-
					-				-			-		-		-
					-				-			-		-		-
					-				-			-		-		-
					-				-			-		-		-
					-				-			-		-		-
Subtotals					27,670		197		5,271			-		-		35,851
Labor Burden on Labor	INCLUDED IN UNITS ABOVE							0%								-
Total Direct Cost		27,670							5,270			2,910		-		35,851
Overhead	10%															3,585
Contractor Fee	10%															3,585
Insurance & Bond	NOT APPLICABLE															
Sales Tax on Material	BY GENERAL CONTRACTOR															
Total Subcontractor Cost																**$43,021**
															COST/SF	$8.60

FIGURE 03410-1

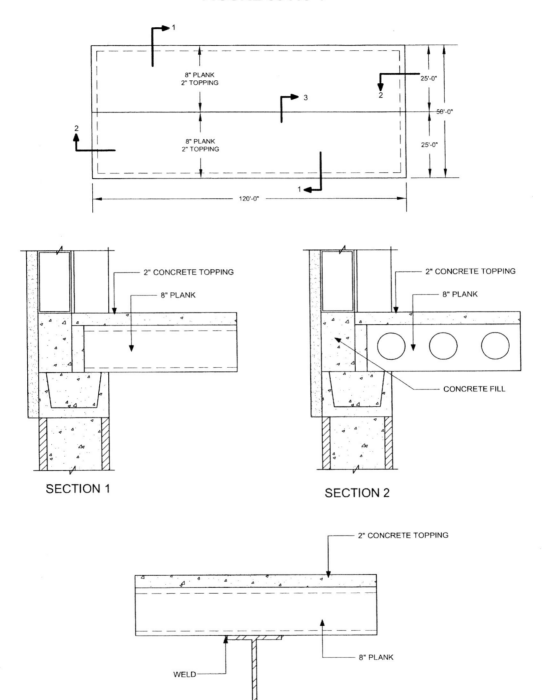

03470 – TILT-UP PRECAST CONCRETE

Introduction

Tilt-up construction is a method of constructing concrete walls in which panels are site cast and cured flat on the building slab or on a casting bed, then lifted up into their final positions with a crane. The panels are braced and commonly connected to each other by the use of an embedded steel weld plate on each side of the panel. The second floor structure or roof structure is applied to the wall panels. Tilt-up precast concrete can be classified as structural concrete (03430) or architectural concrete walls (03470). The determining factor is whether the panels support the structural components. In either case, the estimating procedure is the same. Tilt-up concrete construction is not a new method or a new building system. Tilt-up construction has increased in popularity in the last few decades due to increased creativity in architectural and structural design and advances in engineering and concrete technology. The development of larger mobile cranes and the refinement of methods have made Tilt-up construction competitive with other types of building systems.

Types and Methods of Measurements

The concrete formwork associated with tilt-up wall panels is measured in lineal feet if the height is less than 12 IN or square feet if greater than 12 IN. The concrete reinforcing will consist of wire mesh measured in square feet or squares (100SF) or reinforcing steel measured in pounds or tons. Chamfer and reveals are measured in lineal feet. The quantity of concrete is measured in cubic yards.

The precast concrete items will consist of the foundation embeds, bond breaker, concrete, lifting inserts, bracing inserts, structural embeds, panel & slab patching, concrete finishing and panel base grouting. Foundation embeds, lifting inserts, bracing inserts and structural embeds are measured by counting each one. Panel layout bond breaker, panel & slab patching, and concrete finishing are measured in square feet. Grouting the panel base is measured in lineal feet.

Organization & Specific Factors Affecting the Estimate

The first step in preparing an estimate for site casting concrete wall panels is to prepare a casting plan. This is different from almost all other site cast concrete because the project documents will not dictate where the panels will need to be cast. The design will dictate the width, height and thickness of the panel but the casting location is up to the contractor. The panel might be cast four feet away from its final resting place or it might be cast far away in a remote corner of the site. The casting plan should try to place all panels as close to the final location as possible and minimize panel stacking. Making a scale drawing of the available slab on grade area will be very helpful. The casting plan will need to block out areas of the slab on grade for penetrations (plumbing, fire sprinkler, electrical, etc.); structural steel columns and perimeter pour strips. The available area should be compared to the face area of concrete panels. If the available area is larger than the face area of the concrete panels, then the panels can be cast on the slab on grade. If the face area of the concrete panels is larger than the available area, then some or all of the panels will need to be stacked or cast on a temporary casting slab. The temporary casting slab can be estimated similarly to the slab on grade. This cost and that of removal should be included in the tilt-up panel cost. The estimator should plan casting locations for the concrete panels, while keeping in mind the panel erecting order and the distance to final location. The plan will need to allow for equipment and forming access as well.

If the panels are to be erected from the inside, the estimator needs to verify that the slab on grade (as designed) will support the crane. In some cases modifications to the slab on grade must be made. These modifications must be coordinated with the estimate for the slab on grade.

The completion schedule must be considered in order to determine whether it is realistic. There may be penalties or bonuses associated to the completion date. Compression/acceleration of the schedule not only affects the general conditions, overhead, and profit, it may affect productivity. It may be necessary to alter the concrete cure time, but additional cost must be addressed. Availability of materials such as embeds, inserts, special forms must be determined. Additional cost may be incurred for immediate fabrication, special shipping or delivery. Shop drawings, submittals and other architectural/structural requirements may also affect the schedule. Determine if there is labor available to maintain the schedule and if required, premium time, additional supervision, and/or additional or special tools should be included.

Safety regulations dealing with crane placement while picking up tilt-up panels vary due to the panel design. Some projects or the structural consultant will allow the crane to sit outside the building lines and tilt the panels in place with the panels tilted toward the crane. Other projects and consultants will require the crane to sit inside the building lines and tilt the panels into place so that the panels are tilted away from the crane. This requirement will affect the panel-casting layout because extra room will be needed for crane access. This may mean additional panels will need to be stacked which will have an impact on cost. Stacking panels on top of one another is very efficient space wise but is not very cost effective because the forming labor is more expensive and it extends the time spent constructing panels.

Overview of Labor, Material, and Equipment

Once the panel layout plan has been determined, quantities can be calculated. The gross panel area should be calculated in square feet by multiplying the panel width in feet by the panel height in feet and summing the results for all panels. The net panel area can be calculated by subtracting the square footage of any openings in the concrete panels from the gross panel area.

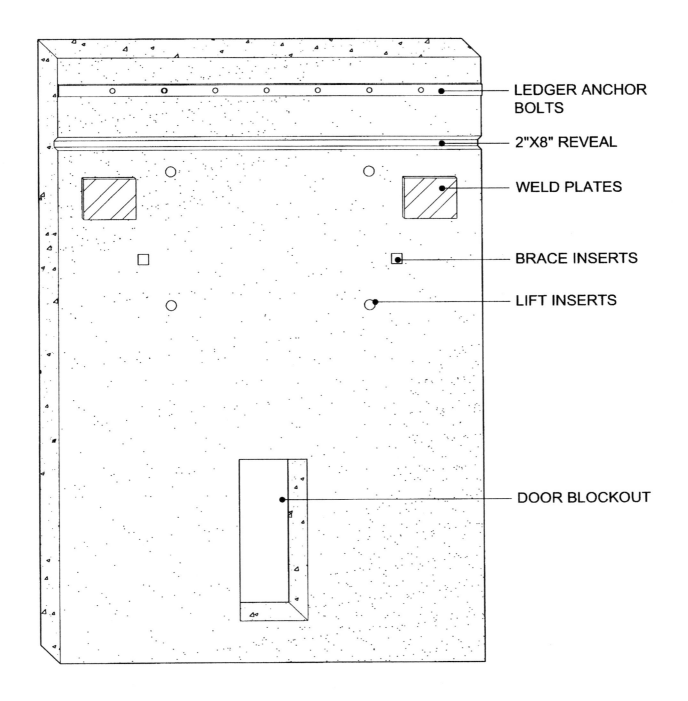

SAMPLE PANEL
NOT TO SCALE

03470 Tilt-up Precast Concrete

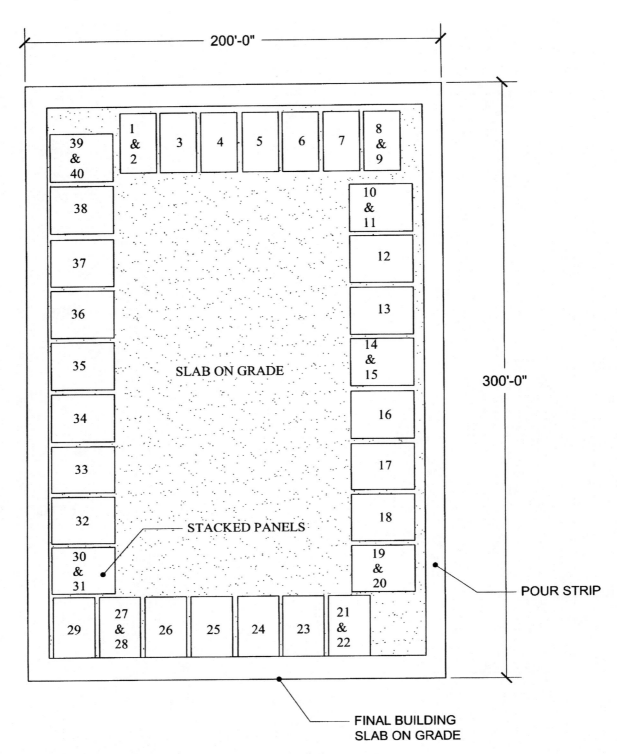

SAMPLE TILT-UP PLAN VIEW
NOT TO SCALE

Recess joints, column footing blockouts and other slab blockouts need to be temporarily surfaced to assure the required panel finishes. The slab on grade must be designed to accommodate the weight of the crane if the panels are to be erected from the inside. The application of curing agents and bond breakers is necessary to prevent the adhesion of the panels to the slab on grade. Casting the panels on the slab on grade may damage the slab. Repairs to the slab should be anticipated and included in the estimate. The slab on grade finish and flatness will affect the finish of the architectural face of the concrete panel and it may impact the concrete yield. The estimator should take into consideration the slab on grade specifications as well as the flatness history for the company placing the slab on grade when determining the waste factor to use for the concrete. If the slab is used as the casting bed, the floor finishes need to be considered. The area under the concrete panels will have bond breaker on it, as well as many screw holes from the formwork. If the area will receive floor coverings or a concrete sealer, the bond breaker will need to be stripped. If the slab will be left exposed, the screw holes will need to be patched. The costs of stripping the bond breaker and/or patching screw holes need to be added to the estimate (if required). Coordinating the placing of the pour strip after the panels are set needs to be considered and usually included in the slab on grade cost. Using the layout plan prepared for the concrete panels, calculate the lineal footage of edge forming around the perimeter of the panel groups. The edge form can be constructed with any type of forming system that works with the panel dimensions and can be easily fastened and unfastened from the slab. Next takeoff the lineal footage of bulkheads between panels within each panel group. Examine the architectural elevations and determine the locations and sizes of any reveal and/or chamfer strips. Calculate the lineal footage of reveal strips unless it is wider than 12 IN. If it is wider than 12 IN, calculate the area of reveal strips in square feet. Chamfer strips should be taken off in lineal feet. In addition to the material for formwork, reveal strips and chamfer strips, the estimator will need to include concrete screws to attach the formwork to the casting slab. The formwork materials can be reused multiple times but the estimator will need to carefully analyze the construction schedule and plan form usage accordingly.

If required the net panel area will be used to determine the square footage of wire mesh reinforcing. Wire mesh is sold in both rolls and flat sheets. It is usually best to use flat sheets. Rolls or wire mesh will retain some of the curvature and can stick out of the concrete face. Flat sheet sizes vary based on mesh size. The required sheets of wire mesh can be calculated by dividing the net panel area by the sheet size. Don't forget to take into account lapping the sheets at the joints. The estimate will also need to cover the cost of wire mesh supports to keep the mesh centered in the concrete panel. The rebar is priced in pounds or tons. Take off the lineal footage of each size of rebar and multiply by the weight per foot and divide by 2000 pounds per ton. The estimator should consider whether the rebar will be tied in place or pre-tied and then placed. If the rebar is going to be pre-tied, the estimator will need to include the cost of a crane to swing the mats into place.

The concrete should be quantified by the cubic yard and is calculated by multiplying the net panel area by the thickness of the concrete in feet and dividing by 27. The estimator needs to include additional material cost for waste. The estimator will need to carefully analyze the placement method of the concrete. If the slab on grade is strong enough to support loaded ready mix trucks and there is good access around the perimeter of the panels you may be able to place the concrete into the forms direct from the concrete truck. If the access is limited a conveyor, pump or crane and bucket may need to be used. The concrete is to be placed, vibrated and finished to the desired thickness and texture.

The extent of the finish will be indicated on the elevations and included in the specification subsection titled "Finishes". Sandblasting the concrete panels is often used to create a variety of textures along with form liners; stamping, brooming, aggregate finishes and colored admixtures being other ways to provide architectural variety. This cost can vary significantly.

Tilt-up panels will need special inserts for lifting and bracing. The inserts and braces should be counted and totaled. The costs of labor to place and secure the inserts, the material, and design fees associated with the design of the lifting inserts and bracing should be included in the estimate. Brace rentals are usually by the month and the length of rental time is determined by the time required for the building to reach safe structural completion. Some lifting and bracing insert manufacturers do not charge design fees if you use their system.

Tilt-up panels will also include structural embeds. At a minimum each panel will contain structural embeds for connection to the foundation and the foundation will contain embeds for connection to the panels. The connection of panels to each other is commonly by the use of an embedded steel weld plate on each side of the panel. These weld plates, are sometimes spot welded to the reinforcing steel within the panel. When the panels have been erected, a third steel plate is welded to the two side-by-side weld plates to join them together. The productivity rate for the welder is based on the number of connections per panel and the method of elevating the welder to the work place using ladders, lifts or scaffolding. Embed plates or ledgers for other structural connections such as trusses, beams and joists may be required. Each type of embed items should be counted and totaled. Ledgers are determined and listed in lineal feet.

The lifting phase of tilt-up construction is the actual erection and placement of the panels in their final wall position. The capacity of the crane that the erector is going to use can impact the cost of the concrete panels. The capacity of the crane might limit the placement of panels because the panels must stay within the safe reach of the crane. This may require more stacking of panels or temporary casting beds. The crane selection should be determined by referencing a crane-lifting chart, which is available from crane service companies or manufacturers. The anticipated crew size will vary depending on the speed of the lift, experience of the crew and the complexity of the panels. Repositioning the crane and crew will increase the time to set the panels. The crane cost should be included as an equipment cost. To determine the crane production time divide the total number of the tilt-up panels by number of panels estimated to be set each day. The total crane time should also allow for the mobilization, set-up, breakdown & demobilization.

Estimate

All of the above items and their associated quantities should be shown on a summary sheet similar to the attached example estimate. The pricing for labor, materials, subcontractor, and equipment are totaled and the appropriate mark-ups should be applied. The labor cost should be marked up by a percentage to cover the labor burden (payroll taxes, insurance, vacation, worker's compensation, and other benefits). The material total should be marked up by a percentage to cover sales tax. If performing the work as a subcontractor, your overhead and profit should be determined and included. The General overhead and profit should not be added until the total of all of the work is calculated. The subcontractor and general contractor or construction manager should reconcile or verify that all markups are included in the total cost.

Sample Checklist and Estimate for a warehouse project of 60,000 square feet.

- ✓ From the schedule review, it has been determined to be realistic and there are no penalties or acceleration required.

- ✓ Standard materials are to be used and the submittal review process is sufficient.

- ✓ Skilled labor is available and no overtime or special equipment is required to maintain the schedule.

- ✓ Prepare a casting plan to determine the panel casting method.

- ✓ Use building slab as the casting slab by stacking several of the panels. See drawing L1.

- ✓ Calculate the gross area and net panel area.

- ✓ Use the layout plan to estimate the formwork for edge forms, bulkheads, reveals and/or chamfer.

- ✓ Use the net panel area to calculate the wire mesh and or reinforcing steel.

- ✓ Count the lifting and bracing inserts, braces, structural embeds and connection embeds. See drawing L2.

- ✓ Concrete quantity is calculated with the net panel area plus 5% waste factor. Place concrete with pump.

- ✓ Use the gross area for the layout, bond breaker, finish & curing, point & patching slab, point & patch panels, and sand blast exterior quantity.

- ✓ A local crane and rigging subcontractor will do the lifting from the outside of the building and take one week.

- ✓ Grouting the base of each panel is equal to the slab on grade perimeter length.

- ✓ Another subcontractor will complete the caulking of the panel joints at exterior walls.

- ✓ The building slab on grade concrete subcontract will place the pour strip.

03470 Tilt-up Precast Concrete — Part Two

Example 034701-1

SAMPLE ESTIMATE SHEET FOR TILT-UP PANEL CONSTRUCTION

60,000 sf Tilt-Up Building. Perimeter walls are 30' h x 25' w x 8" thick. All panels can be cast on the slab with several needing to be stacked. One horizontal 8" wide x 2" deep reveal at exterior. Door, windows and roll-up doors need to be blocked out. Final exterior finish will be medium sandblast.

Item or Description				Material		Labor				Equipment		Sub		Total
	DESCRIPTION	Quantity	Un	Unit $	Cost	Hr/Unit	Hours	Rate/Hr	Cost	Unit $	Cost	Unit $	Cost	Total Cost
1	LAYOUT PANELS	30,000	SF	-	-	0.0032	96.77	31.00	3,000	-	-	-	-	3,000
2	BOND BRAKER	30,000	SF	0.03	900	0.0016	48.39	31.00	1,500	-	-	-	-	2,400
3	2" X 8" PERIMETER FORM	4,400	LF	1.20	5,280	0.0484	212.9	31.00	6,600	-	-	-	-	11,880
4	2' X 8' OPENING FORM	620	LF	1.20	744	0.0645	40.00	31.00	1,240	-	-	-	-	1,984
5	¾" CHAMPFER AT EDGES	8,040	LF	0.25	2,010	0.0161	129.6	31.00	4,020	-	-	-	-	6,030
6	BLOCKOUT & SET WELD PLATES	80	EA	5.00	400	0.5806	46.45	31.00	1,440	-	-	-	-	1,840
7	SET ROOF TRUSS WELD PLATE	36	EA	5.00	180	0.5806	20.90	31.00	648	-	-	-	-	828
8	SET LEDGER ANCHOR BOLTS	667	EA	1.50	1,001	0.0565	37.65	31.00	1,167	-	-	-	-	2,168
9	2" X 8" REVEAL	1,000	LF	1.35	1,350	0.0726	72.58	31.00	2,250	-	-	-	-	3,600
10	SET LIFT EMBEDS	160	EA	3.75	600	0.1613	25.81	31.00	800	-	-	-	-	1,400
11	SET PANEL BRACE INSERTS	80	EA	2.50	200	0.1613	12.90	31.00	400	-	-	-	-	600
12	BRACE RENTAL (TWO MONTHS)	80	EA	13.00	1,040	0.0000	0.00		-	-	-	-	-	1,040
13	CONCRETE = 5% WASTE	781	CY	55.00	42,955	0.226	251.9	31.00	7,810	10.00	7,810	-	-	58,575
14	FINISH & CURE PANEL	30,000	SF	0.05	1,500	0.0052	154.8	31.00	4,800	-	-	-	-	6,300
15	POINT & PATCH PANEL	31,000	SF	0.06	1,800	0.0071	212.9	31.00	6,600	-	-	-	-	8,400
16	POINT & PATCH SLAB	30,000	SF	0.04	1,200	0.0048	145.1	31.00	4,500	-	-	-	-	5,700
17	GROUT PANEL BASE	1,000	LF	2.50	2,500	0.1048	104.8	31.00	3,250.00	-	-	-	-	5,750
18	REINFORCING STEEL	39,000	LB	-	-							0.50	19,500	19,500
19	CRANE AND RIGGERS (SET 8/DAY = 40 PANELS)	1	W	-	-							12,500	12,500	12,500
20	SANDBLAST FINISH EXTERIOR	30,000	SF	-	-							0.85	25,500	25,500
Subtotals					**63,660**		**1,614**		**50,025**		**7,810**		**57,500**	**178,995**
Labor Burden on Labor					%			45%	22,511					22,511
Total Direct Cost									**72,536**		**7,810**		**57,500**	**201,506**
Overhead	10%													20,151
Profit	5%													10,075
Insurance & Bond	2%													4,030
Sales Tax on Material	6%													3,820
Total Subcontractor Cost														**$239,582**
													COST/SF	

Notes:
$7.99
Average labor rate per hour = $31.00

04050 – BASIC MASONRY MATERIALS AND METHODS

Introduction

Masonry is one of the oldest forms of construction. Modern materials such as concrete masonry have increased its popularity because of its adaptability to all types of construction and relative economy.

General Requirements

This section will not attempt to review all aspects of general requirements. It will highlight those major elements of Division 1 which are unique to masonry construction and would affect the bid price for the work.

- ✓ 01300 - Submittals. Shop drawings (requiring approval) for fabrication, bending, and placement of reinforcement material. Certificate of compliance for each type of masonry used. Certificate of compliance for each grout and mortar mix used.

- ✓ 01400 - Quality Control. Outlines inspection and testing required of the contractor, or by third party, and scheduled by the contractor. Specifies masonry mock-up wall requirements. Requirement for mock-up wall affects labor and material requirements for the total construction effort.

- ✓ 01500 - Construction Facilities. A water supply is necessary for mixing mortar and grout. Temporary shoring requirements for support of openings while mortar sets up. Provide chemicals for final cleaning of finished masonry (per specification).

Projects that require masonry construction in the winter (during freezing conditions) take several extra steps. Keep the masonry units, mortar mix and grout mix above freezing (usually 12 hours prior to laying). To achieve this, cover and heat the storage area. Keep the mix water from freezing. Cover and heat the work area while laying CMU. Keep the cover in place several days after laying block to allow mortar to set properly.

- ✓ 01600 - Material and Equipment. Keep concrete masonry units, mortar mix, and grout mix materials dry at all times. Delivery of most of these is on pallets that provide enough elevation above the ground to protect from ground moisture. Provide tarps or plastic shrouds to protect from precipitation.

Typical equipment items usually provided by the masonry contractor and included in the estimate are:

- ✓ Scaffolding
- ✓ Scaffolding boards
- ✓ Forklift
- ✓ Mortar mixer
- ✓ Hod and mortar board
- ✓ Tubs
- ✓ Masonry saw
- ✓ Wheelbarrows
- ✓ Shovels
- ✓ Brick tongs

Note: Masons usually provide their own tool kit. Do not include the cost of these in the estimate. Some of these items are:

- Level (18 IN & 4')
- Plumb bob
- Masonry hammer
- Rubber gloves
- Trowel
- Mason's line
- Folding rule
- Brick set chisel
- Jointing tools
- Chock block
- Stiff bristle brush

04060 – MASONRY MORTAR

Introduction

Mortar is a critical component of masonry construction. It must bond the masonry units into a strong resilient structure. Essential mortar properties include plasticity, adhesion, water retention, and hardening consistency. Mortar is a mixture of Portland cement, fine aggregate (sand), lime and water. The specific proportions, scientifically tested, achieve desired compressive strengths. For a particular project, the specifications will say which mortar mix to use.

Determine mortar quantities by consulting any of many charts available in estimating manuals or from materials suppliers. In addition, many masonry contractors have their own equally reliable data developed from experience. Mortar is quoted in cubic feet per wall area (i.e., 5.6 cubic feet of mortar per 100 square feet of wall for 8 IN × 8 IN × 16 IN block.)

Note: The amount of mortar per square foot is directly related to the block dimensions.

Quantity Survey

Determine the quantity required and refer to the specifications for the specified mix. For example, ASTM C270 type "S" is specified in proportions:

- ✓ 1 CF Portland cement
- ✓ ½ CF Lime
- ✓ 4½ CF Sand

Sand, being the coarser commodity, will equal the desired mortar quantity. The finer materials will not add to the volume when mixed. Therefore, the other ingredients are ratios based on the specified mix. Price these materials separately for delivery to site for mixing. Summarize the quantity take off by total cubic feet of each material.

Grout - Determine the number of vertical reinforcing sections by multiplying the number of sections by the number of block courses. This gives the number of cells for grouting. Each block manufacturer specifies the volume of voids in manufactured block. Multiplying the number of cells for grouting by the specified volume equals the total grout requirements.

Use the same procedure as with mortar to ratio materials according to specification.

04090 – MASONRY ACCESSORIES

Quantity Survey

Typical joint reinforcing is the pre-manufactured ladder or truss type made of heavy wire. These lengths lie horizontally between the courses of block and are imbedded in the mortar. The specifications define the exact spacing. A typical example for 8 IN block height might be every second course or every 16 IN. Load bearing walls (a wall specifically designed and built to support an imposed load in addition to its own weight) require vertical reinforcing. Use de-formed steel bars for this purpose. Install the bars in the block voids and then fill with grout. CMU walls may be double wide (referred to as double wythe) or with brick or brick veneer. Use reinforcing to tie the two wythes together. For additional strength requirements, use bond beams. Bond beams use specially configured block that forms a trough for receiving steel and concrete. There is usually a bond beam at the top of a masonry wall or at each story height in multistory building. Extreme conditions might dictate more frequent use.

Identify wire ladder/truss type reinforcing as typical where appropriate. Use building dimensions to determine total quantities. Divide wall height by specified spacing to determine the number of courses requiring reinforcing. Multiply this total by the horizontal length to get neat total horizontal reinforcing requirements.

Summarize wire ladders by length and bars by size and weight.

Use a control joint (also known as contraction joint) to regulate the location of cracking resulting from expansion and contraction. The rule of thumb in a one or two story structure is one every 40 feet of horizontal wall plus at corners and openings. Place control joints closer together as buildings become taller. Insert rubber or PVC pre-formed gaskets in vertically aligned specially grooved block. Seal this with caulking or sealant, allowing expansion and contraction to occur while remaining water tight. As an alternative, blocks may be tongue and groove design and not require an insert.

Determine the quantity for control joints. Refer to specification Division 04150 Masonry Accessories for type of control joint required.

Other common accessories which are integral parts of a completed concrete masonry wall are:

- ✓ Lintels - required over door or window openings. Frequently lintels are steel shapes. Use steel beams or steel angle depending on the span of the opening. The steel framing plans and structural details will specify the exact materials required. Be sure to measure and review each opening separately as the requirements may vary from opening to opening within a single structure. Summarize the take off in pounds by steel type and size for pricing.

- ✓ Steel angles, built-up steel members, bond beams filled with reinforcing steel and grout, and pre-cast shapes may function as lintels.

- ✓ Anchor bolts - grouted into the masonry or masonry cavity.

- ✓ Insulation - may be in loose form installed in the block voids. For cavity walls, attach rigid board insulation in the cavity between the wythes.

04210 – CLAY MASONRY UNITS – MASONRY

Quantity Survey

Walls

The preparation of a brick takeoff requires the following information:

- ✓ The type of brick selected for this project. This is important since each brick type is a different size. The most common type of brick used today is modular. Dimensions of this type of brick are 8 IN long × 2 2/3 IN high × 4 IN wide. Standard non-modular brick dimensions are 8 IN × 2¼ IN × 3¾ IN. Other brick unit designations are engineer, pavers, roman, norman, economy, jumbo, utility, queen, and king. The above nomenclature is not standard throughout the industry. When the estimator is in doubt, ask the supplier to identify his brick by length, width, and thickness. Remember that manufactured dimensions can vary from specified dimensions. These variances are acceptable as long as they do not exceed the ASTM specifications. (The specifications for face brick are found in ASTM C-216).

- ✓ Identify solid bricks or special shaped bricks. These can be very costly, especially if used in small quantities.

- ✓ Check the width of the mortar joint. The mortar joint is typically 3/8 IN but can easily vary from 1/8 IN to ¾ IN. Example: A common brick is 8 IN long × 2 ¼ IN high. When added to a normal motor joint of 3/8 IN, use the following equation. The dimensions of 8 3/8 IN × 2 5/18 IN = 21.984 or 22 square inches area on the face of each brick. Divide 144 by 22 which equals 6.54 or 6 ½. Six and ½ becomes the multiplier the estimator uses to determine the number of bricks needed, after finding the area.

- ✓ Another consideration is the brick pattern specified. Patterns commonly used are running bond, Dutch bond, English bond, and Flemish bond. A variation to any of these patterns is possible. Example: Quoined corners is a variation commonly used on bank buildings.

- ✓ In addition, consider the position that bricks occupy in the wall. Brick positions include stretcher, header, rowlock, rowlock stretcher, soldier and sailor. The position variations of bricks is not uncommon. For example: A double or triple soldier course at the top of masonry walls.

- ✓ Another consideration is the type of mortar joint specified. Types commonly requested are concave, convex, v-tooled, stripped, rodded, raked out, weathered, struck and flush cut. Of these, the flush cut is the most economical. This joint only requires using the trowel to strike off the excess mortar. The stripped joint is one of the most expensive, requiring wood strips inserted in the joints and removed after the mortar has hardened. The struck joint is the most common since it is the easiest when working from inside a structure.

Once the estimator understands the type and configuration of the units, the takeoff can begin. Use the following procedure:

- ✓ Coordinate the building elevations with the floor plans to check the true length of any exterior surface area. Elevations and sections will show wall height.

- ✓ Mark the plans to show the starting point of the takeoff and the direction in which the takeoff will move.

- ✓ After surveying the exterior of the building, take off the structural portion of the drawings. Structural plans provide information about height of the walls between floors, etc. The height of the wall should correspond to the brick coursing for the type of brick specified.

- ✓ Check interior architectural wall sections. Room finish schedules and interior elevations will show any interior brick work. Note ceiling heights, since brick is usually no more than one course above the finished ceiling height.

- ✓ Using the floor plan for length and the elevation or section for height or width, determine the square feet for each wall.

Site Improvements

The final drawings reviewed are the site improvements usually found in the civil drawings. These drawings show building signs, retaining walls, planter boxes, sidewalks, and manhole work. Signs, depending upon their design, can be a very expensive item. Retaining walls and planter boxes are usually not in detail and do not appear on elevations. Therefore, an experienced estimator must thoroughly review the drawings for reference to details of these items.

Sidewalks and entrances may use bricks called pavers. Square and basket weave patterns are common patterns and require no cutting. The herringbone pattern is very difficult and requires a considerable amount of cutting.

Specialty work also exists. Fireplaces are not uncommon and may include boilers, incinerators, smokestacks, etc.

General

Deduct the area not in brick from the quantities above. Examples of such deductions include doors, windows, louvers, or other building materials. Other areas usually subtracted include accents laid in a different pattern. Example of these are rowlocks, soldier courses, and header courses.

After deducting these areas, determine the thickness of the wall. If the thickness is four inches, deduct the exact width and length. When the thickness is an opening with a twelve-inch reveal, consider this in the deduction. For example, reduce a 5'0 IN × 7'0 IN opening to a 3'0 IN × 7'0 IN opening if the wall thickness is 12 inches.

Use a separate page for each type of construction surveyed. A separate page is also needed for different patterns, as well as different brick types, and mortar joints. Title each page for future reference. Mark each wall after surveying. Use colored pencils for this task. Use a different color to identify each wall type. Make a one-line drawing over the area or actually color in the entire wall. Place a check mark beside each opening counted. List each wall element and opening separate on the takeoff sheet. For clarification, the plans show detail numbers. Use the colored pencils or check marks to identify detail take off.

Give particular attention to drawing details. The details usually have priority over the smaller scale drawings and often give additional insight to design intent.

Keep these details separate on the takeoff sheet and identify by the detail number. Often a general heading is used and can be confusing. Example: "fireplace hearth" would be a detail, but if there are two fireplaces, which detail applies? A better description would be "fireplace #2 IN or fireplace-room g.-21.

After finding the net area, determine the proper multiplier and use it to find the number of bricks needed for each type. Include the mock-up panels for each type. Many jobs require three to four panels before final acceptance.

After determining the net quantity, add a percentage for waste. Breakage, cutting, chipping, or other imperfections are some causes for waste. This percentage varies across the country, but a good average is three to five percent. This number is then rounded to the next highest cube - or 500 bricks.

Example: Standard brick has a multiplier of 6.5 Therefore, a wall $10 \times 100 = 1000$ square feet. The wall cavity is 4 inches. There are two 3'0 IN $\times$ 7'0 IN doors and two 3'0 IN $\times$ 5'0 IN windows. The deduction is $3 \times 7 \times 2 = 42$ square feet plus $3 \times 5 \times 2 = 30$ square feet. 1000 minus 72 equals 928. $928 \times 6.5 = 6,032$. With a waste factor of 5 percent, the number of brick needed equals 6,334. Therefore, the quantity is 6,500 bricks.

One final aspect is scheduling that may dictate returning to the job site several times to do small amounts of work. Building signs or patch work in renovation of existing facilities are two common examples.

Summarize similar brick, combine patterns and price. The estimator should remember that schedule or phasing can drastically alter the unit prices originally thought to be proper.

Extension of quantities for brick masonry is as important as a proper takeoff.

The estimator must show that the quantity takeoff is in the same unit of measure as the company's historical costs.

It is helpful if the company has historical data on each of the items met during the takeoff. If the company has no such data, get pricing from outside sources possibly suppliers, independent estimating services, or one of several books that contain pricing. When using these prices, the estimators must keep in mind that their companies did not perform the work. There is a learning process for each new type of work. When there is a large quantity of the work, repetition will make the task easier time after time.

Estimate the exact amount of area and combine like quantities of work. Be sure to exclude all alternate work from the base bid. Also list phased work. Example: A building sign may have the same pattern and type of brick as the building face brick. However, estimating the sign probably is one of the last items.

04220 – REINFORCED UNIT MASONRY ASSEMBLIES

Quantity Survey

The following is a step-by-step procedure for estimating masonry materials. For discussion purposes, assume a wall or single story structure. Construction is a running bond (stretcher) pattern. This is a pattern where all courses are laid lengthwise. The vertical joint of one course occurs halfway between the joints of the courses above and below. This is a structure with load bearing walls that will require taking off quantities of vertical reinforcing and corresponding grout. The CMU size is 8 IN × 8 IN × 16 IN. The specified mortar is ASTM C270 Type S. Block is ASTM C90.

Survey square feet of walls using the elevation drawings. Note whether walls bear on foundations not shown on elevations. When so, add the depth not shown to the height dimensions. The estimator should make a separate takeoff for each type and size of block. Include an allowance for a mock-up wall in the measurements. Deduct the area of all openings (doors and windows). Based on specified block size, divide the net square footage by the nominal square footage of one block. The exposed surface of an 8 IN × 8 IN × 16 IN block is 128 square inches or 0.89 square feet (128/144) per block.

Note: Estimating manuals may show the inverse of this value (144/128) or 1.125 blocks per square foot. In this case, the estimator should multiply the net square footage by the number of blocks per square foot.

Round totals to the nearest whole number. This gives the total neat quantity of full-size block required.

Next, measure the vertical edges of all openings both sides. Divide this total by the specified block height. 8-inch height is equal to 8/12 or 0.67 feet. Deduct 50 percent as every other course will be ½ size block at the edge.

Use this technique to calculate the number of blocks at the control joints. These blocks are different from those taken off in above operations. They have notches or tongues and grooves to accommodate expansion/contraction control device. Measure the vertical length of all control joints and multiply by 2 for both sides. Divide this by the specified block height converted to feet. Then divide into two equal parts, one part is full size blocks and the other part is half size blocks.

Deduct the total block derived from the calculations above to determine total face blocks required.

Study wall sections and details for type of block and desired pattern for pilaster construction. Count blocks per course and multiply by the number of courses. Determine course number by dividing the wall height by the nominal height of block. Multiply this total by the number of pilasters on the floor plan.

At this point add an allowance for waste, cut off, loss and breakage. Company records or experience will determine what value or factor to add.

The estimator now has the total of the block requirements for this project. Summarize the quantities by type, size, and number of blocks.

05120 STRUCTURAL STEEL

Introduction / Description

Structural steel encompasses the steel load bearing members of a structure. Examples of structural steel are columns, girders, beams, joists, metal deck and steel bracing members. Structural steel is usually fabricated at a shop or yard, and then delivered to the project site for erection. Smaller braces and such are usually field fabricated and installed. A structural steel supplier usually submits a price for the fabricated material delivered to the project. A steel erector may submit a price to install the materials supplied to the project. The structural steel price may also include the fabricated material and erection as a lump sum cost. Since the shop labor is more efficient and less expensive, the more materials fabricated in the shop will lower the combined cost for fabrication and erection. The estimator should be able to estimate the costs of materials (fabricated material) and installation (erection) as separate items.

Types and Methods of Measurements

Structural steel pieces that come in manufactured shapes such as flanged beams, tubular steel, angle, channel, etc. are taken off by the piece in lineal feet and can be converted to pounds or tons. Items such as base plates, flanges, tabs connecting to columns, and shear plates, stiffener plates, connection bolts, at beams are taken off as each, or can be added to the takeoff as a percentage of the total weight.

Open web steel joists or "bar joists" are also taken off by the piece in lineal feet and converted to pounds or tons. The difference in obtaining the weight per lineal foot of joist compared to a standard steel shape is that the manufacturers must be referenced to find the weight per foot of joist.

Structural decking is taken off in square feet and maybe divided by 100 to calculate the squares of metal deck.

Specific Factors Affecting Estimate

Pre-estimate planning should always be performed to contribute to the accuracy of the estimate. A review of the specifications is a good start. Things to identify are who will furnish the temporary utilities; insurance requirements, bonding and safety requirements, working hours and security restrictions, other subcontractors on site, wage rates, and any other unusual requirements. These items are indirect cost and should be included in the overhead or listed as a separate item in the estimate. After the review of the specifications a study of the drawings to determine the site size, location and restrictions is the next step. The erection sequence and process should be planned thoroughly as this effort greatly affects the overall construction time. The erection plan will determine the equipment, manpower loading schedules, and the number of mobilizations required. The fabrication and erection of the structural steel scope is a milestone in every project. The structural steel estimator should carefully review the proposed project schedule for realistic time frames with respect to the detailing, fabrication, and erection process. This will have an impact on productivity and should be taken into account in the erection and equipment cost. You now have a good understanding of the extent, size and general environment of the project.

Shop drawings are detailed drawings of each piece of structural steel to be fabricated. A mistake in height or length of one measurement could affect the entire project. Shop drawings are generated by the fabricator after securing a work order and submitted by the general contractor to the design firm for approval. A qualified and experienced person should accomplish the review of the shop drawings. Fabrication will not begin until the shop drawings are reviewed and approved. The cost of preparing and reviewing the shop drawings, along with the expected schedule impact must be considered and included in the estimate. Any complications, restrictions, requirements or other items listed from the specifications review should be incorporated into the estimate. The construction plan set of drawings should be page checked to ensure that a full set has been received. It is also important to ensure that the architectural as well as structural drawings are included. The structural drawings may not show any or all of the architectural steelwork. The estimator should use past experience & knowledge to interpret these documents and achieve a complete scope of work. Structural steel construction projects are unique by nature, making standardization difficult. A good standard practice is to complete a final review of the estimate for errors, omissions, and completeness. There are no set rules regarding the material, labor, and equipment ratios of a structural steel estimate. Erection of structural steel requires workers to work at dangerous and deadly heights. Proper use of safety equipment is necessary for all projects. An implemented safety plan should be another standard for all projects. A list of items that will affect the estimate developed from above will help calculate your unit prices for material, labor & equipment and should be checked in the final stage of the estimate.

Organization of the Estimate / Quantity Survey and Calculating Quantities

A detail reading of the steel specifications division 5 in the specification book and also on the structural drawings is necessary for a listing of all requirements. The takeoff should be completed in a systematic matter as not to double count any of the members. Takeoff of the structural steel materials begins with the building columns (vertical members). The column lines are shown on the drawings and intersect at each column type. Column sizes and weights are indicated at each column, with lengths (or height) usually indicated by top and bottom elevation marks. The columns are taken off in total lineal feet for each column. The total weight is then calculated by multiplying the column weight by the total lineal feet. Then takeoff all detail column material such as anchor bolts, base and cap plates, gussets, etc. Upon completion of the column quantities, the beams (horizontal members) are taken off in total lineal feet for each beam type. Prevent duplication of beam members by taking off the pieces running north to south, then the east to west pieces. The total weight is then calculated by multiplying the beam weight by the total lineal feet. Secondary framing such as angles, "X" bracing, wing bracing and such are taken off in total lineal feet of each size and type of material, and then converted to total weight. Typically these items are not as plainly shown on the drawings, as are columns and beams. Notes and or detail sections sometimes indicate secondary framing items.

The specification for the structural steel may require shop primed steel. If this is required the surface area of steel to be painted must be calculated. The Manual of Steel Construction, Ninth Edition published by AISC gives the surface area for all steel sections. A unit price is applied to the total surface area of steel to be primed or painted.

Connections between different structural components can include bolts, welds or both. Base plates, flanges, tabs, at columns and shear plates, stiffener plates and connection bolts at beams can be taken off by each and converted to total weight. The basic type of welds are continuous field, partial penetration, and full penetration. The sizes of the different welds will be specified by the design drawings or specifications. These different weld types and sizes should be listed and priced separately. Steel connections can also be added as a percentage of the total weight. Consideration should be given to the detail of the take-off and the steel design to determine the amount of percentage to add when using a percentage. A variable of 3% to 15% markup of the total structural steel weight may be necessary.

Open web steel joists or "bar joists" are taken off by their SJI designation. Joists should be taken off in the same manner as beams. The total weight is then calculated by multiplying the joist weight by the total lineal feet. Then takeoff the bridging & bracing in accordance with the SJI requirements. The following is a take-off procedure for material cost of open web steel joists and bridging:

- ✓ Joist girder at grid lines
- ✓ Girder braces or 4 per girder if not shown on drawings
- ✓ Filler joists between girders
- ✓ Bridging per SJI (Steel Joist Institute)
- ✓ Additional connections & details per drawings
- ✓ Send to joist manufactures for price

Metal decking is formed from steel sheets in different gauges and a standardized cross sections. Typical sections are 30 IN in width and are manufactured in various lengths to suit job conditions. The following is a takeoff procedure for the material cost of metal deck.

- ✓ Deck at floors and roofs
- ✓ Closures & gauge mud stops
- ✓ Additional deck shown on drawings
- ✓ Send to deck manufactures for price
- ✓ Shear studs at composite deck

Items not typically included in a structural steel estimate are cost of field inspections, testing, rebar or mesh, ornamental metals, wrought iron, non-ferrous metals, sheet metal, metal studs, tracks & bridging, anchor bolts for other trades, signage, chain link fence, hardware for other trades, finished metal roofing, structural calculations & engineering. Also see section 05500 for metal fabrications (miscellaneous steel).

Overview of Labor, Material, and Equipment

Estimating the fabricated material costs for structural steel is achieved by applying a unit cost to the total weight of the required items. Shapes and weights of structural steel are indicated by use of common designations indicated on drawings. Examples of common drawing designations for steel sections are:

- ✓ Wide Flanged steel would be designated "W10×30". The "W" is the designation for Wide Flange, the "10" is for nominal depth in inches, and the "30" is the weight per foot.

- ✓ American Standard Beam would be designated "S10×30". The "S" is the designation for American Standard Beam, the "10" is for nominal depth in inches, and the "30" is the weight per foot.

- ✓ American Standard Channel would be designated "C10×30". The "C" is the designation for American Standard Channel, the "10" is for nominal depth in inches, and the "30" is the weight per foot.

- ✓ Miscellaneous Beams would be designated "M10×30". The "M" is the designation for Miscellaneous, the "10" is for nominal depth in inches, and the "30" is the weight per foot.

- ✓ Miscellaneous Channel would be designated "MC 10×30". The "MC" Is the designation for Miscellaneous Channel, the "10" is for nominal depth in inches, and the "30" is the weight per foot.

- ✓ Angle would be designated "L5×4×3/8". The "L" is the designation for angle, the "5" is for the length of one leg in inches, the "4" is for length of the other leg in inches and the "3/8" is the thickness of each leg in inches. Note the longest leg should be listed first.

- ✓ Tee (cut from W) would be "WT10×30". The "WT" is the designation for Tee cut from a Wide Flange, the "10" is for nominal depth in inches, and the "30" is the weight per foot.

- ✓ Tee (cut from S) would be designated "ST10×30". The "ST" is the designation for Tee cut from an American Standard Beam, The "10" is for nominal depth in inches, and the "30" is the weight per foot.

- ✓ Tee (cut from M) would be designated "MT 10×30". The "MT" is the designation for Tee cut from a Miscellaneous Beam, The "10" is for nominal depth in inches, and the "30" is the weight per foot.

The project schedule may require the steel erection to beginning sooner than steel from a mill can be fabricated and delivered. Materials supplied from a steel warehouse may be more expensive and drop offs from cut lengths will have a waste cost added. Structural steel material cost should take these items into account to include shop drawings, shop fabrication, painting and delivery to the site.

Steel installation or erection is the cost of field labor to include unloading, placing steel, field fabrication if required, safety cables, connections including welding and small tools. The labor and equipment cost is based on the number of pieces per day on average for each element. The installation rate (number of pieces per day) will not only vary due to crew size, speed and efficiency, but also is affected by job conditions (relative access or mobility) and the weight of the pieces. A crew may set more smaller and lighter pieces (but less total tonnage) per day than larger or heavier pieces. Crew sizes can vary and will depend on the nature of the project.

Labor for the installation of metal deck is calculated by applying productivity to the total square feet of metal deck and multiplying by an hourly rate. The height of the structure, the weld specification or connection method, the average weight of each member, and site conditions will affect productivity.

Equipment such as cranes, forklifts, scissors lifts, welding machines, generators and stud guns are used for the erection of columns, beams, joists and metal deck. Lifting equipment requirements are dependent on the heaviest piece and longest reach of the steel members to be installed. An experienced equipment or erection estimator should determine the pieces of equipment and rental time required with respect to the weights, and safety requirements. The crane selection should be determined by referencing a crane-lifting chart, which is available from crane service companies or manufacturers. Rental cost of each piece of equipment used including welding machines should be included in the estimate.

Estimate

The narrative above is a directional process of how to take-off and summarize any structural steel project for a the beginner to a Senior Level estimator desiring to learn or develop a cost for structural steel. The prerequisite to utilize this section is the ability to read and understand a Construction Set of Contract Documents for any building. The estimate attached is an example of how to price the summary of categories generated from the narrative explanation above. We assume to have taken off a single story building such as a hospital, or office building, with average site, weight conditions and weld specifications. All of the above items and their associated quantities should be listed, totaled and transferred onto an estimate sheet similar to the attached example. The Structural Steel is calculated by the piece, the connections are counted and listed, and the equipment is priced based on the erection time. Installation procedures and methods are different for each category of material. For this example we will limit the categories to those listed below, and distribute man-hours for erection using an average production factor for each. The distribution of labor productivity should be applied to the total quantities of materials broken down into at least the following categories: Columns, Beams, Joists, Secondary Framing and Decking. Erection equipment is listed separately.

EXAMPLE ESTIMATE FOR STRUCTURAL STEEL BY PIECE

Based on a single story building such as a hospital, or office building, with average site and weight conditions, and weld specifications.

	Item or Description				Material		Labor			Equipment		Sub		Total
	DESCRIPTION	Quantity	Unit	Unit $	Cost	Hr/Unit	Hours	Rate/Hr	Cost	Unit $	Cost	Unit $	Cost	Cost
1	Columns	9.2	Tons	1300.00	11,960	-	-	-	-	-	-	-	-	11,960
		16.0	Pcs		-	5.25	84.00	17.50	280	-	-	-	-	1,470
2	Beams	7.9	Tons	1200.00	9,480	-	-	-	1,188	-	-	-	-	9,480
3	Secondary Framing, Field Fab	2.0	Tons	1000.00	2,000	-	-	-	3,500	-	-	-	-	2,000
4	Open Web Joists	1.85	Tons	1500.00	2,775	-	-	-	-	-	-	-	-	2,775
5	Metal Deck	9,500	SF	1.00	9,500	0.006	57.00	15.50	884	-	-	-	-	10,384
6	Connections				-	-	-	-	-	-	-	-	-	
	Bolted	56.0	Ea	55.00	3,080	0.28	15.68	17.50	274	-	-	-	-	3,354
	Cont. field weld	250.0	Lf	1.00	250	0.20	50.00	17.50	875	-	-	-	-	1,125
	partial pent weld				-	-	-	-	-	-	-	-	-	
	full pent weld				-	-	-	-	-	-	-	-	-	
	Forklift	80.0	Hrs		-	-	40.00	17.50	700	10.00	800	-	-	1,500
	Welding Machine	80.0	Hrs		-	-	-	-	-	7.50	600	-	-	600
	Generator	80.0	Hrs		-	-	-	-	-	4.00	320	-	-	320
	TOTAL TONS	21.0	TONS											
Subtotals					40,755		658		11,301		6,920		-	58,976
Labor Burden on Labor					%			45%	5,086					5,086
Total Direct Cost									16,386		6,920		-	64,062
Overhead	10%													6,406
Profit	5%													3,203
Insurance & Bond	2%													1,281
Sales Tax on Material	6%													2,445
Total Subcontractor Cost														**$77,398**

Notes:
Average labor rate per hour = $17.50 STRUCTURAL & MISC.   COST/SF DECK $8.15 $15.50 JOISTS & DECK
COST/TON $3,694

05120 Structural Steel — Part Two

EXAMPLE ESTIMATE (A) FOR STRUCTURAL STEEL BY TON

For this estimate we will use a single story building such as a hospital, or office building, with average site and weight conditions, and weld specifications.

#	Item or Description / DESCRIPTION	Quantity	Unit	Material Unit $	Material Cost	Labor Hr/Unit	Labor Hours	Labor Rate/Hr	Labor Cost	Equipment Unit $	Equipment Cost	Sub Unit $	Sub Cost	Total Cost
1	Columns	9.2	Tons	1300.00	11,960	9.75	89.70	17.50	1,750	350.00	3,220	-	-	16,750
2	Beams	7.9	Tons	1200.00	9,480	8.89	70.22	17.50	1,229	300.00	2,370	-	-	13,079
3	Secondary Framing, Field Fab	2.0	Tons	1000.00	2,000	110.00	220.00	17.50	3,850	700.00	1,400	-	-	7,250
4	Open Web Joists	1.85	Tons	2500.00	4,625	20.00	37.00	15.50	574	300.00	555	-	-	5,754
5	Metal Deck	9,500.0	SF	1.00	9,500	0.006	57.00	15.50	884	0.02	190	-	-	10,574
6	Connections	3,820	Lbs	1.05	4,011	0.023	87.86	17.50	1,538	0.25	955	-	-	6,504
	TOTAL TONS	22.9	TONS											
	Subtotals				41,576		562		9,643		8,690		-	59,909
	Labor Burden on Labor		%					45%	4,339					4,339
	Total Direct Cost								13,982		6,920		-	64,249
	Overhead 10%													6,425
	Profit 5%													3,212
	Insurance & Bond 2%													1,285
	Sales Tax on Material 6%													2,495
	Total Subcontractor Cost													**$77,665**

COST/SF DECK $8.18 COST/TON $3,397

Notes:
Average labor rate per hour = $17.50 STRUCTURAL & MISC.
$15.50 JOISTS & DECK

06100 – ROUGH CARPENTRY

A building construction project may involve various wood framing elements including, but not limited to interior or exterior load or non-load bearing walls, sheathing, decking, blocking and supports to join the members and anchor the framework to other construction.

Introduction

Wood framed wall systems are found on many types of projects, mostly in the construction of residential buildings, either single or multi-family housing. Estimating prefabricated trusses is included in 06170 Prefabricated Structural Wood. Estimating wood blocking and other supports are similar to estimating the components of wood framing and will not be addressed. Wood framing when estimated is most commonly broken down into "loads", which derives its term from the delivery of a particular package of lumber to the jobsite. Traditionally, loads are packaged and delivered in order of their assembly. Using an example of a wood framed, two story, built-up floor system building, the first load delivered would be the 1st floor load. Just before the completion of the floor framing, the 1st floor wall load will arrive. Soon after, deliveries of the 2nd floor load, the 2nd floor wall load, and finally the roof load arrive.

Contractors systematically order framing loads, lumberyards systematically deliver framing loads, and carpenters systematically assemble framing loads. Given the inherent subdivision of a wood frame system, it is logical that estimators of wood flame construction systematically estimate the wood framing by "loads."

Types and Methods of Measurement

Wood materials are specified by species, grade, seasoning, surfacing, nominal size, and treatment, if required. Lumber sizes in the United States are given as nominal dimensions. A piece nominally 1 by 2 IN in cross section is a 1 × 2(one by two), a piece 2 by 10 IN is a 2 × 10, and so on. Pieces of lumber less than 2 IN in nominal thickness are called boards. Pieces ranging from 2 to 4 IN in nominal thickness are referred to as dimension lumber. Pieces nominally 5 IN or more in thickness are termed timbers. Typically lumber is priced per thousand board foot (M BD FT). Board foot measurement is based on nominal dimensions. A board foot of lumber is defined as 2 SQ IN in nominal cross-sectional area and 1 FT long. For example a 1 × 12 or 2 × 6 - 10 FT long each contains 10 BD FT.

Dimension lumber is usually supplied in two foot increments of length, typically 8, 10, 12, 14 or 16 FT lengths. It is important to note that the length of material will affect the purchase price. Due to the dwindling supply of long straight trees, longer lengths of dimensional lumber and timber will be relatively more expensive.

Some lumberyards may quote dimensional lumber on a lineal foot or board foot basis rather than per thousand board feet. Care should be taken that the quantities and material quotations are in the same unit of measure (LF, BD FT, or M BD FT.)

Wood panel products (plywood, composite panel, waferboard, orientated strand board or particleboard) are specified by grade, span rating, thickness, and exposure durability classification. Plywood panels for use in paneling, furniture, and other uses where appearance is important are graded by the quality of their face veneers. Standard plywood panels are 4 FT by 8FT in surface area and range in thickness from 1/4 to 1 IN. Longer sheets are manufactured for siding and industrial use. Wood panels are usually priced per piece or per square foot.

Factors Affecting Takeoff and Pricing

Architectural floor plans should indicate the locations and dimensions of all the walls, partitions, and openings. Elevations show side views of the building, with vertical framing dimensions indicated as required. Section drawings and/or details should show the dimensional relationships of the various floor levels and roof planes, and the slopes of the roof surfaces. While there are local variations in framing details and techniques, the sizes, spacings, and connections of the framing are regulated by building codes.

The following elements and measurement of wood framing need to be taken off.

§ The sill preferably of treated or naturally decay-resistant wood is usually bolted to the foundation as a base for the wood framing. Installation may require drilling of the sill plate for previously placed anchor bolts or expansion anchors. Note the requirement for sill sealer to be installed under the sill plate. The top of the foundation is usually uneven, requiring the use of a wood shim. Note the lineal feet and the dimensional lumber required for the sill, anchoring requirements, and other miscellaneous items.

§ Floor framing consists of joists, bridging, and subflooring. Prefabricated floor trusses may be furnished and delivered to the jobsite by a supplier. Often prefabricated floor trusses are notched at the top of the truss to accommodate a 2 × 4 band around the perimeter to tie the floor system together. If the floor joist system is to be field fabricated, note the spacing, lengths, and dimensional lumber required. Bridging between joists, where required, may be solid blocks of joist lumber, wood crossbridging, or steel crossbridging. Where ends of joists are butted into supporting headers, as around stair openings and at changes of joist direction for projecting bays, metal joist hangers will be required and should be noted. Adjust the quantities to account for double joists under partitions, laps at bearings, or ledgers. Subflooring should be glued to the joists to prevent squeaking and increase floor stiffness. Subflooring is usually taken off in square feet, noting the type and thickness of the material required.

§ Wall framing consists of three elements: bottom plate(s), studs, and top plate(s). Other items to be taken off may include framing for openings, corners, partition intersections, and supporting studs for the headers over openings.

Typical building code requirements for stud sizes and spacings are listed in the table below. Note that 2 × 4 studs are inadequate for walls that support two floors and a roof above or basement studs below a two story building. Note also that very tall studs may have to be increased in size.

STUD SIZE	SPACING	MAXIMUM STUD HEIGHT	MAXIMUM STORIES SUPPORTED
2 X 4	16 IN	14 FT	One floor plus roof
	24 IN	14 FT	Roof only
2 X 6	16 IN	20 FT	Two floors plus roof
	24 IN	20 FT	One floor plus roof

Framing costs vary considerably simply due to geographic location. Material cost can widely vary due to the availability, distance from resources, local market conditions, and even material handling costs for remote regions. Note the species of lumber required.

Often related to geographical location, the seasonal effects on wood frame construction can be quite significant. In wet climates, additional costs are commonly added for weather protection. Wood is a porous material and is prone to warping. Warping occurs most when wood is subject to moisture and then allowed to air-dry in an uncontrolled environment. Major weather events (hurricanes, tornadoes, etc.) may have a significant effect on both the availability and price of wood products.

Overview of Labor, Material, Equipment, Indirect Costs, Approach, and Mark-ups

The following estimate of an exterior wood framed project is organized in "loads". Both the floor and wall loads will be considered, the roof load will not be discussed within this section.

Lumber pricing will be in the cost per BD FT. The estimate is for a single story, built-up floor system residence. It will include 2 × 12 floor joists, 3/4 IN plywood floor sheathing, and 2 × 4 stud walls 8 FT in height. The floor plan example will measure 20 FT by 60 FT for a total of 1200 SQ FT. Four window openings measuring 4 FT × 4 FT and two door openings measuring 3 FT × 7 FT are included. Labor, material, and equipment will be summarized separately. Application of indirect costs and mark-ups will follow.

Floor Load: Assume the floor joists run in the 20 FT building direction, are spaced 16 IN on center, and rest on a concrete stem wall at the perimeter and midpoint span (see Floor Framing Sketch 1). To determine the quantity of joists divide the building length (perpendicular to the running direction of the joists) by the joist spacing: 16 IN = 1.33 FT, therefore, 60 FT/1.33 FT = 45 joists

This method does not account for the very first joist; therefore, you must add one joist to each calculation. The total joist count is then 46 each at 20 FT long. (If the building width had been 18'–6 IN, a 20 FT piece would be included in the estimate.)

To determine the length of the joists, one might assume it to be 20 FT in overall length. However, a premium is paid for framing lumber greater than 16 FT in length. Inexperience may lead to assuming that 10 FT lengths are adequate and the count is simply doubled. Joists, if broken in overall span must be overlapped a minimum of two feet and the overlap must occur over the center support or mid-span. This would require each joist length to be 11 FT. Another issue occurs since lumber of this type is purchased in two foot increments. If 12 FT lengths are then purchased, one foot of each member is essentially waste. For this example, the waste equals 92 FT. The cost difference between wasting 92 FT of lumber should be compared to the premium paid for 20 FT lengths. Consideration should also be given to the additional labor involved in handling ninety two pieces versus forty six pieces. In this example, assume paying for 20 FT lengths to be the more cost effective. If actual lengths are specified by the design professional, the cost for the specified lengths should be anticipated.

The material takeoff for the floor is as follows:

MATERIAL – FLOOR FRAMING					
Description	Quantity	Unit	BD FT	Unit Cost per BD FT	Total Cost
FRAMING					
2" x 12" x 20" floor joist	46	EA	1840	0.705	1297.20
2" x 12" rim joist	160	LF	320	0.705	225.60
2" x 12" blocking at midspan	60	LF	120	0.075	84.60
Waste factor, 5% of BD FT			114	0.705	80.37
DECKING					
3/4" T&G Plywood	1200	SF		0.70	840.00
Waste 10% of SF		SF	80	0.70	56.00
Subtotal					2,583.77
Hardware, 5% of lumber cost	1	LS			129.19
Subtotal					2,712.96
Tax	7.25%				196.69
TOTAL					**$2,909.65**

To calculate the labor for this portion of the work, a number of methods may be used. First, experience and historical cost data may show that a crew of two carpenters and one laborer is a standard size crew for this portion of the work. This is referred to as the crew size or crew balance. Crew balance is derived from the concept that optimum efficiency is obtained by assembling particular quantities and types of workers together as a team or "crew" to perform particular tasks. Adding more workers to a crew may increase productivity, but may not justify the increased cost. To the contrary, decreasing the number of workers may decrease cost but may also decrease productivity to an unacceptable rate, hence, the term "balance".

Other sources such as estimating reference manuals may be useful in determining crew size. Along with crew balance, the crew output must also be determined. The crew output is the quantity of work the crew can perform in a given amount of time. The output is a unit of measure per given time (I.e. 600 SF/Day, 52 LF/HR, etc.)

Two carpenters and one laborer with a daily output of 600 SF per day will be the crew size and output for this floor framing example.

The next step in determining labor costs is to calculate the "crew rate", or unit cost per hour. Combine the hourly cost of each member thereby establishing the hourly crew rate. This hourly rate is then multiplied by the total number of crew hours. If the crew output is 600 SF/Day to frame and sheath the joists and plywood flooring, then framing 1200 SF will require two days. Two days equals 16 crew hours. The labor calculations follow:

LABOR – FLOOR FRAMING (Crew: 2 Carpenters and 1 Laborer)			
Description	Hourly Rate	Quantity	Total
Carpenters	30.00	2	60.00
Laborer	20.00	1	20.00
Hourly Crew Rate	**Crew Hours**	**Crew Rate**	**Total**
Total	16	$80.00	$1,280.00

Often reference books and historical cost data provide yet another type of unit referred to as a unit rate. This is equivalent to the time it takes to produce one unit of work. If the output of our crew is 600 SF/Day then the output per crew hour equals:

600 SF/Day divided by 8 crew hours/Day = 75 SF per crew hour

To calculate the time it takes to produce one SF of work, simply invert the output rate. If 75 SF can be completed in one crew hour, then one SF can be completed in $1/75^{th}$ of a crew hour. Therefore, 1 SF takes .013 crew hours to complete.

LABOR – FLOOR FRAMING – Unit Rate Method				
Quantity	**Unit**	**Unit Rate**	**Crew Rate**	**Total**
1200	SF	0.013	$80.00	$1,024.00

Notice the total cost differs from the previous calculation. This is due to the rounding error in the unit rate itself. Our entry only accounts for three decimal places as do many reference books. Caution should be taken when using this method to insure that rounding errors do not adversely affect your outcome.

Wall Load: (See Wall Framing Sketch 2) Similar to floor framing, the wall framing material takeoff is calculated in much the same manner. The 8 FT wall height determines the stud length. Standard exterior framing is anticipated with a stud framing of 16 IN on center Several methods are employed throughout the industry in determining the total quantity of studs in typically framed walls, Obviously counting each piece would likely be the most accurate. One method is to measure the lineal feet of wall, taking no allowances for window or door openings, and dividing it by the spacing dimension of the studs (16 IN on center). Only openings for doors larger than 8 FT wide should be deleted from the total footage. This method allows additional stud lumber to be on hand for the cripples and sub-sills required to flame window openings. To this, only trimmers, king studs, blocking, plates, headers, posts and beams must be added. Our takeoff is as follows:

06100 Rough Carpentry

MATERIAL – WALL FRAMING					
Description	Quantity	Unit	BD FT	Unit Cost per BD FT	Total Cost
2" x 4" x 8' studs	120	EA	640	0.545	348.80
2" x 4" x 8' trimmers & king studs	24	EA	128	0.545	69.76
2" x 4" bottom & double top plates	480	LF	320	0.545	174.40
2" x 4" fire blocking	160	LF	107	0.545	58.13
4" x 12" headers at doors & windows	22	LF	88	0.855	75.24
Waste factor, 5% of BD FT			64	0.545	34.95
Subtotal					797.60
Tax	7.25%				57.83
TOTAL					**$855.43**

For this type of wall framing, we will assume a labor output rate of 800 SF per crew day. Our actual quantity of square footage is (60 FT + 60 FT + 20 FT + 20 FT) × 8 FT height = 1280 SF.

Our unit rate per crew hour is 800 SF per 8 hour crew day = 100 SF per hour

1 SF = 1/100 hour = .010 hour

Using the previous crew balance, labor cost calculates as follows:

LABOR – WALL FRAMING (Crew 2 Carpenters and 1 Laborer)			
Description	Hourly Rate	Quantity	Total
Carpenters	30.00	2	60.00
Laborer	20.	1	20.00
Hourly Crew Rate	**Crew Hours**	**Crew Rated**	**Total**
Total	12.8	$80.00	$1,024.00

Part Two 06100 Rough Carpentry

LABOR – WALL FRAMING – Unit Rate Method			
Quantity	Unit	Unit Rate	Crew Rate	Total
1280	SF	0.010	$80.00	$1,024.00

PROJECT SUMMARY				
Description	Materials	Labor	Total	$/SF
Floor Framing	2,910.00	1,280.00	4,190.00	$3.49
Wall Framing	855.00	1,024.00	1,879.00	$1.57
TOTAL – DIRECT COST	3,765.00	2,304000	6,069.00	
General Conditions, overhead, and profit		20%	1,214.00	$.96
Subtotal			7,283.00	
Construction Contingency		5%	364.00	$.29
Escalation to Mid-Point Construction		0%	0.00	
TOTAL CONSTRUCTION COST			$7,647.00 .00	
COST PER SQUARE FOOT (SF of Floor Area)			$6.37	

Generally, the Project Summary is followed by detailed quantity takeoff and pricing sheets. Specific mark-ups have been added as percentages of Direct Cost. General condition costs are discussed in other parts of this manual. Construction contingency may be added if the estimator believes there are unforeseen costs relating to issues not fully developed but may be the responsibility of the contractor. Escalation to mid-point of construction is commonly added for projects with extended periods of duration to cover inflationary costs.

06170 - PREFABRICATED STRUCTURAL WOOD

INTRODUCTION

A truss roof system is comprised of a series of prefabricated "trusses" which span between load bearing walls and/or beams and includes the roof decking (06170 Prefabricated Structural Wood and 06150 Wood Decking). Spans for trusses can range from 10 feet to 200 feet or more. Trusses can be manufactured in a variety of styles to suit both architectural and structural requirements. These styles include gable, hip, shed, flat and mansard. The structural design must meet code requirements that may include live, dead or snow loads, etc. This will require a brief discussion of 06150 Wood Decking, but will not expand on any other elements of that section.

TYPES AND METHODS OF MEASUREMENT

In order to accurately estimate the truss roof system, the estimator must examine its two major components: the individual roof trusses and the roof decking system. Each of these components will need a separate estimate and require different measurement calculations.

The body of the truss system is comprised of individual trusses and grouped by length of span, roof pitch and configuration. This categorization is applied to all truss roof systems regardless of architectural style and/or structural requirements.

In situations where truss roof systems intersect or are complicated by roof features, such as dormers, some stick or loose framing will be required. This framing will complete the roof system in areas where it is not cost effective to manufacture and set a truss.

In order to account for these pieces, a linear foot takeoff per lumber size must be generated and may include ridge board, rafters and bottom plates.

The roof decking system is measured in square feet which is calculated by multiplying the overall length of the roof by the rafter length or slant height.

FACTORS WHICH MAY EFFECT QUANTITY TAKEOFF AND PRICING

Effect of Small Quantities vs. Large Quantities

The set up costs associated with the manufacture of trusses (specific span, pitch and style requirements), are evenly distributed over the number of pieces manufactured. A smaller job has a higher per truss manufacturing cost than a larger project, because it has fewer pieces to absorb the set-up costs. As with the manufacturer's labor costs, the labor to install the trusses is greatly impacted by the truss count. Not all large jobs have lower unit labor costs. Extremely complex roofing systems, large or small, will require additional labor for layout and assembly.

Effect of Geographical Location

In certain regions of the country, the "on center" span between trusses is adjusted to allow for code requirements. Due to these requirements, the overall count of trusses may increase, thereby impacting the total cost for the material and installation labor. In the northeast United States, roof pitches are extreme to prevent the collection of heavy snow. Because of these extreme pitches, more material per truss is required thus, the cost of the individual truss is higher, as is the cost of the entire truss system. As the roof pitch increases both the amount of plywood decking and the difficulty of installation are affected. Geographic variations can affect the installation cost.

Seasonal Effect on the Work

Seasonal effects that impact the cost of building a truss roof system include lost time due to extreme weather conditions. Scheduling with these conditions in mind is critical to containing costs.

TRUSS ROOF SYSTEM

There are many factors that affect the labor costs to install this system. In order to estimate labor costs, the estimator must consider the roof design and the installation method.

Installation considerations include:

A. Complexity of system
B. Large vs. small project
C. Field assembly of trusses if the trusses need to be shipped in pieces
D. How the trusses will be installed, i.e. by hand or equipment
E. Size and experience of the installation crew
F. Miscellaneous framing

Once these issues are resolved labor hours can be calculated. If necessary, include the expense of hoisting equipment to set the trusses. The use of hoisting equipment allows a crew to set a higher volume of units. This generally outweighs the cost of "handing up" the trusses. Equipment is usually a per day/week/month cost which includes an operator, transportation of the equipment to and from the job site and fuel.

The manufacturer will generate a detailed takeoff of the trusses for each project, complete an estimate and provide a quotation for either a total cost or unit cost to furnish the material.

This quotation should include:

- ✓ All truss units per the contract documents
- ✓ The cost of the material
- ✓ Delivery charges
- ✓ Special handling and transportation costs (oversized trusses)

Complex roof systems, which include intersecting truss systems, dormers, gables and other architectural features require stick or loose framing. This framing addresses areas in the truss system where it is cost prohibitive to manufacture a truss and is not included in the truss material quotation.

Wind and temporary bracing will need to be determined and carried as a line item expense. Fasteners, clips, straps, spacers, and blocking, etc. are costs calculated as a percentage of the material cost.

Roof Decking

The labor required to install roof decking is based on a per sheet or per square foot cost. This cost includes the labor to unload and stock the material, as well as the labor to install the material.

A variety of material is used in roof decks and is usually supplied in sheets. The quantity required is determined by dividing the total square feet of roof area by 32 square feet, assuming 4 FT × 8 FT sheets. A factor should be included to account for roof design, complexity and waste.

Additional material costs include H-clips, fasteners and glue and may be carried as a percentage of the material cost.

RATIO AND ANALYSIS

An effective way to test the final bid price is through comparison to historic data. Price checks, such as square foot costs or material to labor ratios, must address variables such as volatile lumber costs, truss, design, roof design and labor costs. These can significantly impact the value of historic data.

MISCELLANEOUS PERTINENT INFORMATION

While an estimator is generally not a licensed structural engineer, one should understand how a truss system is structurally supported. The truss layout plan, as prepared by the design professional, should be cross referenced with the structural plan. If there are any inconsistencies, the design professional should be notified.

SAMPLE ESTIMATE

The attached sample estimate is for a gable roof, with dormers. The main truss span is 67'– 0 IN, trusses are shipped in three pieces and will be lifted into place using an 80 ton crane. The General Contractor's plan has one carpentry crew assembling the large trusses on the ground and another crew installing the trusses.

Conversion Factors for Sloped Surfaces:	
Pitch of Roof*	Multiply Flat Area by
2:12	1,014
3:12	1,031
4:12	1,054
5:12	1,083
6:12	1,118
7:12	1,158
8:12	1,202
9:12	1,250
10:12	1,032
11:12	1,357
12:12	1,413
13:12	1,474
14:12	1,537
15:12	1,601
16:12	1,667
17:12	1,734
18:12	1,803
19:12	1,875
20:12	1,948
21:12	2,010
22:12	2,083
23:12	2,167
24:12	2,240

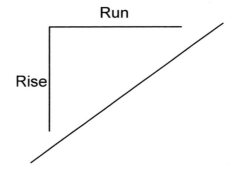

* Commonly designated on drawings as Rise/Run

06170 Prefabricated Structural Wood

	ITEM DESCRIPTION	U/m	Quantity	Labor u/p	Material u/p	Subcontr.	LABOR	MATERIAL	SUBCONTR.	TOTAL
1	**TRUSSES**									
2	24" c.c., fire treated lumber									
3										
4	67'-0" span truss	ea	91	78.50	381.00		7,144	34,671	0	41,815
5										
6	10'-0" span truss	ea	18	27.50	150.00		495	2,700	0	3,195
7										
8	**STICK FRAMING**									
9										
10	Main roof intersection	l.f.	1,000	0.60	1.50		600	1,500	0	2,100
11	2 x 6 fire treated									
12	Dormer	l.f.	140	0.60	1.50		84	210	0	2,100
13	2 x 6 fire treated									
14	Dormer front & side Walls	l.f.	862	0.75	0.75		647	647	0	1,293
15	2 x 4 x 8' SYP @16" o.c.									
16	(bearing partitions)									
17										
18	End gable wind bracing	l.f.	115	0.71	0.90		82	104	0	185
19	2 x 4 x 16', fire treated									
20										
21	Blocking	l.f.	356	0.50	0.50		178	178	0	356
22										
23	Pre-staging/Assembly	ea	91	55.00			5,005	0	0	5,005
24										
25	Crane – 80 ton	l.s.	1			1,000.00	0	0	1,000	1,000
26										
27	Fasteners/Hardware	l.s.	1		400.00		0	400	0	400
28										
29	**SUBTOTAL – TRUSSES**						$14,234	$40,409	$1,000	$55,643
30										
31										
32	**ROOF DECK**									
33										
34	7/16" t&g PLYWOOD	sht	417	9.75	16.50		4,066	6,881	0	10,946
35										
36	Material waste 5%	sht	21	9.75	16.50		203	344	0	547
37										
38	Fasteners/Hardware	l.s.	1		250.00		0	250	0	250
39										
40	**SUBTOTAL-ROOF DECK**						$4,269	$7,475	$0	$11,744
41										
42										
43	**SUBTOTAL DIRECT COST**						$18,503	$47,884	$1,000	$67,386
44	Labor Burden 35%									6,476
45	Sales Tax 6.5%									3,112
46	**TOTAL DIRECT COST**									$76,975
47	Gen. Conditions, Bond 15%									11,546
48	Home office Overhead & Profit 10%									8,852
49	**TOTAL ESTIMATE**									$97,373

TRUSS ROOF SYSTEM

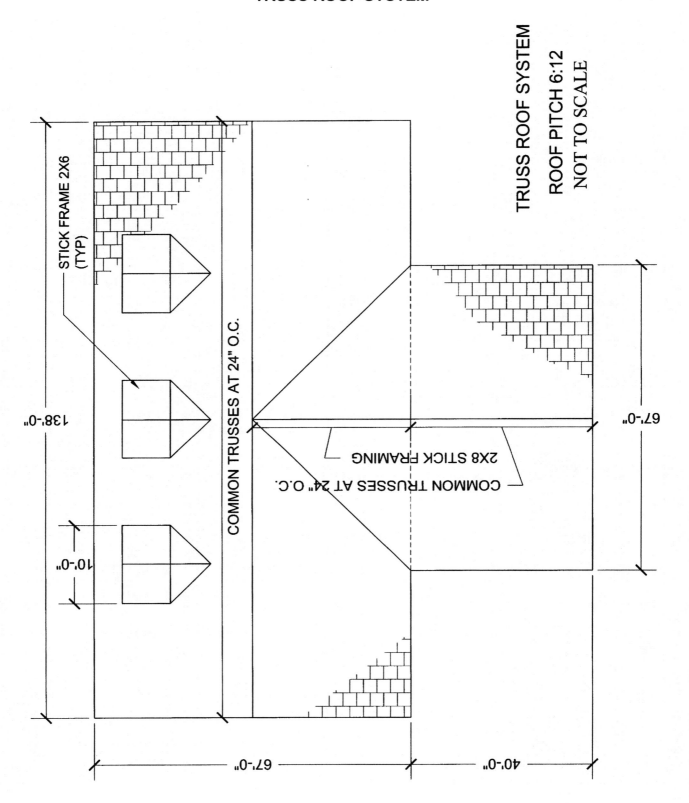

TRUSS ROOF SYSTEM
ROOF PITCH 6:12
NOT TO SCALE

Figure 06200-1

300-G-5

Identification of Standing and Running Trim and Rail Parts

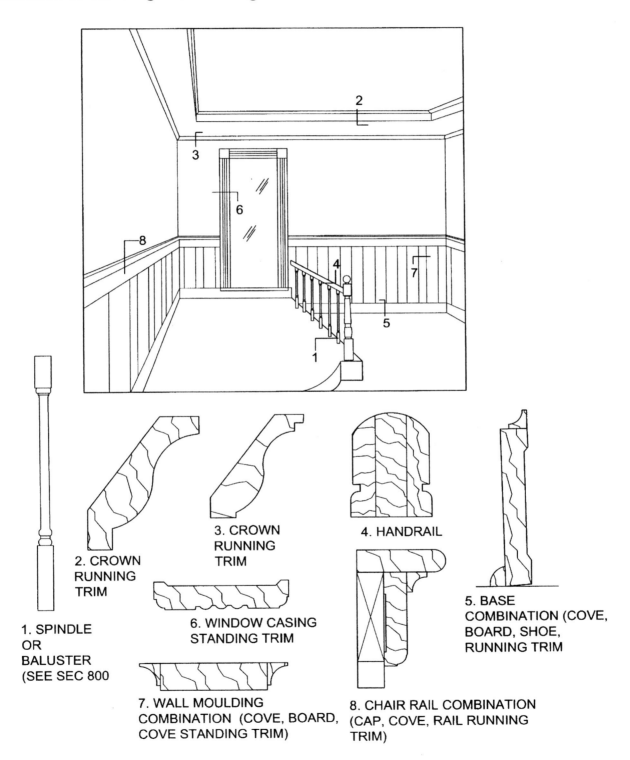

1. SPINDLE OR BALUSTER (SEE SEC 800
2. CROWN RUNNING TRIM
3. CROWN RUNNING TRIM
4. HANDRAIL
5. BASE COMBINATION (COVE, BOARD, SHOE, RUNNING TRIM)
6. WINDOW CASING STANDING TRIM
7. WALL MOULDING COMBINATION (COVE, BOARD, COVE STANDING TRIM)
8. CHAIR RAIL COMBINATION (CAP, COVE, RAIL RUNNING TRIM)

Figure 06200-2

OFFICE FLOOR PLAN

Part Two 06170 Prefabricated Structural Wood

"C"

BUILDING: SAMPLE
LOCATION:
ARCHITECT:
SUBJECT: EXAMPLE No. 1 – WOOD BASE

SHEET No.: 1 OF 2
ESTIMATOR: #0500056
DATE: MAY 2000

DESCRIPTION OF WORK		
PARAMETERS -		
CUSTOM GRADE		
3PC BASE		
PRE-FINISHED OAK		
5TH FLOOR		
QUANTITY	Q	
MANAGER	50 LF	
OFFICE	48 LF	
GENERAL OFFICE	100 LF	
	198 LF	
MANPOWER		TOTAL MH
UNLOADING MATERIAL		1 MH
SETUP & BRAKDOWN EQUIPMENT		1 MH
INSTALL - BOARD		9 MH
" - SHOE		4.5 MH
" - COVE		4.5 MH
TOUCHUP		2 MH
		22 MH
EQUIPMENT & FASTENERS		
COMPOUND MITER RENTAL		3 DAYS
NAILS		1 BOX
MISC. SMALL TOOLS		1 LS

} LABOR RATE BASED ON PREDRILLING OAL

Part Two 06170 Prefabricated Structural Wood

BUILDING: **SAMPLE**
LOCATION: _____
ARCHITECT: _____
SUBJECT: **EXAMPLE No. 1 – WOOD BASE**

SHEET No.: **2 OF 2**
ESTIMATOR: **#0500056**
DATE: **MAY 2000**

DESCRIPTION OF WORK	Q	UNIT $	T S
SUMMARY			
MANPOWER (CARPENTER)	22 MH	$40 / HR	880
COMPOUND MITER (RENTAL)	3 DAYS	$10 / DAY	30
NAILS	1 BOX	$2	2
MISC. SMALL TOOLS	1 LS	$50	50
			$ 962
OVERHEAD 5%			48
			$1,010
PROFIT 10%			101
TOTAL			$1,111

Standard Estimating Practice

"D"

BUILDING: __SAMPLE__
LOCATION: _____
ARCHITECT: _____
SUBJECT: __EXAMPLE No. 2 – CASE WORK__

SHEET No.: __1 OF 2__
ESTIMATOR: __#0500056__
DATE: __MAY 2000__

DESCRIPTION OF WORK				
PARAMETERS				
PLASTIC LAMINATE CABINETS				
P. L. COUNTER W/INTEGRAL BACKSPLASH				
QUANTITY				
BASE CABINETS		4 UNITS		
WALL CABINETS		4 UNITS		
COUNTERTOP		10 LF		
SINK CUT OUT		1		
MANPOWER				
UNLOAD CASEWORK TO 5TH FL	2 MEN		2 HRS =	4 MH
SETUP & BREAKDOWN EQUIP.				1 MH
INSTALL WALL CABINETS		4 UNITS	2 MEN X 1HR/UNIT	8 MH
" BASE CABINETS		4 UNITS	2 MEN X .75HR/UNIT	6 MH
" COUNTER TOP			2 MEN X 1 HR	2 MH
SINK CUT OUT				1 MH
				22 MH
EQUIPMENT & FASTENERS				1 LS
SMALL TOOLS				1 BOX
SCREWS				

06200 FINISH CARPENTRY

Finish Carpentry covers two types of work: (1) job built and installed carpentry work and (2) the installation of many different pre-manufactured products specified in different divisions and sections. 06400 Architectural Woodwork installation is typically included in this section. Both interior and exterior work are covered by this section and can be quoted labor only or both labor and materials. Since so many different sections are included, the quotation requires a more detailed description of your inclusions and exclusions. The scope may include 06220 Millwork, 06250 Pre-finished Paneling, 06260 Board Paneling, 06270 Closet and Utility Wood Shelving, 06400 Architectural Woodwork, 06500 Structural Plastics, 06600 Plastic Fabrications, 07400 Siding Panels, 08000 Doors and Windows, 10000 Specialties, 11000 Equipment, 12000 Furnishings.

Knowledge of wood products is necessary when estimating the cost of finish carpentry.

TYPES AND METHODS OF MEASUREMENT

Millwork includes all wood interior finish components of a building and is produced from much higher quality wood than that used for framing. Running and standing wood trims are measured in lineal feet and each corner or return, which requires a field cut, is counted individually. Figure 06200-1 shows examples of standing and running trim.

Cabinets can either be measured by the lineal foot of cabinet or by counting each individual cabinet.

Countertops, which normally sit on base cabinets, are measured in lineal feet. Countertops, supported by panels, are also taken off in lineal feet but also must have an additional takeoff of the supports, which carry an additional labor factor.

Paneling should be taken off by the piece, since it costs the same to install a 2 IN × 2 IN panel as it does a 2 IN × 6 IN panel.

The panel pieces are converted to square feet and compared to purchase sizes available of the product.

Closet shelving, utility shelving and coat rods are taken off by the lineal foot of running shelf.

Storage shelving is also taken off by the running foot of shelf and multiplied by the number of shelves.

It is necessary to quantify the related fasteners, adhesives, treatments, fittings or hardware required for product installation. The miscellaneous pre-manufactured products are individually counted, then details applied to the count regarding size, weight and components. When manually completing a quantity survey, using colored pencils or highlighters to mark items as you count them will keep you from counting them twice.

ORGANIZATION AND FACTORS EFFECTING TAKEOFF AND PRICING

Start with an outline of the items or work to be included. This is obtained from reviewing both the plans and specifications. In addition to the above listed sections, it is important to review sections for allowances and the instructions to bidders regarding any owner furnished and contractor installed items. List all items of work, obtain and set up columnar sheet or a spreadsheet to list material by room number.

Note the length, width, height, quantity and any special information. Obtain current component and installation instructions by calling or emailing the manufacturer; or search on line for the product information. Determine the substrate surface usually furnished by others upon which you will install the materials of your estimate. Different substrates or structures can require different installation procedures. Determine to what extent the substrate will be ready for the installation. Include any substrata preparation in the estimate. Determine if the materials furnished by others will be staged at the location of installation or include unloading from transportation at the street. Temporary storage may be required and include the associated costs, if necessary. Include clean up and disposal of package waste if it is to be included in your work. While establishing the estimate outline, list your estimate assumptions and clarifications; specifically any similar type work that you do not plan to include in your estimate and proposal. The following is an example of your outline headings:

Item #	Item Description	Room #
Elevation #	Qty	EA / LF / SF
Height	Width	Length
#Cut-outs	Comments	

The following is a short list of some associated work, but commonly excluded: furniture, fixtures, stone or tile countertops, steel, painting or staining, glass/glazing and sealant or caulking. The following is a short list of work that might be included in this type of estimate and typically would be quoted with both labor and materials included, for a complete proposal: wood trim, millwork, wood finish material, paneling, shelving and job built counters. Additional work items are specified in other sections. Typically contractors furnish both material and labor, this example includes installation labor only. Although the term labor only is used commonly, the proposal includes the necessary equipment, supervision and fastening materials. The following list of items are possible to have in this estimate: cabinets, cases, stairs, railings, wood veneer solid lumber panels, wood ornaments, simulated wood ornaments/trim, any wood trim, wood frames, screens, blinds, shutters, plastic lumber, cultured marble, glass fiber reinforced plastic, plastic laminates, plastic handrails, plastic paneling, solid surfacing, exterior siding, wall panels, walk doors, door hardware, accordion doors, access doors, folding doors, grilles, double acting doors, wood framed windows, chalk boards, marker boards, tack boards, toilet partitions, toilet / bath / laundry accessories, access flooring, lockers, fire extinguisher cabinets/brackets, telephone enclosures, coat hooks, projection screens, laboratory furniture/casework, dock bumpers, library shelving, retractable stairs, unit kitchens, and any other item that doesn't fall clearly in another common trade discipline.

OVERVIEW OF LABOR, MATERIAL, EQUIPMENT, INDIRECT COSTS, AND APPROACH TO MARKUPS

The following examples will show how to properly takeoff, quantify and price various types of finish carpentry installation. Renovations to an existing facility requires a site visit to determine access to the work area and existing substrates and conditions. The takeoff sheet should be organized in such a way so that each area of work can be identified. Quantity takeoffs organized by room number or elevation designation is the preferred method. First, identify the quality level of material and workmanship that is specified for the particular project.

Quality Standards The Architectural Woodwork Institute (AWI), Section 1700-G-2 defines three grades of woodwork and the estimator should be familiar with each as defined below:

Premium Grade - The grade specified when the highest degree of control over the quality of workmanship, material, installation and execution of the design intent is required. This grade is usually reserved for special projects, or feature areas within a project.

Custom Grade - The Grade specified for most conventional architectural woodwork. This grade provides a well defined degree of control over the quality of workmanship, materials and installation of a project. The vast majority of all work produced is Custom Grade.

Economy Grade - The grade, which defines the minimum expectation of quality, workmanship, materials, and installation within the scope of AWI standards.

The Architectural Woodworking Institute, Section 1700-G-5 also offers the following installation recommendations:

The methods and skill involved in the installation of woodwork in large measure determine the final appearance of the project. Architectural woodwork should be allowed to come to equilibrium on site prior to installation. Factory finished woodwork may require a week or more to acclimatize. The design, detailing and fabrication should be directed toward achieving installation with a minimum of exposed face fastening. The use of interlocking wood cleats or metal hanging clips combined with accurate furring and shimming will accomplish this. Such hanging of woodwork has the additional advantage of permitting movement that results from humidity changes or building movement. AWI guidelines are commonly referenced in specifications and this knowledge is particularly important to the estimator. It is important to understand the full scope of the project by reviewing the drawings, access to the work area, phasing and scheduling. All of these issues play an important role in determining installation productivity.

Estimate Example 06200-1 Installation of Wood Base

This simple example will illustrate the procedure for quantifying and pricing the installation of wood base. See Figure 06200-2 for floor plan layout. The project specification indicates a custom level of quality, which is most common in commercial type construction. Determine the type of wood base, such as single or multiple applications, wood species and the substrate to which the base will be attached. Then determine the proper fastening method and the quality level desired.

The wood base detail indicates a three-piece base consisting of a cove piece, a flat board and a shoe. See Figure 06200-3 for details. The installation of a three-piece base will require three individual installations. The most costly application is the flat board piece. Once laid out and installed the cove and shoe pieces can be easily attached and usually do not carry as high a labor factor. The wood species indicated in the project specifications is a pre-finished oak. Because oak is a hard wood and should pre-drilled before nailing or screwing to eliminate the potential for splitting, productivity should be adjusted accordingly.

The project conditions indicate that the work is to take place in an existing office space, which is being renovated and upgraded. Since the existing space is climate controlled, maintaining, the relative humidity and moisture content should not be a concern but must still be verified prior to installation. Since the work is on the fifth floor of an office building, unloading the finish material, as well as tools and equipment, and loading it into the work area will have to be factored into the price for installation. This can be accomplished in several ways. The most accurate would be to determine the quantity of material being installed and calculate the amount of labor that would be required to load this material. A major factor that would influence this cost is the availability of an elevator. This would obviously be quicker than having to walk the material up five flights of stairs. Another less accurate but widely used method would be to add a percentage of the total labor cost to the bottom line.

Our next major concern is the substrate to which the new wood base will be applied. Some of the walls are new construction made up of metal studs and drywall. The project details indicate that wood blocking be installed in between the metal studs at all locations where new wood base is being installed. The finish carpentry contractor should coordinate with the general contractor and/or other trade contractors to insure the wood blocking is installed prior to the gypsum wallboard. Since wood blocking is being installed as a substrate we will be able to drill and nail the three-piece base into place at all new walls. Sometimes it is necessary to glue, in lieu of nailing, the wood base if the proper substrate has not been installed in order to provide a secure installation.

The next step would be to perform the actual quantity survey. The first thing we want to do is establish the new walls and the existing walls. Often this is easier if the designer has shaded the different wall types. Once we have established new and existing walls, perform a lineal foot takeoff. Next count the number of cuts, meaning inside and outside corners as well as doorjambs. The number of cuts and miters will significantly affect the productivity and the cost of installation.

The equipment used for this type of installation usually consists of a compound chop/miter saw, drill and hammer.

Once the installation is complete it is the responsibility of the installer to patch the nail or screw holes and provide any miscellaneous touchup to pre-finished wood. The material supplier should provide the installer with touchup materials such as stains. Some installations require that edges of standing and running trim be caulked with a clear sealant to provide a tight, clean joint. This is usually a requirement in hospitals and in premium and custom quality installations.

The methodology described in the first example can be applied to virtually any standing and running trim installation.

Estimate Example 06200-2 Installation of Manufactured Casework

This example illustrates the procedure for quantifying and pricing the installation of finished casework.

First determine the type of casework, plastic laminate or pre-finished wood and type of countertop such as plastic laminate, polymer solid surface, stone or stained wood. For this example the project specifications indicate a custom level of quality, which is most common in commercial type construction. Figure 06200-3 illustrates the installation of a plastic laminate countertop with wall and base cabinets in a lunchroom.

There are several methods for quantifying and pricing manufactured casework. The most accurate involves counting each individual piece of cabinetry, as the takeoff illustrates. This would require the estimator to determine the quantity of wall and base cabinets individually. Wall cabinets usually require two workers to install where as one person could install base cabinets. Two workers are usually required for loading and unloading of the cabinets.

Another method of quantifying casework is to measure the total linear footage. Only those firms with accurate historical cost should use this pricing technique.

Base cabinets almost always have countertops. Countertops are taken off and the installation is priced by the lineal foot. Determine how the back splash is attached if it is required. Determine whether the top requires a sink cut out or whether it will be delivered with cut out already made. Most of the time countertops will come precut to length and width, however field adjustments may be necessary. Cutting a countertop to fit existing conditions is known as scribing. In this example the countertop is a one-piece unit with an integral back splash and the sink cut out was made in the shop.

The lunchroom is on the fifth floor and will require the use of an elevator in order to load the units onto the floor. On projects where an elevator is not available it is sometimes necessary to bring in a crane and remove a window in order to load the material. It is important to know where the project is and what kind of access there is to the site. Determining the substrate to which the cabinets are attached is very important. The general contractor should provide wood blocking or kitchen metal, (16ga sheet metal), attached to the studs. If the wall exists it should be stripped out with wood furring run horizontally at the top and bottom of the cabinet on which to fasten the cabinet. The methodology used in this example can be used to install most types of casework.

SPECIAL RISK CONSIDERATIONS

The cost of finish carpentry is affected by the project schedule, access to the work area, climate control and acclimatization of material, and the availability of onsite storage of material.

06410 – CUSTOM CABINETS

Introduction

This section addresses the installation only of custom fabricated cabinets items. As more fully described below, there are groups of items having similar installation procedures. This is a subsection of 06400 Architectural Woodwork, which is not describe fully because the previous section 06200 Finish Carpentry includes estimating this work. This section further clarifies custom cabinets.

This work is closely related to, and often done by same contractor, the following sections:

06600 Solid surfacing.

12300 Manufactured casework

This section generally includes the following cabinets:

- ✓ Wood
- ✓ Restaurant
- ✓ Medical and laboratory
- ✓ Laboratory
- ✓ Optical
- ✓ Ecclesiastical
- ✓ Custom fixtures
- ✓ Library
- ✓ Educational
- ✓ Hospital
- ✓ Veterinary
- ✓ Display
- ✓ Bank fixtures
- ✓ Dormitory
- ✓ Dental
- ✓ Hotel and motel
- ✓ Residential
- ✓ Pharmacy

The following items are likely to be found in the above list.

- ✓ Cupboards — Cabinet Hardware
- ✓ Cabinets and vanities — Laminate clad surfaces
- ✓ Counters/cases —Opaque finish Wood Cabinets

Installation of fabricated cabinets can be accomplished under a number of different arrangements as follows:

- ✓ A general contractor's own forces with the material purchased from a supplier
- ✓ A subcontractor who both manufacturers and installs the finished product for a general contractor

06410 Custom Cabinets

Types of Measurement

Usually, measure items in this particular section as follows:

✓ Total lineal length of fixtures for the entire project by category

Description	Unit	Quantity
Low level cupboards	LF	
Counter tops	LF	
High level cabinets	LF	
Vanity counter	LF	
Vanity cupboard	LF	

✓ Enumerated - EA (Each), NO (Number of), or AVG (Average)

Description	Unit	Quantity
Low level c'b'ds avg 4'0" long	EA	
Counter tops avg 5'0" long	EA	
High level cabinet avg 5'0" long	EA	
Vanity counter avg 5'0" long	EA	
Vanity cupboard avg 4'0"	EA	

Determine the choice of measurement by the size and type of project. For large projects, it is usually easier and faster to measure the items in lineal measure. For smaller projects with one only type items it is usually more practical to measure these by number or each.

Factors Affecting Takeoff and Pricing

Custom Cabinets items are normally indicated on both floor plans and interior elevation drawings. Consider the following when taking off quantities for a project:

On large projects when measuring items linearly it is useful to use a wide columnar pad. Write the headings of the different items at the top of each column. Then start on a room-by-room basis following the room schedule. Schedules usually go from the basement to the uppermost floor. This is an orderly procedure that provides a good audit trail. Watch for apartment blocks where there is one floor description typical for additional floors. Penthouse suites are usually different from the typical floor. It is always wise to double check each floor as a precautionary measure.

It is good practice to state the room number along with the relevant elevation number (i.e., Room 2012 plan drawing A3 and north elevation Room 2102 drawing A54 detail 1/3/54).

Estimators should always make a habit of cross referencing the plan drawings with the elevation drawings. Occasionally draftsmen forget to draw both parts. This leaves a floor plan with no corresponding elevation or an elevation shown with no corresponding floor plan. When estimators refer only to floor plans or elevations, it is easy to be short on the quantity survey.

Check both the plans and elevations to be sure the survey is complete. Bring discrepancies to the attention of the architect for correction.

These problems usually compound on larger projects. As an example, use a hospital with many patient rooms in different wings of the building. There is a typical detail describing the rooms to which the detail applies. Sometimes there is omission of an entire wing, so estimators must be alert.

Does the project require any unusual or unique pieces of millwork? There is more expense involved when the project includes plumbing fixtures mounted in casework because of requirements for notches, cut outs or holes for piping.

A similar problem arises with computer millwork and its requirements for multiple cables and plugs. These items may have to come on site in a knock down condition because of transportation problems. This will entail more field assembly.

If the project is phased, segment the quantity survey by phase to determine how this will affect the overall project. Determine the millwork installation concentration over the project duration. Estimators must determine this schedule to know how to estimate the work.

Building configuration can increase installation costs. It costs more to install fixtures on the sixteenth floor of a building than it will to install them on the ground floor.

Occasionally some projects may have all the custom casework built in place which requires a considerable amount of site labor.

Consider the custom cabinet supplier's time frame for making the products. When there is a long lead time for special material, there could be an impact on the project schedule. One method to solve this production time schedule is to increase the project time. There may be additional costs if there is a completion deadline with installation by multiple shifts or by working weekends.

Sometimes a general contractor will not take off the custom cabinets because they expect to receive a furnished supply and installed price from a vendor. Generally, it is good practice to take off these items in case no bids are received, to quantify rough blocking and for off loading and placing the units. Handling of units require care to prevent damage and costly replacements.

Productivity can be effected by availability of skilled tradesmen, project complexity, and if it is a retrofit project within an occupied facility.

Special Risk Considerations

These considerations may include remote locations where delivery of manufactured items is in a knockdown form for ease of handling and shipping. This usually creates a cost saving in transportation charges. However, this is offset in part by the extra cost of assembly at the destination point.

There may be a requirement for a larger than normal amount of field installation work such as applying plastic laminate to counter tops on the job site as opposed to shop applied. There may be an unusual amount of scribing and fitting units to irregular or uneven wall finishes. This occurs in old buildings that may have plaster wall finishes. Quite often these older buildings have a lot of walls out of square and installation becomes much more time consuming.

Interfacing with other trades on the site may create some difficulty when installing woodwork. This is particularly true with mechanical and electrical trades.

Phased projects may create problems when trying to optimize crews.

One problem that can occur on a site is the provision for adequate ventilation when applying contact adhesive on plastic laminates. When there are applications of large areas of plastic laminate, there can be a strong build up of vapors. This can cause respiratory and health problems for installers. There is also a risk of fire or explosion if sufficient ventilation is not in place.

Post Project Analysis and Recording of Historical Data

A system for recording and analyzing past projects is necessary for addressing weaknesses in estimating and creating a database for future projects which should address labor productivity, location, complexity, type of project, and other information deemed valuable. To avoid duplication, refer to other related sections.

07100 – DAMPPROOFING AND WATERPROOFING

Introduction

These guidelines and procedures provide the estimator with methods that will result in a complete and accurate survey of job costs. This will help the company's managers to bid and manage a project to its successful and profitable completion.

Designers, lenders and general contractors usually do not have the means to allow for waste and degrees of difficulty when preparing budgets and preliminary estimates. The guides and procedures for waterproofing estimators are for use on specific projects where plans and specifications are complete and available to the estimator as a package. They include: general and supplementary conditions of the contract, soil reports, civil, architectural, structural, and mechanical plans.

The key word in the definition of the estimate is reliable. To be reliable, the estimate must be complete and accurate. Every item and condition in the plans and specifications that affects the waterproofing appears as a cost in the estimate summary.

The estimate should reflect the impact of allowances, alternates, addenda, and the project schedule.

Quantity Survey

The specifications for waterproofing should be consistent with other divisions of the work. The work is affected by, but not limited to, the following items:

- ✓ Soil reports
- ✓ Concrete mix and curing time
- ✓ Excavation and backfill procedures
- ✓ Location/types of control and expansion joints

Identify, locate, and include the specifications for these items in the survey.

Architectural drawings may not include sprinkler systems, drainage systems, plumbing and mechanical projections. Get a complete set of drawings for the project. Make a complete list of all items furnished and installed by other trades that rest on or pass through the waterproofing membranes. Include the list of all drawings in the bid package and the dates of the drawings with the latest revision numbers, if any. For change orders or disputes concerning scope, these notations will be critical.

The difficulty in estimating waterproofing is very often defining the scope of the work. Draw the foundation plan with all elevator and special equipment pits. Include the site retaining walls whether or not requiring waterproofing because a clarification may expand the scope of the work.

The drawing should also show the length and height of all walls and the width of the footings. Note the type and location of all joints. The estimate must indicate who provides the joint sealant. Refer to the sealants, elastomeric coatings, and tile flooring sections to determine the type and extent of this work. The drawing should also include floor and room numbers as well as room names. Determine the responsibility for the final cleaning, finishing, and preparation of substrates.

Be sure that all project specifications and details are consistent. Also determine whether or not they agree with the manufacturer's requirements for warranty.

Materials

The actual coverage rates for waterproofing materials may vary the manufacturer's published data depending on a variety of conditions. Accurate records of previous projects can be used to establish the material requirements for future projects.

Labor costs for waterproofing can be job and system specific. The estimate should include accessory items, site equipment, and safety requirements.

07190 – WATER REPELLENTS

Introduction

Carefully review plans and specifications to insure accuracy and completeness of the estimate. Read the specification and note type of water repellent, length of warranty, and approved manufacturers. Cross-reference specifications of substrates to insure that no unusual conditions exist which may increase material or labor costs. Raked joints on CMU or sandblasted (architectural) concrete are examples. Review plans and classify treated surfaces by manufacturer's recommended rate of application necessary to insure the compliance with warranty requirements. Estimates for water repellent application should not be based solely on other quantity surveys such as masonry or concrete takeoffs. Measurement of these surfaces is usually in square feet. Missing from these surveys are surface areas such as exposed parapets, wraps, and concrete stems. Such "estimates" are only rough approximations. Due to the high cost of some water repellents, this may result in a serious shortfall. Manufacturer's recommendations for material application rates may vary due to porosity and profile of the substrate.

Quantity Survey

Labor production is proportionate with increased or decreased material application rates. Therefore, the quantity survey should be grouped into varying substrates that require water repellents. Itemize any specific variation of a substrate which significantly affects material consumption and labor production rates. For example, split-face block, wire-cut brick, and architectural concrete require increased material and labor. This is due to porosity and profile compared to common CMU, brick and concrete. When estimating water-repellent stain, group substrates according to both color, and opacity (semi-transparent or opaque). Determine lineal dimensions of exterior wall surfaces from dimensional floor plans. Determine heights by elevation and section. Add for extra surface created by offsets of rowlocks or other coursing to the elevation height. Heights also should include the height of any exposed stem wall added to the finished floor height dimension. Survey additional surface area created by popouts, indentations, and wraps.

Determine all surface areas for clear water repellents in SF. When estimating water repellent stains, the estimator must determine LF of any color banding or color separation, to determine additional labor difficulty.

Complex structures usually require an elevation-by-elevation takeoff. Define a starting point and proceed in a pre-determined direction away from and back to the starting point until each wall survey is complete. Take special care to account for hidden elevations by carefully cross-referencing elevations with floor plans. Define true height of building walls partially hidden by other walls. The estimator should survey all parapets and fire walls not covered by other materials such as roofing or waterproofing.

In most instances, do not deduct door and window openings from gross square footage. Deduct openings only when there are enough in size and count to affect the estimate. Typically do not deduct openings less than 100 SF in size from gross surface area. The basis of this standard is:

- ✓ labor production seldom diminishes and usually increases because of an opening (due to protective requirements).
- ✓ not deducting openings usually accounts for material waste due to method of spray application.

07190 Water Repellents

- ✓ the material cost saved may offset some of the additional cost of protective materials.
- ✓ the additional accuracy is not worth the necessary time needed to get it.

However, when deducting a large opening, add the wrap, before deducting the opening. Although openings may be deducted, labor is still determined by gross square footage with an additional percentage on material to account for waste.

Multipliers for Developed Surface Area of Concrete Masonry Units

Accuracy is always preferable to approximation. Developing surface areas for water repellents cannot always be simple methods of multiplying height times length of wall. In certain instances, development of multipliers for additional surface area is necessary to account for the third dimension of profile. Manufacturers usually recommend specific application rates based on judgment and experience. Material application rates for fluted masonry are usually twice that of common masonry since the surface area is usually double.

Whenever possible, determine the additional area created by flutes per block and add to the modular block size. Divide this sum by the modular block size to determine a multiplier. Raked joints add considerable surface area due to additional joint depth. Measure the nominal perimeter of the block module in inches and multiply by the depth of the rake (if depth of rake is not known, use 3/8 IN as a nominal dimension). Add this to the surface of the block. Divide this sum by the surface area of the block to determine the multiplier.

Example: Determine the actual surface area of an 8 IN × 16 IN CMU with a 3/8 IN raked joint in a wall 15 feet high and 100 feet long. Flat Surface Area - 15 FT × 100 FT = 1,500 SF. Multiplier - [(8 IN × 16 IN) + (3/8 IN × 48 IN)] (8 IN × 16 IN) = 1.14. Developed Surface Area = 1,710 SF.

Figure multipliers for sandblasted joints similarly, adjusting depth of joint according to degree of blast.

For 8 IN × 16 IN block with nominal 3/8 IN raked joints, use these multipliers to determine developed surface area:

Standard -	Multiply flat surface area by 1.14
Single-Score -	Multiply flat surface area by 1.19
Double-Score -	Multiply flat surface area by 1.25
Triple Score -	Multiply flat surface area by 1.28

The multiplier for 8 IN × 8 IN block with 3/8 IN raked joints is the same as single score 8 IN × 16 IN CMU.

The multiplier for 4 IN × 16 IN slumpblock, adobe, or other similar units with 3/8 IN raked joints is 1.07.

Multipliers for Developed Surface Area of Brick Masonry

Multipliers for Raked Joints for Common Brick (2 3/8 IN × 8 IN):

1/4 IN rake -	Flat surface area times 1.27
3/8 IN rake -	Flat surface area times 1.41

Since brick size varies widely by manufacturer and geographic location, figure other multipliers for brick masonry similar to the CMU example.

Multipliers for Formed Concrete Walls

Use judgment and experience when developing surface area of sandblasted, exposed aggregate, roped, fractured or grained effect concrete. For ripple-rib forms (semi-circular), multiply surface area times 1.57 to determine developed surface area. For triangular forms with right-angles, multiply by 1.41 for developed surface area. For triangular forms composed of equilateral triangles, multiple by 2 to determine developed surface area. Trapezoidal forms vary greatly in depth and on-center dimension (OCD). For practical approximation of a multiplier for developed surface area, add twice the depth to the OCD and divide the sum by the QCD to determine the multiplier. The more accurate method is to determine the actual additional surface area added by trapezoidal forming by viewing the forms in profile.

The length of the form before corrugation is divided by length of the form after corrugation usually between on-center dimensions. The following multipliers are based on Burke, Greenstreak and Fitzgerald trapezoidal formliners. These multipliers are less than the approximation method since the sloped sides actually decrease surface area by comparison with such approximations.

OCD	Depth	Multiplier	OCD	Depth	Multiplier
½"	¼"	1.78	2"	1"	1.78
5/8"	¼"	1.80	2"	1 ¼"	1.93
5/8"	½"	2.25	2"	1 ½"	2.08
1"	½"	1.87	3"	1 ½"	1.72
1"	5/8"	2.13	4"	¾"	1.27
1 ¼"	½"	1.48	4"	1 ½"	1.54
1 ½"	½"	1.41	4"	2"	1.78
1 ½"	¾"	1.72	6"	¾"	1.18
2"	½"	1.39	6	1 ½"	1.36
2"	¾"	1.54	8"	¼"	1.11

Note: The above multipliers are also useful in determining surface area for concrete release agents, retarders, sandblasting, and painting.

Extension of Quantities

After completing the quantity survey, the estimator is ready to extend quantities. Determine additional surface area added by a third dimension of profile and use an appropriate multiplier to determine actual surface area.

The estimator, or another person, should cross check all quantities. The estimate detail sheet should list, from left to right, the following columns: SURFACE, SF, SF/HR, HRS, SF/GAL, GALLONS. Add columns for SF, HRS, and GALLONS to determine total square feet, total labor hours, and total gallons required for the job.

All material application rates should be according to manufacturer's recommendations to insure compliance with warranty requirements.

For multistory buildings, include the additional costs for access, labor productivity and equipment. Adjust labor production rates. Account for the additional cost of access equipment as a specific item.

07240 EXTERIOR INSULATION AND FINISH SYSTEMS

Exterior Insulation and Finish Systems; commonly known as EIFS are classified in CSI Division 7. It is further defined in Level Two Classification of 07200, Thermal Protection; and in Level Three Classification as 07240, EIFS.

EIFS is a multi-layered system comprised of insulation board, adhesive/base coat, reinforcing mesh, and finish coat. It can be panelized or installed in place, over a variety of substrates including gypsum sheathing, cement board, concrete, plywood, or unit masonry. The components can be combined in many different ways and the insulation board fabricated in an endless number of ways to accomplish a wide variety of Architectural needs and desires. The Estimator should be careful to review construction documents to assure special shapes, thickness, or specification options are included as required.

Types and Methods of Measurement:

EIFS is typically installed on flat wall surfaces with reveals or build outs around windows, doors or other features. Reveals and build outs can also be used to add accent colors or shapes to the features of the exterior of the building. The flat surfaces will be measured in square foot of surface. Window details will typically be measured by lineal foot; as will reveals. Build outs or areas of thicker insulation will be measured in square feet or lineal feet depending on the shape designed. There may also be special shapes, such as simulated keystones, that would be quantified by each required. EIFS is very flexible in that it can take on almost any shape. Therefore the estimator must also know how to calculate the surface area of convex and concave shapes by utilizing circumference formulas.

The take off quantities must be converted to the purchase size of each product in the system.

Insulation Board is sold in 2'×4 FT sheets and is available in varying thicknesses. The square footage of each thickness of material must be divided by 8 square feet to achieve the number of insulation boards required for the project, plus a waste factor. With experience the estimator will be able to judge from the size and shape of the building what would be a proper waste factor to be applied when pricing the insulation board.

The adhesive/base coat is polymer based product packaged in its dry form and sold by the bag. When mixed with water in the proper ratio each bag will provide enough material to cover a certain square footage of surface area at the required thickness. The manufacturer's literature must be referred to for the exact factor to use. Dividing the square footage of surface area to be covered by the coverage factor for each bag will result in the number of bags needs.

Reinforcing mesh is typically supplied in rolls that cover a set area. The estimator must be careful to research the mesh requirements for the project very carefully. There are various mesh products that may be specified for different areas such as high impact mesh around areas of high traffic. Or there may be a requirement for two layers of mesh. The estimator must be aware of the possible scenarios and identify the square footage of each option specified. Once the area of each is calculated simply reply divide the square footage of surface area by the square footage per roll to determine the number of rolls required, plus a waste factor.

The finish coat is available in an array of colors and textures. The estimator must identify each in the design and calculate the square foot area of each specified option. The finish materials are typically provided in factory mixed containers or dry and in bags ready for mixing at the point of application. Consult the manufacturer's literature for the exact coverage per bags. Utilize that factor to determine the number of bags required.

Special shapes and reveals will be measured by lineal feet and an amount for each product required for these architectural features must be determined by the estimator.

ORGANIZATION / FACTORS AFFECTING THE ESTIMATE:

The estimator must identify each product combination incorporated in the design and prepare a quantity survey to properly account for each. In addition to the different combinations of products the estimator must also identify different areas in the building that should be analyzed on its own. The cost of field applied EIFS may be different for a first floor area than a third floor area, or a fascia, or back of parapet. The following is a brief sample of organizing a project takeoff:

Walls 1st Floor	_____	SF
Walls 2nd Floor	_____	SF
Columns	_____	SF
Parapet rear side	_____	SF
Fascias	_____	SF
Soffits	_____	SF
Reveals	_____	LF
Window Frames	_____	LF
Control Joints	_____	LF
Caulking	_____	LF

The estimate must be organized to easily identify the items that will be priced differently. The height of walls will determine the amount of scaffold or machines required to reach all the surfaces. Applying the system from the ground is easier and less costly than if the material is placed on the third level of scaffold and the applicator is working around scaffold frames and has limited mobility. Therefore work preformed from the ground should be separated from areas to be worked from scaffold or machines. The backs of parapets may be worked from a roof area, which may allow for mixing of product at the work area, but you still have to factor getting the men and material to the location.

Soffits and Fascia tend to be areas that are worked from scaffold or lifts; the production rate for applying product to those areas will be different than that of a wall surface. Through tracking of actual field productions a company can determine the production rate for the different areas that may be are encountered.

The finish chosen by the designer may affect the cost of work. Different textures may require different production rates. Also some finishes require spray and trowel application versus straight trowel applications. The number of color changes in the finish system may add cost.

Overview of Labor, Material, and Equipment:

Labor:

EIFS typically falls under the jurisdiction of the Plasterers. Crew sizes and ratios of plasterers to laborers may vary depending on the region the work is in. The manufacturers of EIFS Systems require applicators to be certified in order for warranties to honored.

Material:

1. Insulation Board is available in 2×4 sheets of various thicknesses of polystyrene or polyisocyanurate foam. Insulation board is either mechanically fastened utilizing screws and special washers; or it is applied with adhesive. The construction documents must be thoroughly reviewed and understood to determine the method of application.

2. Special Foam Shapes may be bought or fabricated by the applicator. The insulation board is very easily out and shaped as desired.

3. Adhesive/Base Coat is a polymer-based product mixed with water to achieve the desired consistency. It can be used to adhere the insulation board to the substrate. It is used to hold the mesh in place. It also serves to build up the product thickness to achieve the desired protection.

4. Reinforcing Mesh is typically an open-weave glass fiber mesh available in a variety of types to meet different impact resistance requirements at different location of the project.

5. Finish Coat is a polymer based product provided in a factory mixed or dry form mixed with water to achieve the required consistent. It is available in a variety of textures and colors that are typically trowel applied to the surface.

Equipment:

1. Drill and paddle are used for mixing of most base and adhesive coats. The finish coats that are factory prepared must be mixed using a drill and paddle on the site.

2. Hand Trowels are used for applying the base and finish coats. Smooth trowels and notched trowels are used in differing applications. Consult the manufacturer's literature for the proper trowel.

3. Hot wire foam cutting knifes are used to cut the insulation board to size and to create reveals in the surface.

4. Scaffold and lifts may be required to allow access to the work.

Sample Estimate

07240 Exterior Insulation and Finish Systems — Part Two

The sample estimate will be for one bay of a commercial building with fixed aluminum windows. There will be architectural features around each window and a cornice at the top of the parapet wall. The system specified will be for adhesive applied 1 IN foam, one base coat with standard mesh, and a finish coat. The architectural features around the windows will be 1 IN × 6 IN protrusions beyond the face of the finished wall. The cornice will be 12 IN high with a 6 IN 'step FT in the face. See the section A-A and elevation 1 in exhibit 07240-1 for more details.

1. The following data shall be used to determine the costs for this sample estimate.

   a. Labor rate (unloaded) = $20.00 / hr
   b. Insulation Board 1 IN thick = $0.50 / sf
   c. 2 IN × 6 IN insulation Board = $0.65 / lf
   d. 2 IN × 12 IN Insulation Board = $1.20 / sf
   e. Reinforcing Mesh Material = $0.18 / sf
   f. Base coat material = $0.25 / sf
   g. Finish Coat Material = $0.35 / sf

2. The following production rates shall apply to the sample contractor.

   a. Insulation board is installed at a rate of 600 square feet per day per worker.
   b. 2"×6 IN and 2"×12 IN Insulation Boards are installed at a rate of 200 lineal feet per day per worker.
   c. Base coat with mesh on flats is completed at a rate of 400 square feet per day per worker.
   d. Base coat and mesh is completed at a rate of 200 lineal feet per day per worker on window wraps and cornice work.
   e. The Finish Coat is completed at a rate of 200 square feet per day per worker.
   f. Finish coat is completed at a rate of 100 lineal feet per day per worker on window wraps and cornice work.
   g. Production rates take into account support labor and working up to 40 feet above grade.

3. All equipment including, scaffold, mixers, tarps, etc. are owned by the installing contractor.

Part Two — 07240 Exterior Insulation and Finish Systems

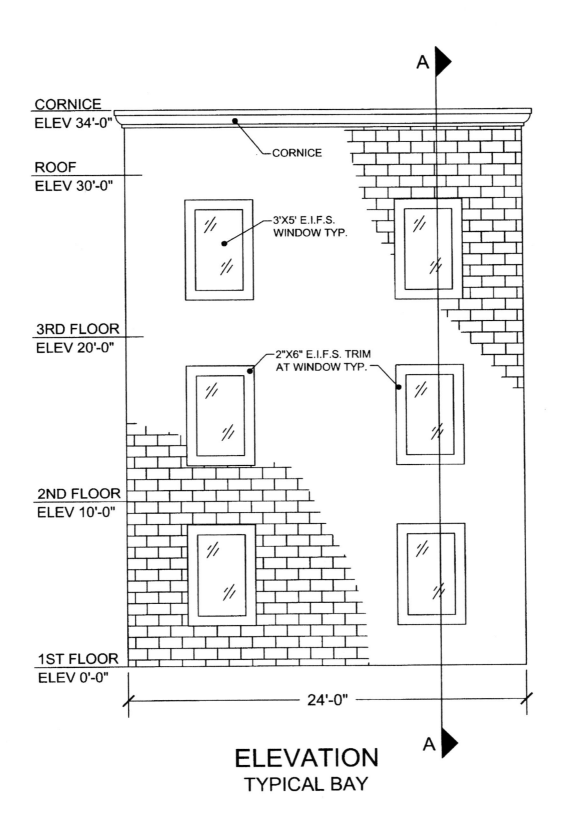

ELEVATION
TYPICAL BAY

Standard Estimating Practice

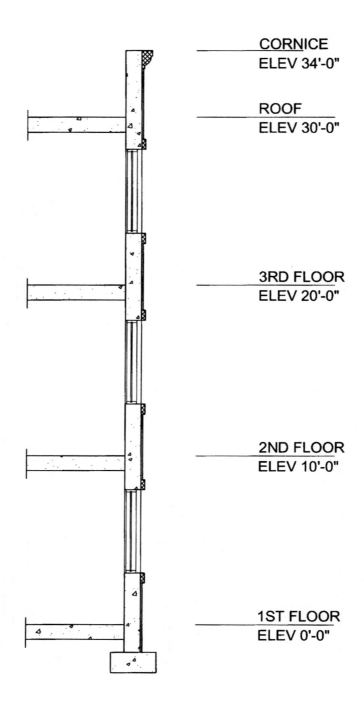

Part Two — 07240 Exterior Insulation and Finish Systems

Estimate Detail

Company Name: American Society of Professional Estimators
Project: Standard Estimating Practice, Sixth Edition
Address: Wheaton, MD 20902
System/Work Area: Standards Committee
CSI Section: 03470
Estimate #: 6th Edition
Bid Date: July 2004

Estimator: Standard Board — Date: January 2003
Pricing: Standard Board — Date: January 2003
Extended: Standard Board — Date: January 2003
Summary: Standard Board — Date: January 2003
Verified: Standard Board

Item or Description			Material				Labor		Equipment		Sub		Total
DESCRIPTION	QUANTITY	UNIT	UNIT $	COST	HR/UNIT	HOURS	UNIT $	COST	UNIT $	COST	UNIT $	COST	COST
Insulation Board				-				-		-		-	-
Flats – 1" Insulation Board	816	SF	0.50	408	0.01	11	0.27	218					626
Window 'Trim' Boards 2" x 6" wraps	96	LF	0.65	62	0.04	4	0.80	77					139
Cornice 2" x 12"	24	LF	1.20	29	0.04	1	0.80	19					48
Cornice 2" x 6"	24	LF	0.65	16	0.04	1	0.80	19					35
Base Coat				-				-		-		-	-
Flats	816	SF	0.25	204	0.02	16	0.40	326					530
Window wraprs	96	LF	0.25	24	0.04	4	0.80	77					101
Cornice total	24	LF	0.25	6	0.04	1	0.80	19					25
Mesh	816	SF	0.18	147		0		-					147
Finish Coat				-				-		-		-	-
Flats	816	SF	0.35	286	0.04	33	0.80	6531					938
Window wraprs	96	LF	0.35	34	0.08	8	1.60	154					187
Cornic total	24	LF	0.35	8	0.08	2	1.60	38					47
				-		0		-		-		-	-
Scaffold	3	tiers		-	6.00	16	120.00	360	180.00	540		-	900
				-		0						-	-
Subtotals				1,223		96		1,960		540		-	3,723
Labor Burden on labor	50%							960					960
Small Tools	5%							96					96
Total Direct Cost								3,038		540		-	4,801
Overhead	10%												480
Profit	5%												264
Insurance & Bond	2%												111
Sales Tax on Material	6%												73
Total Subcontractor Cost													**$5,730**

Notes: Labor Rate Unburdened $20 per hour

Standard Estimating Practice — 375

07500 MEMBRANE ROOFING

Introduction

Membrane roofing includes roofing assemblies applied to the structural substrate, over insulation, or protected with insulation, (protected membrane). Roofs can be covered with many different materials including, but not limited to the following: built-up bituminous roofing; cold-applied bituminous roofing; elastomeric membrane roofing; thermoplastic membrane roofing; modified bituminous membrane roofing; fluid-applied roofing; foam roofing; and roll roofing. Roofing can be organized into two groups: steep roofs and low-slope roofs. Typically membrane roofing is used on low-slope roofs, but with special construction most types of membrane roofing can be applied to steep roofs also. Knowledge and experience in the specific roof systems being estimated are important for a complete and accurate estimate.

The various components of a roof system include: the *deck* which is the structural surface that supports the roof; the *thermal insulation* which slows the passage of heat into and out of the building; a *vapor barrier* that is sometimes used in colder climates and on buildings with high interior humidity to prevent moisture from accumulating within the insulation; the *membrane* which keeps water out of the building; *drainage* components to remove the water that runs off the membrane.

Other specification divisions, which may affect the roofing are, but are not limited to:

- ✓ Division 2 Site Construction — Requirements may affect material storage and building access.

- ✓ Division 3 Concrete — The type of concrete roof decks (structural vs. lightweight) may determine the application method of the new roof system.

- ✓ Division 5 Metals — The type, gauge, configuration of the metal roof deck may determine the minimum thickness and attachment methods for the roof insulation. Acoustical roof deck usually requires the installation of insulation in the flutes of the metal roof deck and the roofing contractor may be required to complete the installation.

- ✓ Division 6 Wood and Plastics — Wood nailers may be required as a part of the roof system and may be specified in this section.

- ✓ Division 07600 Flashing and Sheet Metal — Often the sheet metal flashings have to be covered by the roof warranty. Configuration of the metal flashings and how they are installed in the roof system affects the watertightness of the system.

- ✓ Division 13 Special Construction — Removal of any existing roofing may involve asbestos containing roofing material (ACRM) and would be covered by this section.

- ✓ Division 15 Mechanical — The design of the roof top equipment (welded watertight curbs, factory insulation, etc.) may affect the roofing estimate.

Determining Material Costs

Even though the type of roof system varies, the same information is required to prepare a list of material is:

1. Approved manufacturers for the membrane, roof insulation, system components and accessories

2. Specific roof type (built-up roofing, elastomeric, modified bitumen, etc.)

3. Specific type of roof membrane (.060 IN non-reinforced EPDM vs. .045 IN reinforced EPDM.)

4. Manufacturer's specification number or system (Fully Adhered, Ballasted, Mechanically Fastened, Johns Manville Specification #3GIS, Tamko Specification #201C, Awaplan 170, etc.)

5. Specified application guidelines which may or may not be overridden by the manufacturer's specifications

6. Insulation type, thickness (tapered or flat), number of layers, fastening requirements, etc.

7. Warranty type and duration (five, ten, fifteen, and/or total system, limited, prorated, etc.)

8. Uplift and Fire Rating requirements

9. Special products — roof drain inserts, special pipe boots/supports, walkpads, concrete pavers, roof expansion joints, etc.

Assessing the Scope of the Job

Determine the physical parameters of the job (takeoff), measuring the roof area in square feet (SF) or squares (SQ = 100 SF), various linear calculations (LF), each individual items (EA), and noting any special job characteristics.

Roof Plan

A roof plan if included in the drawings is the best place to start. If possible make a copy of the roof plan or at the very least make a sketch or tracing. The sketch does not have to be drawn to scale but should closely approximate the shape of the building. The sketch should locate the roof drains, roof penetrations, and walkways. Be sure to note the different roof deck type(s) and locations. Use printed dimensions where available. Use inside parapet dimensions in calculating the roof area without eliminating area for roof top units or skylights. Another option uses outside dimensions and reducing the area for large roof top units or skylights. Either method is appropriate as long as the method is consistently followed. There are numerous computerized estimating programs available to assist roofing estimators but the essential items included are the same.

Perimeter Details

Determine which details are utilized at the perimeter: parapet details, terminations, metal edges, expansion joints, etc. In addition to the lineal feet (LF), you will need the heights of the various parapets/flashings, and any specific details, especially if they vary from the manufacturer's standard details.

Roof Penetrations

Items which need to be surveyed are listed below include:

1. Roof drains
2. Overflow roof drains
3. Pipe penetrations/supports
4. Penetration or pitch pockets
5. Roof top equipment
6. Skylights
7. Scuppers
8. Overflow Scuppers
9. Walkways (may be required even if not shown on the roof plan)
10. Lightning protection penetrations
11. Tie-ins to any existing roof systems

Alternates

Alternates may be shown on the roof plan, but are often only mentioned in the bid form or in the specifications. If alternates affect the roofing, be sure to survey and price them accordingly.

Determining Labor Costs

Labor costs and production rates should be determined by historical costs for each system and type of roofing. Some items which may be included in determining labor costs are wages, benefits, taxes, unemployment taxes, insurance costs (worker's compensation and liability), small tools, safety, and expendable materials. Other items which may affect labor costs are hoisting, travel, seasonal work, building height, building and roof access, inspection requirements, difficulty of the project, and job schedule.

Equipment Costs

Specialized equipment may be necessary to complete a particular phase of a roofing project. Such equipment may include cranes for hoisting material, specialized "tear-off" equipment, spray equipment for adhesive or coatings, or heat welding equipment. Costs for purchasing and/or rental should be included in the estimate.

Finalizing the Estimate

After all the direct material and labor costs are determined, overhead and profit needs to be added to the job. Company management should determine overhead and profit and a range of percentages should be provided to the estimators for their use. Senior estimators may have the authority to alter the mark-ups for overhead and profit based on specific job conditions and other factors. Comparing the square foot cost to similar projects should indicate whether the project is within the appropriate range.

07590 – ROOF MAINTENANCE AND REPAIRS

Introduction - An Ounce of Prevention

The prudent roofing contractor is in a continuous process of identifying and controlling the risks involved in a commercial re-roofing project.

The estimator does most of the work identifying potential liabilities before submitting the estimate for management review. The invitation to bid will reveal the owner's intention to require performance and payment bonds. Bonds are not usually major cost times. Seven year periods of liability are not uncommon. The general conditions of the contract will specify the types and limits of insurance required. This includes hold harmless clauses indemnifying the owner and the designer. There also are requests to name the owner and designer as additionally insured parties. These clauses extend the roofing contractor's limits of liability.

Quantity Survey

The general conditions will, or should, specify the dates for starting and finishing the work. They also define penalties for late completion. The owner's schedule will often determine the crew size and equipment required. Commercial re-roofing estimators should not assume that it will be possible to work double shifts or weekends. The owner's production or security requirements may affect the contractor's ability to ease the constraints of the schedule.

Look at the terms of payments. Is there provision for progressive billings? Is there a retention clause? Are there provisions to pay for materials properly stored and protected? Some states have lien laws requiring the owner to get lien releases from the contractors and suppliers. These payment terms may affect the capital requirements of the project.

A visit to the job site should include the compilation of a brief history of the building including:

- ✓ Age
- ✓ Type of construction
- ✓ Record of leaks
- ✓ Roof maintenance programs

The estimator's survey of the project will include:

- ✓ Type of roof deck
- ✓ Existing roof system
- ✓ Means of attachment
- ✓ Projections

Take core samples of the membrane and the insulation. Carefully review the specifications for errors, omissions, or inconsistencies. Note the current use of the building, the estimated value of its contents, and the area with the most potential liability. The commercial roofing estimator will arrive at the job site prepared to ask questions. Where is the mainframe computer? Where is the equipment or inventory? What is under the area of roof deck replacement? Is it the employees' cafeteria that is open 24 hours a day and 6- feet below the roof? What precautions are necessary to protect the occupants and the roofers? What does the exhaust system exhaust onto the roof? When fumes have corroded the metal flashings and roof deck, what will they do to the roofers?

The maintenance supervisor of the building should conduct the tour. The supervisor will know what lead time is necessary for the interruption of services. These include heating, air conditioning, electrical, and other services. Determine the maximum permissible down time. The person responsible for maintenance will know the names of the maintenance contractors who service the building. This person will often give the contractor information not contained in the contract documents that will determine the roofing contractor's method of performance. A tour of the area surrounding the project will reveal working space available to the contractor. Investigate the chance of moving employees' automobiles away from the building. This may eliminate re-painting costs.

Include all information collected during the site visit in an addendum issued by the owner. The addendum also will answer any questions about the specifications. What if the owner is unable or unwilling to issue the addendum? The prudent roofing contractor will then attach a letter to the bid document. It would state the bid includes information issued by the owner during the visit to the site. List all the information in the letter.

Management should set up risk control procedures that are specific to the project. Include these into the estimated cost of the job. Not all procedures will be the responsibility of field supervisors.

Contractors often have difficulties paying their bills because payments are late or not approved. Agree on credit terms before the sale not after delivery of the product or service. The careful roofing contractor, before submitting the bid, will know the name of the person who approves payment. Also known will be the method of getting approval and the proper application for payment. Record the name, address, and phone number of the person who issues the check and the date of issuance. Arrange the work to coincide with the owner's payment schedule. For example, the 20th of the month is the owner's cut off date. The 25th of the month is the supplier's cut off date. Do not ship material to the project between those dates.

Many architectural and consulting firms do not have liability insurance. Eliminate clauses in the general conditions that require the roofing contractor to indemnify the designer. Management may decide to buy a few rolls of sheet plastic to protect the contents of the building. They also could insist the owner move the most expensive or critical items away from the work area.

07920 – JOINT SEALANTS

Introduction

Carefully review plans and specifications to insure accuracy and completeness of the estimate. Read the specifications and note extras of sealants and caulking, length of warranty, and approved manufacturers. Cross-reference specifications to insure that no unusual conditions exist which may increase material or labor costs. Manufacturer's recommendations for material application rates may vary due to joint details.

Quantity Survey

Determine the location of caulking by cross-referencing specification with drawings. Pay particular attention to expansion, control, construction, or key joints. Note sections showing the intersection of dissimilar materials.

List each line item by detail number, description of joint type, joint backing, and type of caulking material required. Note whether joint is rectangular or triangular in nature. Survey all caulking by joint dimensions and length in lineal feet (LF).

Base waste allowances on historical data or as determined by estimator judgment and experience.

Waste may be the result of tooling masked joints or overfilling and shaving horizontal joints. Catalyzing multicomponent bulk material that is then leftover or unused is another cause for waste. Make allowances in the estimate for masking materials, rags, and solvent for cleanup.

Extension of Quantities

Extend quantities by multiplying and adding similar joint types.

Express labor production rates LF/HR. Productivity rates should account for setup, cleanup, and mixing. Also include joint preparation such as cleaning, backing, masking, or priming. For each joint type, determine quantity of caulking material required. When figuring bulk material, list quantities in gallons. When figuring tube materials, list quantities in number of tubes required. Also determine the amount of backing rod, rags, cleanup solvent, and masking required.

Determining Lineal Feet Per Gallon

One U.S. standard gallon contains 231 cubic inches of material. Consider caulking as 100% solids by volume. Therefore, one gallon will yield 19.25 LF/GAL of material for a 1 IN by 1 IN rectangular joint.

Method 1 - To determine the length of rectangular joint that one U.S. standard gallon will fill, divide the joint length by the product of the reciprocals of the joint dimensions times 19.25 LF. Example: A particular joint type is ¼ IN deep and ½ IN wide by 940 LF.

Gallons required = 940 LF / (4 × 2 × 19.25 LF/GAL) = 940 LF ÷ 154 LF/GAL = 6.1 GAL

Method 2 - Multiply the joint dimension by .052 LF/GAL.

Gallons required = 940 LF × ¼ × ½ × .052 LF/GAL = 6.1 GAL

Knowing either of these formulas, one can determine gallons of caulking materials using either a calculator or a computer spreadsheet.

For a triangular or V-grooved joint, double the LF/GAL using the above formulas.

To determine tubes required once gallons are known, divide 128 ounces/gallon by the volume contents of the tube for a multiplier.

08100 METAL DOORS AND FRAMES

INTRODUCTION

The following estimating process will focus on metal doors and frames. Related items may be round throughout the plans and specifications, so it is important that all sections of the documents be reviewed in addition to the specific door and frame sections.

Door installation shall mean the process of hanging the door only. This will require a brief discussion of door hanging devices from Section 08700 Hardware but will not expand on any other elements of Section 08700 or 08800 Glazing. These sections are closely related and must be considered in the estimating of an overall construction project.

DESCRIPTION

The purposes of the metal door and frame assembly are to provide:

1. Security
2. Entrance and Egress
3. Control of traffic flow
4. Fire rated openings in rated wall assemblies
5. Privacy (audio and visual separation)
6. Environmental control (between interior and exterior, or between HVAC zones)

Doors and frames are an integral part of overall construction project costs. They are also one of the most visible and often abused elements of a building. It is especially important that proper consideration is given to the estimating of materials and installation of doors and frames. There are several types of door and frame materials available on the market. The most often used types are wood and metal. Metal doors and frames are typically considered more often for commercial and industrial uses, rather than residential uses. *(See Example A, "Basic Door Grades and Applications", courtesy of the Steel Door Institute, for more detailed information on typical metal door usage.)*

Metal doors are usually constructed of a steel framework with a steel sheet face attached to each side of the framework. Inside this framework and skin, the majority of the space is usually empty, thus the term hollow metal is applied. There may be other items inside this space that will be discussed later.

Metal frames, frequently called "hollow metal frames", do not really have an enclosed space, but are typically three (3) sided and roll formed to the desired shape.

TYPES AND METHODS OF MEASUREMENT

The unit of measure for metal doors and frames is typically EACH. LEAF is also a common measurement for doors. Doors and frames should be counted separately because they do not always have a 1:1 ratio. Often, there are multiple doors within a single framed opening, and it is not uncommon to have a framed opening that has no door.

Example A — Basic Door Grades and Applications

BUILDING TYPES	Standard Steel Door Grades			Door Thickness		Door Design Nomenclature					
	Grade 1 Standard Duty 1 ¾" 1 3/8"	Grade 2 Heavy Duty 1 ¾" only	Grade 3 Extra Heavy 1 ¾" only	1 ¾"	1 ¾" or 1 7/8"	(F)	(G)	(V)	(FG)	(N)	(L)
APARTMENT											
Main Entrance		•		•		•		•	•		
Unit Entrance	•	•		•		•					
Bedroom	•				•	•					
Bathroom	•				•	•					
Closet	•				•	•					
Stairwell		•	•	•			•				•
DORMITORY											
Main Entrance		•		•		•		•	•		
Unit Entrance	•	•		•		•					
Bedroom	•			•		•					
Bathroom	•			•		•					
Closet	•				•	•					•
Stairwell		•	•	•			•				
HOTEL-MOTEL											
Unit Entrance	•	•		•		•					
Bathroom	•					•					
Closet	•				•	•					•
Stairwell		•	•	•			•				
Storage & Utility	•	•		•		•					•
HOSPITAL – NURSING HOME											
Main Entrance		•		•		•		•	•		
Patent Room		•		•		•					
Stairwell		•	•	•			•				
Operating & Exam		•		•		•					
Bathroom	•			•		•					
Closet	•				•	•					•
Recreation		•		•		•		•			
Kitchen		•	•	•			•				
INDUSTRIAL											
Entrance & Exit			•	•		•		•	•		
Office	•	•		•		•	•				
Production		•	•	•		•					
Toilet		•	•	•		•					•
Tool		•	•	•		•					
Trucking		•	•	•		•					
Monorail		•	•	•		•					
OFFICE											
Entrance			•	•		•		•	•		
Individual Office	•			•	•	•	•				
Closet	•				•	•					•
Toilet		•	•	•		•					
Stairwell		•	•	•			•				
Equipment		•	•	•		•					
Boiler		•	•	•		•					•
SCHOOL											
Entrance & Exit			•	•		•		•	•		
Classroom		•		•		•			•		
Toilet		•	•	•		•					
Gymnasium		•	•	•		•		•			
Cafeteria		•	•	•			•				
Stairwell		•	•	•			•				
Closet	•			•		•					•

• (F) Flush   • (G) Half Glass   • (V) Vision Lite   • (FG) Full Glass   • (N) Narrow Lite   • (L) Louver

The method used to quantify the doors and frames should be systematic. When a door and frame schedule *(see Example B)* has been prepared by the architect, quantities should be taken off based on the schedule, which is usually grouped by the floor, building or wing.

Additionally, it is good practice to count the doors and frames individually on the architectural floor plan(s) to verify the schedule. Determine a logical starting point, such as the lower left hand corner of the plan, and start at that same point each time, working from bottom to top. Mark each item as it is counted, using a different colored pencil or hi-liter for doors, frames and hanging devices. Many computerized estimating software packages also have the capability to determine quantities through the use of a digitizer. When a door and frame schedule has not been provided, prepare a simple schedule to document and clarify the items for which labor and material estimates have been generated. Documentation is always important and often becomes critical where the documents are not complete.

Example B — Door Schedule

	DOOR				FRAME			REMARKS
No.	Size	Type	Material	Glass	Type	Material	Label	
A	3'-0" x 6'-8" x 1 ¾"	2	WD		A	HM		
B	2'-8" X 6'-8" X 1 3/8"	2	WD		C	Drywall		Bi-fold
C	5'-0" X 6'-8" X 1 3/8"	5	WD		PM	Drywall		Bi-fold
D	2'-8" X 6'-8" X 1 3/8"	3	WD		PM	WD		Louvered
E	2'-4" X 6'-8" X 1 3/8"	2	WD		PM	WD		
G	3'-0" x 6'-8" x 1 3/8"	2	WD		PM	Wd		
J	4'-0" X 6'-8" X 1 3/8"	5	WD		C	Drywall		Bi-fold
K	1'-8" X 6'-8" X 1 3/8"	5	WD		C	Drywall		Bi-fold
100	3'-0" x 6'-8" x 1 ¾"	1	HM		A	HM	20 min.	
101A	3'-0" x 6'-8" x 1 ¾"	1	HM		A	HM	20 min.	
101B	3'-0" x 6'-8" x 1 ¾"	6	HM	¼" Safety	A	HM	20 min.	½ View
102A	3'-0" x 6'-8" x 1 ¾"	1	HM		A	HM	20 min.	
102B	3'-0" x 6'-8" x 1 ¾"	1	HM		A	HM	20 min.	
103A	PR 3'-0" x 6'-8" x 1 ¾"	1	HM		B	HM	20 min.	Pair
103B	PR 3'-0" x 6'-8" x 1 ¾"	1	HM		B	HM	20 min.	Pair
104	3'-0" x 6'-8" x 1 ¾"	2	WD		A	WD	20 min.	
105	3'-0" x 6'-8" x 1 ¾"	2	WD		PM	WD	20 min.	
106	3'-0" x 6'-8" x 1 ¾"	6	HM	¼" Safety	A	HM	20 min.	½ View
107A	3'-0" x 6'-8" x 1 ¾"	1	HM		A	HM	20 min.	
107B	3'-0" x 6'-8" x 1 ¾"	1	HM		A	HM	20 min.	
109	3'-0" x 6'-8" x 1 ¾"	6	HM	¼" Safety	A	HM	20 min.	½ View
110A	PR 3'-0" x 6'-8" x 1 ¾"	4	HM	¼" Safety	B	HM	20 min.	Pair
110B	PR 3'-0" x 6'-8" x 1 ¾"	4	HM	¼" Safety	B	HM	20 min.	Pair
111A	3'-0" x 6'-8" x 1 ¾"	6	HM	¼" Safety	A	HM	20 min.	½ View
112A	3'-0" x 6'-8" x 1 ¾"	1	HM		A	HM	20 min.	
112B	3'-0" x 6'-8" x 1 ¾"	1	HM		A	HM	20 min.	
113	3'-0" x 6'-8" x 1 ¾"	2	WD		A	HM	20 min.	

FACTORS THAT MAY AFFECT QUANTITY TAKEOFF AND PRICING

Fire Rating Requirements

The cost of metal doors and frames can be significantly affected by fire rating requirements. When a door and frame assembly is located within a fire rated wall assembly, it must be fire rated in order for the entire assembly to meet fire code requirements. Door and frame fire ratings range from 20 and 30 minute fire doors, in a one (1) hour wall assembly, to three (3) hour fire doors used to separate buildings or fire zones within a building. Fire rated door and frame assemblies also have restrictions on the type of hardware which may be used as well as the amount and type of glass which they may contain.

The labor required to install fire rated door and frame assemblies will be greater than that required to install an assembly of the same materials without a fire rating. Fire rated doors and frames typically require special hardware such as mandatory use of smoke seals, closers and latching devices which further increase overall costs of both material and labor.

Security and Electronics

The manufacturing of security doors is significantly different than that of standard metal doors and frames. Security doors and frames are often constructed of heavier gauge materials.

Fourteen gauge doors and twelve gauge frame materials may be used in lieu of more typical eighteen or twenty gauge doors and sixteen or eighteen gauge frames for non-security applications.

Metal frames in security and any other heavy-duty application may be grouted solid. The door's interior core is filled with reinforcing steel channels. There are often bullet resistant materials such as Kevlar and soundproofing materials manufactured into the core of the door. Bullet resistant door and frame assemblies must go through a certification process.

Often security doors and frames have integrated electronic devices such as electrical raceways, electronic transfer hinges, electric hold opens, power operators, electric strikes, electromagnetic locks and actuators such as magnetic card readers and electric push button coded locks. The doors and frames themselves must be machined and reinforced to accept these types of electrical devices. Without the actual electronic hardware devices, material costs of the security door and frame may be two to four times of the cost of the same size and configuration assembly without security provisions. Shipping costs per unit will be increased due to the additional weight and labor costs will increase because it may take two (2) craftsmen to install the heavier door. There may also be additional alignment requirements for electrical conduits and other devices in the frame and it will take longer to drill and fasten through heavier gauge materials.

Special Coatings and Material Types

Special coatings, such as galvanizing, bituminous coatings, lead lining, and special materials such as stainless steel are used to address specific conditions where regular steel doors and frames would not withstand the environment or offer the desired protection. Galvanizing is often used where excessive moisture will be present or where chemicals are commonly used, (i.e. wastewater treatment facilities and car washes.) Bituminous coatings are often used in frames that will remain in contact with the moisture. Lead lined doors and frames are commonly used in hospitals and medical facilities as a means of controlling radiation. Stainless steel doors and frames may be used where cleanliness is critical, such as in food preparation or storage areas where any rust or corrosion would cause an unsanitary condition.

All of these special coatings and material types will increase the material costs due to additional and specialized manufacturing processes. Shipping costs will increase for additional weight, especially with lead lined doors and frames. Installation costs for lead lined doors are increased significantly due to handling the additional weight and special anchoring required to support the weight. Craftsmen may require special hazardous material training and equipment if field patching of the lead barrier is required.

Schedule

Metal doors and frames are typically a long lead time item because the shop drawing preparation and approval process is time consuming. It is critical that metal doors and frames be ordered in a timely fashion in order to take delivery when needed in the overall project schedule. Typically the more customized a door and frame assembly is with special jamb depths, glass opening sizes, coatings and other options, the longer it will take to receive. It may be possible to rush delivery of an order for an additional cost.

OVERVIEW OF LABOR, MATERIAL, EQUIPMENT AND INDIRECT COSTS

Labor

The ability to establish future labor cost and productivity is tied to the collection of data developed under specific working conditions on past projects. The total hours required to install metal doors and frames shall take into consideration factors such as:

✓ Material receiving and handling

✓ Proximity of storage/staging areas to installation point (it is most efficient if material deliveries can be scheduled and packaged so materials are handled only once)

Frame attachment/installation methods differ depending on wall construction and frame type. (*See Attachment C, "Anchor Details", and Attachment D, "Common Wall Conditions" courtesy of the Steel Door Institute for anchor types, frame preparations, and common wall conditions.*) Frames may be welded one-piece frames or knocked down (KD) frames which are fabricated in three (3) separate pieces and installed after the surrounding construction is complete. Temporary bracing is required for welded frames, which are typically set before the surrounding construction is in place. Spreaders, used to maintain the same width of the frame opening from the head of the frame to floor, are required for both welded frames and for KD frames.

Metal door types also vary (See Attachment E) but most require a typical installation on some type of hinges or pivots.

Attachment F is an example of "Quantity Takeoff and Pricing Sheets". This example has been generated using Timberline Precision Estimating for Windows with labor rates, productivity rates and material costs from R.S. Means. Doors have been taken off separately from frames and like items have been grouped. The column heading "Location" can be specified for each item to sort and track the estimate by building, floor or wing. The item descriptions should note as much as possible about the materials, such as size, type, gauge, fire rating, door thickness and KD or welded frames. The estimating totals section at the end of the takeoff and pricing sheets summarizes labor, material and equipment costs. The add-on items that follow the cost summary are general conditions/requirements and sales tax, etc. The contractor's fee may then be calculated as a percentage of estimated cost totals or individual category totals. General conditions/requirements may also be included as individual line items within the estimate rather than as a percentage of costs.

Material

Many suppliers of metal doors and frames will provide a valuable service by performing their own detailed takeoff of plans and specifications, from which they will provide a lump sum quotation to furnish these materials for the project. An advantage to this is the opportunity to compare your takeoff with theirs to check for errors and omissions.

Indirect Costs

Storage of the materials, time spent coordinating deliveries, special training and handling precautions are examples of indirect cost factors and should be addressed in the estimate.

RISK CONSIDERATIONS

Issues that could adversely affect the cost of a project:

- ✓ Poor material quality

- ✓ Schedules requiring unanticipated overtime or rush material delivery

- ✓ Material delivery delays

- ✓ Non-standard wall types requiring special jamb depths

- ✓ Special/custom door and frame configurations

- ✓ Availability of skilled and experienced craftsmen

- ✓ Availability of proper tools and equipment

- ✓ Lack of craftsman continuity

RATIO AND ANALYSIS TOOLS

Metal door and frame costs as a percentage of overall building costs do not provide a consistent ratio.

A company's historical data may be the most accurate gauge of costs in comparison to overall project cost. In order to make this analysis it is critical that the scope of work be a direct comparison. The type of building construction, usage and door and frame types must all be known. Another analysis would be the ratio of labor hours to material costs and would be most accurate if related to projects of the same type and usage. Cost guides such as Means Building Construction Cost Data, and Walkers Estimating Manual can be used for comparative analysis of estimated productivities. Some very basic material quantity checks can be performed as follows:

- ✓ If all doors have three (3) hinges then the number of hinges estimated should be three (3) times the number of doors.

- ✓ If all frames are for single doors, and all framed openings have doors in them, then the number of frames should be the same as the number of doors.

Historical cost tracking will show trends and averages and it is important to use them only as a reference, realizing that no two projects are identical.

MISCELLANEOUS PERTINENT INFORMATION

Preliminary Estimating

Often professional estimators are asked to generate cost estimates based on preliminary information. It is important that the scope of work be clarified by the estimator. When estimating at the preliminary stage of a project, the methods used to quantify costs should be modified to reflect the level of detail known.

At the earliest stage of a building's design and estimating it may be necessary to use a chart such as the "Door Density Guide" found in the Means Building Construction Cost Data. This chart can be used when no interior floor plan exists to arrive at the approximate number of door openings to be estimated per square foot of floor space based on building usage. The material and labor costs are often shown combined as a lump sum cost per door opening based on historical information. Many of today's computerized estimating software products allow the estimator to make assumptions about the number and type of door openings and then calculate a detailed cost breakdown of both labor and material for all items within an assembly, with costs based on a company's historical database of similar assemblies.

Example C

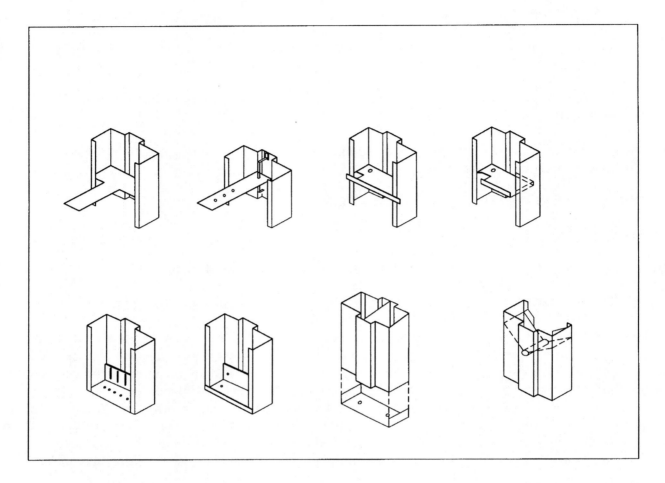

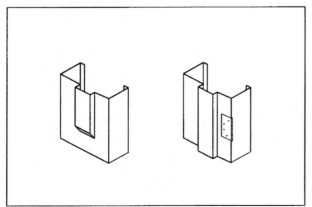

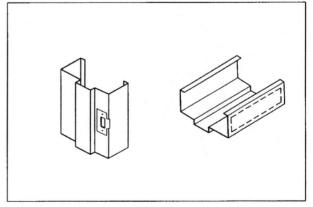

Example D — Common Wall Conditions

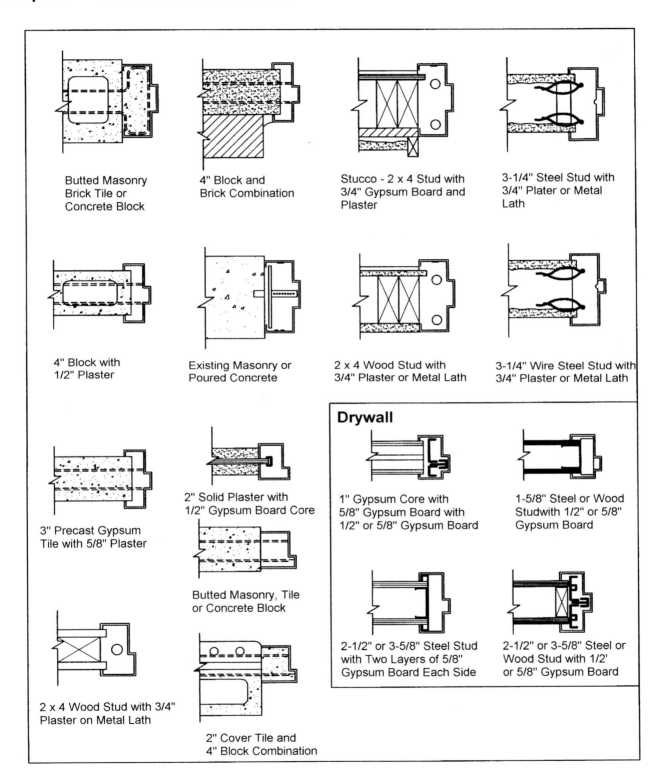

08100 Metal Doors and Frames — Part Two

Example E — Nomenclature for Door Designs

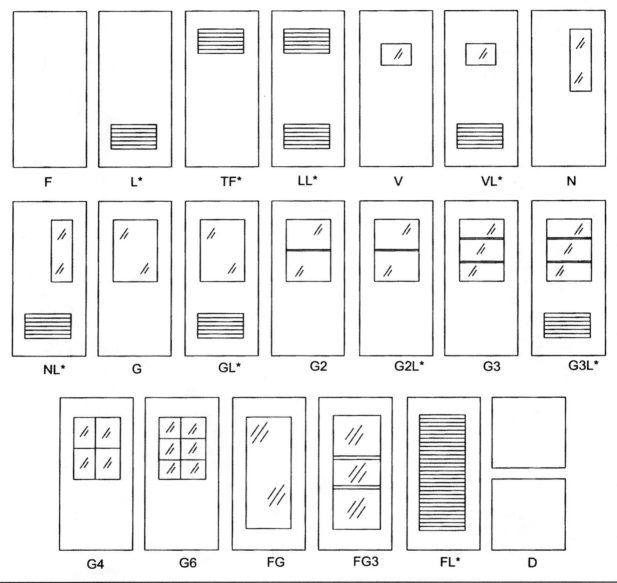

Nomenclature Letter Symbols			
F	Flush	**N**	Narrow Lite
L	Louvered (bottom)	**NL***	Narrow lite and Louvered
TL	Louvered (top)	**G**	Half Glass (options G2, G3, G4 and G6)
LL	Louvered (top and bottom)	**GL**	Half Glass and Louvered (options G2L and G3L)
V	Vision Lite	**FG**	Full Glass (option FG3)
VL	Vision Lite and Louvered	**FL**	Full Louver
		D	Dutch Door

*Louvered door designs: specify design, louver size and/or free area requirements
• Add suffix "I" to indicate inserted louver • Add suffix "P" to indicate punched louver
• Add suffix "A" to indicate air conditioning grille

Example F – Estimate Details Report

METAL DOORS AND FRAMES

Item	Description	Location	Quantity		W%	Conversion	Order Qty	Unit Price	Amount
10	HM Door/Frame HM Door Type 1 3'-0" x 6'-8" x 1 ¾"		10.00 ea.	Lab Mat. Eqp.	0	1.000 hr/ea 1.000 hr/ea	10.00 hr. 10.00 ea. 10.00 hr.	26.00 189.00 2.00	260 1,890 20
10	18 GA, Flush Door HM Door Type 1 3'-0" x 6'8" x 1 ¾" 20 min.		2.00 ea.	Lab Mat. Eqp.	0	1.000 hr/ea 1.000 hr/ea	2.00 hr. 2.00 ea. 2.00 hr.	26.00 204.00 2.00	52 408 4
10	18 GA, Flush Door HM Door Type 1 3'-0" x 7'-0" 1 ¾"		25.00 ea.	Lab Mat. Eqp.	0	1.000 hr/ea 1.000 hr/ea	25.00 hr. 25.00 ea. 25.00 hr.	26.00 197.00 2.00	650 4,925 50
10	18 GA, Flush Door HM Door Type 1 3'-0" x 7'-0" x 1 ¾" 20 min.		5.00 ea.	Lab Mat. Eqp.	0	1.000 hr/ea 1.000 hr/ea	5.00 hr. 5.00 ea. 5.00 hr.	26.00 210.00 2.00	130 1,050 10
10	18 GA, Flush Door HM Door Type 1 4'-0" x 7'-8" x 1 ¾"		12.00 ea.	Lab Mat. Eqp.	0	1.14300 hr/ea. 1.14300 hr/ea	13.72 hr. 12.00 ea. 13.72 hr.	26.00 249.00 2.00	357 2,988 27
10	18 GA, Flush Door HM Door Type 1 4'-0" x 7'-8" x 1 ¾" 20 min		3.00 ea.	Lab Mat. Eqp.	0	1.06700 hr/ea. 1.06700 hr/ea.	3.20 hr. 3.00 ea. 3.01 3.20	26.00 268.00 2.00	83 804 6
10	18 GA, Flush Door HM Door Type 2 3'-0" x 6'-8" x 1 ¾" –1/2 Glass		10.00 ea.	Lab Mat. Eqp.	0	0.94100 hr/ea. 0.94100 hr/ea.	9.41 hr. 10.00 ea. 9.41 hr.	26.00 258.00 2.00	245 2,580 19
10	18 GA, Flush Door HM Door Type 2 3'-0" x7'-0" x 1 ¾" –1/2 Glass		10.00 ea.	Lab Mat. Eqp.	0	0.94100 hr/ea. 0.94100 hr/ea.	9.41 hr. 10.00 ea. 9.41 hr.	26.00 268.00 2.00	245 2,680 19
40	18 GA, Flush Door, ½ Vision Glass HM Frame Type A 3'-0" x 6'-8"		10.00 ea.	Lab Mat. Eqp.	0	1.06700 hr/ea. 1.06700 hr/ea.	10.67 hr. 10.00 ea. 10.67 hr.	26.00 112.00 2.00	277 1,120 21
40	16 GA, Welded Hm Frame Type A 3'-0"x 6'-8" – 20 min.		2.00 ea.	Lab Mat. Eqp.	0	1.06700 hr/ea. 1.06700 hr/ea.	2.13 hr. 2.00 ea. 2.13 hr.	26.00 122.00 2.00	55 244 4

Example F (cont.) — Estimate Details Report

Item	Description	Location	Quantity		W%	Conversion	Order Qty	Unit Price	Amount
40	16 GA, Welded Hm Frame Type A 3'-0"x 7'-0"		19.00 ea.	Lab. Mat. Eqp.	0	1.06700 hr/ea. 106.700 hr/ea.	20.27 hr 19.00 ea. 20.27 hr.	26.00 117.00 2.00	527 2,223 41
40	16 GA, Welded Hm Frame Type A 3'-0"x 7'-0" – 20 min.		5.00 ea.	Lab. Mat. Eqp	0	1.06700 hr/ea. 106.700 hr/ea.	5.34 hr. 5.00 ea. 5.34 hr.	26.00 128.00 2.00	139 640 11
40	16 GA, Welded Hm Frame Type A 4'-0"x 7'-0"		12.00 ea.	Lab. Mat. Eqp	0	1.10000 hr/ea. 1.10000 hr/ea.	13.20 hr. 12.00 ea. 13.20 hr.	26.00 128.00 2.00	343 1,536 26
40	16 GA, Welded Hm Frame Type A 4'-0"x 7'-0" – 20 min.		3.00 ea.	Lab. Mat. Eqp	0	1.10000 hr/ea. 1.10000 hr/ea.	3.30 hr. 3.00 ea. 3.30 hr.	26.00 139.00 2.00	86 417 7
40	16 GA, Welded Hm Frame Type A 6'-0"x 7'-0"		3.00 ea	Lab. Mat. Eqp	0	1.33300 hr/ea. 1.33300 hr/ea	4.00 hr. 3.00 ea. 4.00 hr.	26.00 164.00 2.00	104 492 8

Estimate Totals

Labor	3,553.00
Material	23,997.00
Equipment	$273.00
Subtotal	$27,823.00
General Conditions	2,226.00 (8% x Subtotal)
Sales Tax	1,200.00 (5% x Material)
Contractor's Fee	1,855.00 (6% x (Subtotal + GC's + Tax
Total	$33,1043.00

08710 - DOOR HARDWARE

INTRODUCTION

A building construction project may involve a wide array of hardware. The term door hardware in the broadest definition of the term refers to hardware that is visible and typically requires a special finish regardless of the application. This discussion will be limited to the above referenced Subdivision which refers to those components installed on doors and frames for the purpose of providing the desired performance of the door and frame whether or not made from metal. This work involves many various components ranging from the most sophisticated electronic door control devices to the most basic metal door hinge or plastic weather stripping.

TYPES AND METHODS OF MEASUREMENT USED

The measurement of door hardware is performed by the most basic and tedious of all types of measurement — each component must be listed and counted. The takeoff is usually performed in conjunction with the takeoff of the doors and flames, which are usually identified together on the drawings. Experience has shown that there are nearly as many different ways of identifying door hardware on the drawings as there are people who specify it. It is worth taking the time to study how the drawings identify the hardware, because the approach to takeoff may be greatly facilitated by clearly understanding how the information is being presented. Typically, hardware will be identified in one of two basic methods.

In the first method, each door opening on the floor plan will be give an alphanumeric designation of type. There may be several alphanumeric designations. The plans and specifications will identify for each designated type of opening how the frame, door and hardware are to be configured. In this first method, the estimator must determine the total count of each type of opening designated on the floor plan. By establishing the total of each type of opening that contains different types of hardware items, a total count of each kind of hardware may be determined for estimating costs; or the cost of each opening may be estimated based upon a summation of the cost of the hardware items included in each opening. (See Door Schedule - Example 1)

DOOR SCHEDULE – Example 1									
Opening Type	Opening No.	DOOR				FRAME			Hardware Set
		Material	Width	Height	Thickness	Material	Head	Jamb	
A	1	s.c. wd	3'-0"	7'-0"	1-3/4"	metal	6L/7.1	6L/7.1	10
B	1	metal	4'-0"	8'-0"	1-3/4"	metal	9L/7.1	9J/7.1	13
C	2	s.c. wd	2'-6"	7'-0"	1-3/4"	metal	18L/7.1	18J/7.1	11
D	1	metal	3'-0"	7'-0"	1-3/4"	metal	6L/7.2	6J/7.1	9
E	2	metal	2'-"	7'-0"	1-3/4"	metal	6L/7.1	6`J/7.1	7

DOOR SCHEDULE – Example 2

Door No.	Door Material	Frame Material	Hardware Set No.
1	Wood	Hollow Metal	2
2	Wood	Hollow Metal	2
3	Wood	Hollow Metal	2
4	Wood	Hollow Metal	2
5	Wood	Hollow Metal	3
6	Wood	Hollow Metal	3
7	Wood	Hollow Metal	2
8	Wood	Hollow Metal	2
9	Hollow Metal	Hollow Metal	4
10	Wood	Hollow Metal	1

The second basic method of identifying hardware is where each opening is assigned an individual alphanumeric designation on the floor plan. In this method, some form of a door, frame and hardware schedule is commonly utilized. It is good practice to confirm coordination of the floor plan with the door schedule. (See Door Schedule - Example 2)

Again, there are two common forms of hardware schedules. The first method will establish a table with the openings listed in one direction (e.g., down the first column) and hardware items listed in the other direction (e.g., across the top row). The hardware required for each opening is indicated by marking the cell below each hardware item across from each opening. Each hardware item may easily be counted by going down each column and counting the cells that are marked. (See Hardware Schedule - Example 1)

HARDWARE SCHEDULE – Example 1

Opening No.	Hinges (Pair)	Lever Handle, Lockset	Lever Handle, Passage	Closer	Kickplate	Threshold	Weatherstrip	Floor Stop	Wall Stop
1	1-1/2	X							X
2	1-1/2	X							X
3	1-1/2	X							X
4	1-1/2	X							X
5	1-1/2		X						X
6	1-1/2		X						X
7	1-1/2	X							X
8	1-1/2	X							X
9	1-1/2		X	X	X	X	X	X	
10	1-1/2		X	X	X	X	X		X

HARDWARE SCHEDULE Example 2	
Set No. 1	1-1/2 pair hinges
	lever handle lockset
	closer
	kickplate
	threshold
	weatherstripping
	floor-mounted door stop
Set No. 2	1-1/2 pair hinges
	lever handle passage set
	wall-mounted door stop
Set No. 3	1-1/2 pair hinges
	lever handle lockset
	wall-mounted door stop
Set No. 4	1-1/2 pair hinges
	lever handle lockset
	closer
	threshold
	weatherstripping
	wall-mounted door stop

The second method will designate an alphanumeric hardware set designation to each opening. There will then be a listing of which hardware items are to be included in each hardware set. With this method the estimator must count the number of hardware sets and then, based on the composition of each hardware set, determine the count of each item of hardware. (See Hardware Schedule - Example 2)

SPECIFIC FACTORS WHICH AFFECT THE TAKEOFF, PRICING, ETC.

Pricing of hardware installation is highly dependent on the availability of carpenters experienced in the installation of the hardware type on the project. On a large project, if the openings are similar and situated close enough together, it is possible to significantly increase productivity by working into a system of repetitively installing identical items of hardware on all doors and then coming back through the floor performing the next item of hardware, similar to an assembly line approach. This method also lends itself well to the formation of crews with less expensive individuals to perform the repetitious tasks together with experienced, highly skilled workers who can direct their efforts and perform the more critical activities. In this manner everyone benefits because abilities are utilized to their greatest extent and it maximizes the opportunity to improve skills.

Another, often neglected, aspect of hardware installation is the required investment in quality tools. The more varied and sophisticated the hardware, the higher the investment. The speed of installation is highly dependent on having the proper tools with sufficient power and in good working order.

On small jobs the time involved in getting the tools to the location, laying them out, finding power hookups, moving the tools etc. can be significant.

Many parts of installing hardware are intricately related to the installation of the doors and frames. For example, the installation of a hinge is a part of the installation of a door, therefore, it is common for the installation of some hardware, such as hinges, locksets and strike plates to be included in the estimated cost of the installation of the door or frame. If the doors and frames are not installed properly, then the installation of the hardware will be more difficult than anticipated.

Care should be exercised in coordinating the purchase of doors and frames. Many methods have been developed which can eliminate work required for field installation of hardware. If the doors and frames are to be prehung, the hinges and door are already installed in the frame when it is delivered to the project. If the doors are prefit and premachined, then cutouts for hinges, locksets and other hardware items are prepared prior to being delivered to the project. The advantages to these approaches include a reduction of the time required in the field to install the hardware which reduces manpower and scheduling requirements; a more controlled and repetitious shop environment in which to perform these operations; a reduction in tools required in the field. It does, however, require additional coordination with vendor and/or manufacturer prior to delivery and reduces control and flexibility in the field. Prefinished doors and frames present additional difficulties such as protection, damage and repairs due to construction.

OVERVIEW OF LABOR, MATERIAL, EQUIPMENT, INDIRECT COSTS AND MARKUPS

Many suppliers of hardware will provide a valuable service by performing their own detailed takeoff of plans and specifications, from which they will provide a lump sum quotation to furnish the materials for the project. For estimating purposes, a complete list of all material, with quantities, may be given to the supplier for pricing of material costs. An advantage to this is the opportunity to compare your takeoff with theirs to check for errors and omissions.

Labor is priced by estimating the time required to install each item of hardware and taking into consideration the various conditions previously discussed. Take care to include in the estimate additional time required for supervision, setup and cleanup, receiving, handling, inventory, storage, purchasing and maintaining tools, adjustments and installation errors.

SPECIAL RISK CONSIDERATIONS

An often overlooked but potentially costly consideration is that hardware is not always delivered as ordered. Any delivery must be checked for damage and to verify that all required items are received. Any deficiencies
must be immediately identified and reported. All hardware should be securely stored and readily available when scheduled for installation.
Hardware is an item that often requires readjustment depending on its level of usage and quality of components. With experience this can be anticipated and estimated but a reasonable warranty period must be defined at the end of which the owner will assume responsibility for the maintenance. Factors that can affect the degree of adjustments include wearing of inferior parts, sagging and settlement, loosening of parts after initial use and actual breakage of components.

RATIOS, RULES OF THUMB AND ANALYSIS TOOLS

Hardware quantities relate directly to the quantity of doors and frames. Door quantity can provide a cross check of the hardware quantity.

On similar types of projects, an average cost per door can be a good check. The validity of this check will depend on how similar the current project is to the standard to which it is being compared. For example, a patient wing of a hospital might have one value while a high rise office building with several typical floors might represent another value.

QUANTITY SURVEY					
	Set 1	Set 2	Set 3	Set 4	
Quantity of Each	1	6	2	1	
Item Description	**Quantity of Each Item**				**Total**
Lever handle lockset	1	0	2	1	4
Lever handle passage set	0	6	0	0	6
Closer	1	0	0	1	2
Wall door stop	0	6	2	1	9
Floor door stop	1	0	0	0	1
Kickplate	1	0	0	0	1
Threshold	1	0	0	1	2
Weatherstripping	1	0	0	1	2
Hinges	1-1/2	9	3	1-1/2	15

Another approach would be to figure the cost of a basic average configuration of hardware multiplied by the number of doors and then add the cost of significant additional hardware not included in the average. Although some may use the cost per square foot of building area to determine hardware costs as a check or for preliminary estimates, it should be remembered that this method is not very accurate due to the high variability of costs and because it depends heavily on the density of the doors within the building. It is, however, the easiest and quickest method to estimate if trying to budget costs for the total building. The speed of estimating must be balanced against the desired accuracy when determining the method of estimating to be used.

ESTIMATE								
A	B	C	D	E	F	G	H	I
Item Description	Quantity	Production Rate Quantity/HR	Labor Hours (B)/(C)	Labor $/Hr.	Material Unit $	Total ($) Labor (D)*(E)	Total ($) Material (B)*(F)	Total ($) Cost (G)*(H)
Hinges (Pairs)	15	2	7.50	35.42	15.00	266.00	225.00	491.00
Lever Handle Lockset	4	1	4.00	35.42	100.00	142.00	400.00	542.00
Lever Handle Passage Set	6	1	6.00	35.42	90.00	213.00	540.00	753.00
Closer	2	1	2.00	35.42	75.00	71.00	150.00	221.00
Wall Door Stop	9	4	2.25	35.42	15.00	80.00	135.00	215.00
Floor Door Stop	1	2	0.50	35.42	15.00	18.00	15.00	33.00

Kickplate	1	2	0.50	35.42	50.00	18.00	50.00	68.00
Threshold	2	2	1.00	35.42	30.00	35.00	60.00	95.00
Weatherstripping	2	2	1.00	35.42	25.00	35.00	50.00	85.00
TOTAL			24.75			$877.00	$1,625.00	**$2,502.00**
Payroll Taxes, Insurance (35% of Labor)								307.00
Sales Tax (6.5% of Material)								106.00
Freight (12.5% of Material)								203.00
SUBTOTAL								**$3,177.00**
Overhead and Profit (20%)								623.00
TOTAL QUOTE								**$3,741.00**

09250 GYPSUM BOARD

INTRODUCTION

Typically, partitions in office buildings are comprised of metal stud framed walls with gypsum board attached to one or both sides, taped and spackled ready for final finishing. Ceilings and soffits are often comprised of metal stud framing with gypsum board attached. Larger ceiling areas may utilize suspended cold rolled channels and 7/8 IN hat channel attached perpendicular and tied together with wire. Ceilings and soffits also require the same finish treatment as partitions. Addressing each area is the best way to determine actual costs. Since walls and ceilings are usually configured differently, recognizing the variations is the key to a good estimate.

TYPES AND METHODS OF MEASUREMENT

Metal framing is taken-off and purchased in lineal foot quantities. Gypsum board is taken-off and purchased in square foot quantities. Standard size of wall board is four feet wide in standard lengths, of eight, ten and twelve feet. As with steel studs, gypsum board can be ordered in custom lengths to accommodate special projects. Waste factors must be considered carefully to avoid under estimating. Typically, partitions and soffits are taken-off by the lineal foot. Ceilings are taken-off by the square foot.

FACTORS THAT AFFECT TAKEOFF and PRICING

Office partitions range from standard construction to radius configurations along with decorative trims, accents, etc. Wall heights, ceiling shapes, soffits, coffers, all differ in size which will require modification of average production factors for these components. It takes just as much time, or more, to build and finish a two-foot wall than and eight-foot wall, yet, a two-foot wall has only 25% of the square feet found in an eight-foot wall. The same holds true for ceilings. Large areas allow better production, but small areas are time consuming, cut up, and less productive.

SEASONAL VARIATIONS

Cold temperature and wet weather will slow production, drying times, effect stocking and clean up operations. Temperatures falling below fifty-five degrees will require temporary heat to keep finish trades working. Make sure to consider the cost for heat, or exclude it from your scope of work.

OVERVIEW OF TAKEOFF, LABOR, MATERIAL, EQUIPMENT OF METAL FRAMING FOR WALLS, SOFFITS AND CEILINGS

Wall framing is in lineal feet by each individual wall height. Soffit framing should be addressed in the same. This is essential since you will use this to determine total square feet of wall later for final pricing.

Determine each partition type or ceiling by schedules or details. Many of the walls may be fire-rated assemblies and differ by design. For example, a (1) hour rated demising wall is built from floor slab to bottom of structure with one layer of 5/8 IN type "x" gypsum board applied to each side from floor to structure above. A (2) hour-rated stair or demising wall would be similar except there would be two layers of gypsum board each side. Shaft walls are special design since their configuration entails 1 IN shaft-lineal board on one side of C-H studs, and one to two layers of gypsum board on the other side, to the structure above. Furred and plumbing walls usually extend from 3 IN to 6 IN above finished ceiling heights. Special features such as columns, pilasters, pedestals, etc. should be calculated separately. Interior partitions are usually to bottom of ceiling or 3 IN to 6 IN above finished ceiling height. Always takeoff walls straight through doorways, windows, and small openings, since framing is carried above and studs at the jambs are typically doubled up. The waste from openings allows for mechanical and electrical penetrations that require blocking or backing. Depending on details, an inclusion of additional track may be required for head and sill support. Record the results of your quantity survey in a columnar format with the different partition types across the top of the page and the drawings used listed down the left side of the page. Since individual floors or areas are usually shown on separate drawings; the takeoff can be used in the future for such things as ordering materials per floor or area, or to make comparisons if design changes are made. The total project quantities for each partition type can then be tallied and the estimate performed for that value.

Suspended ceilings are taken off in square feet then the components are converted into lineal feet. The framing components are made up of carrier channels, hat channels, and suspension wires; or metal studs for smaller areas. Suspended ceilings are converted to lineal components. Dividing the area, square feet, by four calculates the lineal feet of carrier channel. Multiply the square feet area by .75 for 16 IN on center or by .5 for 24 IN on center to calculate the lineal footage of hat channel. Suspension wire quantities are found by dividing the total lineal feet of carrier by four (or the area by sixteen). The gypsum board required is estimated using the square foot total.

Soffit and fascias are taken-off by lineal foot of section. Soffits and fascia are usually shown in section view in the detail drawings. Use a columnar layout like that used for partitions to record the quantities for each section or soffit detail on a per drawing or per area basis. The sections are estimated individually quantifying the individual components required to construct the detail.

Gypsum board applications for fire rated assemblies are typically taped and spackled to an average ceiling height, with the balance fire-taped. Multiplying the lineal feet of wall by the required height of gypsum board will determine the total square footage of taping, surface finish, floated and fire-taped, regardless of layers. Total square footage of ceiling area is determined by length times width.

High work should be kept separate for evaluating equipment requirements. If the floors are typical; and takeoff is based on typical floor; it is imperative that the correct number of additional floors be determined and calculated.

LABOR

Each component of the individual partition types; soffit and fascia sections and ceiling must be estimated individually to account for different product rates. Partitions are typically constructed using a top track and bottom track attached to the structure with vertical studs (spaced 16 IN or 24 IN on center) installed inside the tracks and fastened with screws. Gypsum panels are attached in one or multiple layers to one or both faces of the partition framing; panels are attached to the structure above, ceiling height or slightly higher using full height sheets or 'stand-ups'. Gypsum panels required above the stand-ups are called top-fill. In multiple layer applications, layers under the outside or 'Face' layer are called base layers. Insulation for sound or thermal reasons may be required to be placed in the stud cavities. This is usually done after a layer of gypsum board is installed on one side of the partition. Each component; top track, bottom track, studs, face layers of gypsum, base layers of gypsum, stand-ups, top fills, and insulation must be listed separately in the estimate as they apply to the each partition type. Refer to the sample estimate to see a typical office partition type and the components listing as it should appear in the estimate. Production rates are applied to each component of the different partition types individually.

In typical commercial office applications the gypsum board is install with the long edge vertical; this is called 'stand-ups'. Most commercial ceilings heights are 9'-0 IN or lower allowing the use of stand-ups. This helps eliminate a continuous horizontal joint 4'-0 IN above the floor that will need to be taped and finished. Gypsum board required above the stand-ups; such as in a rated partition; is usually installed horizontally. This is because most situations require 4"-0 IN or less of top fill; thus horizontal applications can increase productivity. Stand-ups can be installed at a daily rate of 32 sheets a day or 16 lineal feet per hour. Top fill board production starts at about 12 lineal feet per hour, if minimal above ceiling obstructions are expected. If it is expected that the gypsum board installer will have to work around a great number of items penetrating the wall, production rates must be adjusted accordingly.

Soffits should be figured as an independent unit and include horizontal and vertical flaming numbers. For vertical application of the gypsum board a two-man crew is used for heights up to 10 FT – 0 IN. As with smaller framing, production rates are adjusted for the soffit drops. Special attention should be paid to the exterior wall construction. Usually, these are designed using structural materials, making installation of drywall more difficult. Screw attachment requires self tapping screws which take long to penetrate.

For joint treatment the surface area of all walls are combined by their respective heights. Walls and ceilings to 10'-0 IN are based on one man crew. Finished walls and fire-taped walls are kept separate. Fire-taped areas above ceilings are determined the same way as the third man was calculated for gypsum board installation except that additional layers are not included in the total. Corner and soffit treatment are calculated by lineal feet. Edge trims and expansion joints are also figured by lineal feet. If high or difficult work is anticipated, production should be reduced.

Production rates are best based on prior project experience. Recording past projects is the best tool in future estimating. If this is not available or this type of project has not been completed in the past, these using similar rates and adjusting factors can achieve a reasonable result. Rates should be adjusted for similar but smaller in size. Rates should also be adjusted for high or difficult work. Make note of these areas for final pricing. Stocking labor may be determined using variety of methods such as percent of total labor or material, or actual calculation based on job specific requirements. Drywall is determined in a similar manner.

Laborer production rates can be determined by dividing the total man hours by thirty-two Representing one laborer required for every four man days. This will cover scrap-out and clean up.

Finally, supervision is figured. This is done by again dividing the total production man hours by twenty. This represents one hour for every twenty hours of production. For example, if you had a two-man crew working, a supervisor will check the project for a total of 4 hours each week.

As the crew size increases, so will the supervision hours. Some projects require full time supervision. If this is necessary, determining the total production hours divided by the crew days, multiplied by eight, will account for on site continuous supervision.

09250 Gypsum Board

MATERIALS

Pricing of materials is mandated by current availability of raw materials and special size configurations. Local suppliers will have standard pricing or will provide special pricing for specific projects. When figuring number of metal studs required, multiply length of wall by the spacing factor required, 0.80 for 16 IN on center; 0.75 for 24 IN (these will provide additional studs as needed for corners and doorways.). Then multiply the number of studs by the height of wall to get the length of material for pricing. Track is figured by the lineal feet of wall, times the amount of runs required. Ceiling carriers, hat channels and angles, are also priced by lineal feet. These are separate because they generally are more expensive than standard components.

Gypsum board material is priced by the square foot. Pricing is relatively simple, since all material is priced in square feet. Some materials will be more expensive, for example water resistant at the restroom walls. Joint treatment is determined by square feet also. Finish taping and fire-taping have different coverage rates for material. However, most fire taping in offices is above the ceiling where items penetrate the wall requiring the finisher to 'seal' to the object, which takes extra time. The total cost per unit tends to be the same for each. Corner and edge trims are priced by the piece, depending on their length. Specialty materials are usually kept separate to estimate higher pricing schedules. Set up costs and freight need to be included if applicable.

EQUIPMENT

Equipment cost pricing is calculated two ways. If the equipment is owned, a replacement cost for the equipment is used. If the equipment is to be rented, rental costs are determined for each type. These rates are available from a local rental company. Just like material pricing, quotes for equipment can be obtained for individual projects. Additional cost for stocking equipment, crane use, or scaffolding for special conditions need to be included if applicable.

SPECIAL RISK CONSIDERATIONS

There are a variety of adverse conditions that will affect the final estimate. Scheduling, differing heights of work, weather and labor requirements, all play a part in determining actual costs. These items are generally kept aside for individual evaluation. Adjustments to each vary depending on their composition.

RATIOS AND ANALYSIS

Partition and ceiling framing is basically labor dominate compared to material. The ratio is approximately 70/30 to 60/40, labor and material, respectively. This, of course, changes by design, and special conditions. Cross-checking total costs can be achieved by using common square footage cost multiplied by the total square feet of wall area.

MISCELLANEOUS PERTINENT INFORMATION

With reference to the sample sketches, we have assumed standard trade practices. Not only is each project unique on its design, so are the components. There should be a thorough familiarity with the project before any takeoff or pricing is attempted. Furthermore, items such as, sound and fire caulking, sound attenuation batts, mineral wool at top of walls, mechanical, plumbing and electrical penetrations will also have impact on partition framing and finishing. Scheduling of trades, night work due to existing facilities in operation, and return trips where work was not ready, will also impact pricing.

SAMPLE ESTIMATE

For simplicity reasons a drawings used have been separated into 'shell' and 'tenant improvement' floor plans and reflected ceiling plans. The scope to be bid shall be both the shell work and tenant fit out work together as one contract; to be performed at one time. The takeoff is shown in exhibit 09250- 1. Each partition type; soffit detail; and the area of ceilings is listed in its own column. Partitions are listed by height. Descriptions of each partition type and soffit have been included on the respective drawings. Deck to deck heights shall be 12'-0"; except the fifth floor deck to structure above shall be 14 FT - 0 IN. The exterior wall framing shall be considered to be installed by others; 6 IN insulation and a layer of 5/8 IN gypsum panels shall be installed by this contractor. Note the plans provided are for the first floor of a five story office building; all floors shall be considered the same. Include all floors in the estimate.

The following production rates and hourly wage rates shall be used this sample estimate

1. Carpenters labor is based on $36.00 per hour, burdened.

2. Finishers or Tapers labor is based on $32.00 per hour, burdened.

3. Labors are based on $26.00 per hour, burdened.

4. Burdened rates do not include overhead and profit mark ups.

5. Top track is installed at a rate of 30 lineal feet per hour.

6. Bottom track is installed at a rate of 60 lineal feet per hour.

7. Metal Studs are installed at a rate of 12 lineal feet per hour when spaced 24 IN on center.

8. Stand-up gypsum panels are installed at 16 lineal feet per hour.

9. Top fill productions are 12 lf per hour for horizontal applications up to 4'-0 IN above the stand-ups.

10. Sound insulation batts can be installed at a rate of 150 square feet per hour.

11. Thermal insulation for exterior walls shall be 6 IN foil faced at $0.21 per square foot. Installation rate is 150 square feet per labor hour.

12. Taping production is 150 square feet per day for finished work, sanded and ready for painting.

13. Beads and trims can be installed at a rate of 60 lineal feet per hour.

14. Other rates of production based on historical data have been used in the development of costs for differing walls and soffits. Refer to the sample Estimate.

09250 Gypsum Board

The following material costs shall apply to the sample estimate. Costs are based on purchasing from a supply house that will place the materials in the building at no additional cost:

1. 3 5/8 IN metal studs cost $0.23 per lineal foot.

2. 3 5/8 IN metal track costs $0.21 per lineal foot

3. 4 IN J-Runner for shaft wall construction costs $0.36 per lineal foot.

4. 4 IN CH Studs cost $0.40 per lineal foot.

5. 5/8 IN Gypsum Panels cost $0.24 per square foot

6. 1 IN Gypsum Shaft Wall Liner Board is $0.32 per square foot.

7. 2 ½ IN Sound Insulation Batts cost $0.18 per square foot.

8. Joint compound and tape cost an average of $0.06 per square foot.

9. Corner Beads and trims cost $0.12 per lineal foot.

10. The cost of fasteners, screws, and shots and pins shall be calculated at $0.05 per square foot of gypsum panel.

09250 Gypsum Board

Sample Office Estimate ASPE

		WALL TYPE 1- LF	HGT	WALL TYPE 1A- LF	HGT	WALL TYPE 1B- LF	HGT	WALL TYPE 2- LF	HGT	WALL TYPE 3- LF	HGT	WALL TYPE 4- LF	HGT	EXT WALL-INSUL & SHEATH- LF	HGT	SOFFIT TYPE C3- gypsum @ 10" EF- LF	HGT	SOFFIT TYPE C3- gypsum @ 1'-0" one side- LF	HGT	SOFFIT TYPE C3- gypsum @ 1'-0" one side & 2'-0" other side- LF	HGT	CEILING TYPE C2- SF
	HGT																					
Sketch A	1st floor – SHELL	440	9.5	11	9.5	40	12	81	12	50	5	n/a	12	443	12	n/a		n/a		n/a		n/a
Sketch A	1ST floor – SHELL columns	440	9.5	11	12	40	44															
Sketch A	2nd floor – SHELL	440	9.5	11	9.5	40	12	81	12	50	5	n/a	12	443	12	n/a		n/a		n/a		n/a
Sketch A	2nd floor – SHELL columns	440	9.5	11	12	40	44															
Sketch A	3rd floor – SHELL	440	9.5	11	9.5	40	12	81	12	50	5	n/a	12	443	12	n/a		n/a		n/a		n/a
Sketch A	3rd floor – SHELL columns	440	9.5	11	12	40	44															
Sketch A	4th floor SHELL	440	9.5	11	9.5	40	12	81	12	50	5	n/a	12	443	12	n/a		n/a		n/a		n/a
Sketch A	4th floor – SHELL columns	440	9.5	11	12	40	44															
Sketch A	5th floor – SHELL	440	9.5	11	9.5	40	14	81	14	50	5	n/a	14	443	14	n/a		n/a		n/a		n/a
Sketch A	5th floor – SHELL columns			11	14	40	44															
Sketch C	1st Floor – Fit Out	n/a		494	9.5	n/a		n/a		n/a		468		n/a		n/a		n/a		n/a		n/a
Sketch C	2nd Floor – Fit Out	n/a		494	9.5	n/a		n/a		n/a		468		n/a		n/a		n/a		n/a		n/a
Sketch C	3rd Floor – Fit Out	n/a		494	9.5	n/a		n/a		n/a		468		n/a		n/a		n/a		n/a		n/a
Sketch C	4th Floor – Fit Out	n/a		494	9.5	n/a		n/a		n/a		468		n/a		n/a		n/a		n/a		n/a
Sketch C	5th Floor – Fit Out	n/a		494	9.5	n/a		n/a		n/a		468		n/a		n/a		n/a		n/a		n/a
Sketch B	1st Floor – Ceilings – Shell	n/a		n/a		n/a		n/a		n/a		n/a		3.5		17	3	45	3.5	n/a		940
Sketch B	2nd Floor – Ceilings – Shell	n/a		n/a		n/a		n/a		n/a		n/a		3.5		17	3	45	3.5	n/a		940
Sketch B	3rd Floor – Ceilings – Shell	n/a		n/a		n/a		n/a		n/a		n/a		3.5		17	3	45	3.5	n/a		940
Sketch B	4th Floor – Ceilings – Shell	n/a		n/a		n/a		n/a		n/a		n/a		3.5		17	3	45	3.5	n/a		940
Sketch B	5th Floor – Ceilings – Shell	n/a		n/a		n/a		n/a		n/a		n/a		5.5		17	5	45	5.5	n/a		940
Sketch D	1st Floor – Clg's – Fit Out	n/a		n/a		n/a		n/a		n/a		n/a		n/a		n/a		n/a		341	3.5	204
Sketch D	2nd Floor – Clg's – Fit Out	n/a		n/a		n/a		n/a		n/a		n/a		n/a		n/a		n/a		341	3.5	204
Sketch D	3rd Floor – Clg's – Fit Out	n/a		n/a		n/a		n/a		n/a		n/a		n/a		n/a		n/a		341	3.5	204
Sketch D	4th Floor – Clg's – Fit Out	n/a		n/a		n/a		n/a		n/a		n/a		n/a		n/a		n/a		341	3.5	204
Sketch D	5th Floor – Clg's – Fit Out	n/a		n/a		n/a		n/a		n/a		n/a		n/a		n/a		n/a		341	5.5	204
	SUMMARY	2201	9.5	2523	9.5	199	12	324	12	200	5	2338	12	1770	3.5	68	3	181	3.5	1363	3.5	5717
		440			12	176	14	81	14	50			14	443	5.5	17	5	45	5.5	341	5.5	

PARTITIONS / SOFFIT TYPES

09250 Gypsum Board

	Quantity/ Typ. Rm	Total # Typ. Rm	Quantity/ Plr A401	Total # Plr A401	Quantity/ Plr A402	Total # Plr A402	Quantity/Plr A403	Total # Plr A403	Project Total	
Drapery Pocket – lf	12	221	0	14	0	7	0	7	2652	LF
S/R ceilings	84.5	221	119	14	72	7	88	7	21460.5	SF
S/R Ceilings	318.5	33	674	2	345	1	362	1	12565.5	SF
OTS Ceilings	234	188	555	12	273	6	274	6	53934	SF
Bullhead Above cabinets	0	221	11	14	0	7	0	7	154	LF
½ Durock @ Showers	72	221	72	14	72	7	72	7	17928	SF
2 ½" framing @ tub Head	3	221	0	0	0	0	0	0	663	LF

Type A1 Partitions
Type C1 Partitions
Type C2 Partitions
Type D1 Partitions
Type D2 Partitions
Type D3 Partitions
Type D4 Partitions
Type D5 Partitions
Type E1 Partitions
Type E2 Partitions
Type E3 Partitions
Type E4 Partitions
Type F1 Partitions
Type F2 Partitions
Type F3 Partitions
Type F4 Partitions
Type F5 Partitions
Type G1 Partitions
Type G2 Partitions
Type G3 Partitions
Type G4 Partitions
Type C1 Framing @ Showerhead Locations
Added Layer ½" Cement Board – (Tile Backer) @ Showers & Tubs
Exterior Wall – 6" Insulation w/ 5/8" Drywall
Guestroom Suspended Drywall Ceilings
Guestroom Drapery Pockets
Guestrooms – Flat Tape Slabs @ Textured Paint Areas
Parlor Suite Bulkheads above Cabinets
Drywall Ceilings @ Health Club Level 3
Corridors Suspended Drywall Ceilings
Corridors Drywall Soffits

	Flat							Stepped	
	S/R	Soffits x4'/lF	Soffits 5sf/lf	2x2 ACT	2x4 ACT	2x4 vinyl roof	2x4 Scrub.	S/R	tectum
HOTEL 1ST LEVEL									
Entrance Lobby	4851	375	97						
Coats				178					
Lounge	935	206							
Prefunction	2120	274		291					
Ballroom	3365	923		2269					
Restaurant	3059	134	137	553					
Toilet rooms	1564								
Kitchen Areas	118						4759		
Front desk / offices				954					
Employee Entrance					5752				
Holding/Loading areas						231			
Service Corr's/ Storage/Backhouse				Incl in employee entr/					
HOTEL 2ND LEVEL									
Auditorium	22817	865		958					
Prefunction/break/corridors	Incl	Incl		Incl	976				
Conference Rooms				2975					
Breakout Rooms				1934					
Toilet Rooms	Incl								
Hotel Offices					3495				
Service Corridor/Black of House							4327		
level 3 health club/pool				3697					1132
1ST LEVEL									
entrance	1010	73							

Part Two 09250 Gypsum Board

ASPE SAMPLE DRYWALL ESTIMATE

ASPE SAMPLE DRYWALL ESTIMATE

PARTITION SUMMARY

#	TITLE	PART. TYP	HT	THCK	GA	TOTAL LGTH					UNIT MATL			UNIT MATL COST	UNIT /HR		HRS	LABOR RATE	MATL COST	LABOR COST	MATL COST/SF	LABOR COST/SF	TOTAL COST	TOTAL COST/SF
1	Type 1	WALL	12	4 7/8"	25	2201	1	EA	3 5/8"		TOP TRACK	2201	LF	0.21	30	O C	74	36.00	462.21	2,864.00	0.99	1.77	72,903.82	2.76
							1	EA	3 5/8"		BTM TRACK	2201	LF	0.21	60	O C	37	36.00	462.21	1,332.00				
							1	EA	3 5/8"		STUD	1,651	PC	0.23	12	O C	184	36.00	4,556.07	6,624.00				
		to 10'					2	sd	5/8"	level	FINISH	4,402	HT	0.24	16	O C	276	36.00	10,564.80	9,936.00				
		10 to 14					2	sd	5/8"	level	FINISH	4,402	HT	0.24	12	O C	367	36.00	2,112.96	13,212.00				
							2	ls			T & S	52,824	SF	0.06	150	1 T	353	32.00	3,169.44	11,298.00				
											S.P.S.	52,824	SF	0.06					3,169.44	0.00				
							2	ln			CAULK	4,402	LF	0.11	200	O C	23	36.00	484.22	828.00				
							1	ln			FIRE SAFE	2,201	LF	0.47	100	O C	23	36.00	1,034.47	828.00				
							20	EA			CRNR BD	200	LF	0.12	60	O C	4	36.00	24.00	144.00				
2	Type 1	WALL	14	4 7/8"	25	440	1	EA	3 5/8"		TOP TRACK	440	LF	0.21	30	O C	15	36.00	92.40	540.00	0.97	1.60	15,848.20	2.57
							1	EA	3 5/8"		BTM TRACK	440	LF	0.21	60	O C	8	36.00	92.40	288.00				
							1	EA	3 5/8"		STUD	330	PC	0.23	12	O C	37	36.00	1,062.60	1,332.00				
		to 10'					2	sd	5/8"	level	FINISH	880	HT	0.24	16	O C	55	36.00	2,112.00	1,980.00				
		10 to 14					2	sd	5/8"	level	FINISH	880	HT	0.24	12	O C	74	36.00	844.80	2,664.00				
							2	ls			T & S	12,320	SF	0.06	150	1 T	83	32.00	739.20	2,656.00				
											S.P.S.	12,320	SF	0.06					739.20	0.00				
							2	ln			CAULK	880	LF	0.11	200	O C	5	36.00	96.80	180.00				
							1	ln			FIRE SAFE	440	LF	0.47	100	O C	5	36.00	206.80	180.00				
							5	EA			CRNR BD	50	LF	0.12	60	O C	1	36.00	6.00	36.00				
3	Type 1A	WALL	9.5	4 7/8"	25	2523	1	EA	3 5/8"		TOP TRACK	2523	LF	0.21	30	O C	85	36.00	529.83	3,060.00	0.94	1.42	56,439.55	2.35
							1	EA	3 5/8"		BTM TRACK	2523	LF	0.21	60	O C	43	36.00	529.83	1,548.00				
							1	EA	3 5/8"		STUD	1,892	PC	0.23	12	O C	211	36.00	4,134.57	7,596.00				
		to 10'					2	sd	5/8"		FINISH	5,046	HT	0.24	16	O C	316	36.00	11,504.88	11,378.00				
							2	ls			T & S	47,937	HT	0.06	150	1 T	320	32.00	2,876.22	10,240.00				
											S.P.S.	47,937	SF	0.06			0	36.00	2,876.22	0.00				
							0	ln			CAULK	0	LF	0.11	200	O C	0	36.00	0.00	0.00				
							0	ln			FIRE SAFE	0	LF	0.47	100	O C	0	36.00	0.00	0.00				
							20	EA			CRNR BD	200	LF	0.12	60	O C	4	36.00	24.00	144.00				
4	Type 1B	WALL	9.5	4 7/8"	25	199	1	EA	3 5/8"		TOP TRACK	199	LF	0.21	30	O C	7	36.00	41.79	252.00	0.58	1.04	3,063.87	1.62
							1	EA	3 5/8"		BTM TRACK	199	LF	0.21	60	O C	4	36.00	41.79	144.00				
							1	EA	3 5/8"		STUD	149	PC	0.23	12	O C	17	36.00	326.11	612.00				
		to 10'					1	sd	5/8"	level	FINISH	199	HT	0.24	16	O C	13	36.00	453.72	468.00				
							1	ls			T & S	1,891	SF	0.06	12	O C	13	32.00	113.43	416.00				
											S.P.S.	1,891	SF	0.06	150	1 T			113.43	0.00				
							8	EA			CRNR BD	80	LF	0.12	60	O C	2	36.00	9.60	72.00				
5	Type 1B	WALL	12	4 7/8"	25	178	1	EA	3 5/8"		TOP TRACK	176	LF	0.21	30	O C	6	36.00	36.96	216.00	0.59	1.18	3,732.96	1.77
							1	EA	3 5/8"		BTM TRACK	176	LF	0.21	60	O C	3	36.00	36.96	108.00				
							1	EA	3 5/8"		STUD	132	PC	0.23	12	O C	15	36.00	364.32	540.00				
		to 10'					1	sd	5/8"	level	FINISH	178	LF	0.24	16	O C	11	36.00	422.40	396.00				
		10 to 14					1	sd	5/8"	level	FINISH	176	HT	0.24	12	O C	15	36.00	84.48	540.00				
							1	ls			T & S	2,112	SF	0.06	150	1 T	15	32.00	126.72	480.00				
											S.P.S.	2,112	SF	0.06					126.72	0.00				
							32	EA			CRNR BD	320	LF	0.12	60	O C	6	36.00	38.40	216.00				

Standard Estimating Practice

09250 Gypsum Board — Part Two

PART. TYP.	HT	THCK	GA	TOTAL LGTH					UNIT MAT'L		UNIT MAT'L COST	UNIT /HR			HRS	LABOR RATE	MAT'L COST	LABOR COST	MAT'L COST/SF	LABOR COST/SF	TOTAL COST	TOTAL COST/SF
6 Type 1B	WALL	14 4 7/8"	25	44	1 EA	3 5/8"	TOP TRACK	44	LF	10 LF/PC	0.21	30	O	C	2	36.00	9.24	72.00	0.94	1.81	1,693.86	2.75
					1 EA	3 5/8"	BTM TRACK	44	LF	10 LF/PC	0.21	60	O	C	1	36.00	9.24	36.00				
					1 EA	3 5/8"	STUD	33	PC	14 LF/PC	0.23	12	O	C	4	36.00	106.26	144.00				
	to 10'				2 sd	5/8"	FINISH	88	HT	10 HT	0.24	16	O	C	6	36.00	211.20	216.00				
	10 to 14				2 sd	5/8"	FINISH	88	HT	4 HT	0.24	12	O	C	8	36.00	84.48	288.00				
					2 ls		T & S	1,232	SF	14 HT	0.06	150	1	T	9	32.00	73.92	288.00				
							S.P.S.	1,232	SF		0.00						0.00	0.00				
					8 EA		CRNR BD.	80	LF	10 LF/PC	0.12	80	O	C	2	36.00	9.60	72.00				
7 Type 2	WALL	12 4 7/8"	25	324	1 EA	3 5/8"	TOP TRACK	324	LF	10 LF/PC	0.21	30	O	C	11	36.00	68.04	396.00	1.59	2.69	16,620.08	4.27
					1 EA	3 5/8"	BTM TRACK	324	LF	10 LF/PC	0.21	60	O	C	6	36.00	68.04	216.00				
					1 EA	3 5/8"	STUD	243	PC	12 LF/PC	0.23	12	O	C	27	36.00	670.68	972.00				
	to 10'	level			2 sd	5/8"	BASE	648	HT	10 HT	0.24	16	O	C	41	36.00	1,555.20	1,476.00				
	10 to 14	level			2 sd	5/8"	BASE	648	HT	2 HT	0.24	12	O	C	54	36.00	311.04	1,944.00				
	to 10'	level			2 sd	5/8"	FINISH	648	HT	10 HT	0.24	16	O	C	41	36.00	1,555.20	1,476.00				
	10 to 14	level			2 sd	5/8"	FINISH	648	HT	2 HT	0.24	12	O	C	54	36.00	311.04	1,944.00				
					2 ls		T & S	7,776	HT	12 HT	0.06	150	1	T	52	32.00	466.56	1,664.00				
							S.P.S.	15,552	SF		0.06						933.12	0.00				
					2 ln		CAULK	648	LF		0.11	200	O	C	4	36.00	71.28	144.00				
					1 ln		FIRE SAFE	324	LF		0.47	100	O	C	4	36.00	152.28	144.00				
					8 EA		CRNR BD.	80	LF	10 LF/PC	0.12	80	O	C	2	36.00	9.60	72.00				
8 Type 2	WALL	14 4 7/8"	25	81	1 EA	3 5/8"	TOP TRACK	81	LF	10 LF/PC	0.21	30	O	C	3	36.00	17.01	108.00	1.57	2.51	4,636.81	4.09
					1 EA	3 5/8"	BTM TRACK	81	LF	10 LF/PC	0.21	60	O	C	2	36.00	17.01	72.00				
					1 EA	3 5/8"	STUD	61	PC	14 LF/PC	0.23	12	O	C	7	36.00	195.62	252.00				
	to 10'	level			2 sd	5/8"	BASE	162	HT	10 HT	0.24	16	O	C	11	36.00	388.80	396.00				
	10 to 14	level			2 sd	5/8"	BASE	162	HT	4 HT	0.24	12	O	C	14	36.00	155.52	504.00				
	to 10'	level			2 sd	5/8"	FINISH	162	HT	10 HT	0.24	16	O	C	11	36.00	388.80	396.00				
	10 to 14	level			2 sd	5/8"	FINISH	162	HT	4 HT	0.24	12	O	C	14	36.00	155.52	504.00				
					2 ls		T & S	2,268	SF	14 HT	0.06	150	1	T	16	32.00	136.08	512.00				
							S.P.S.	4,536	SF		0.06						272.16	0.00				
					2 ln		CAULK	162	LF		0.11	200	O	C	1	36.00	17.82	36.00				
					1 ln		FIRE SAFE	81	LF		0.47	100	O	C	1	36.00	38.07	36.00				
					2 EA		CRNR BD.	20	LF	10 LF/PC	0.12	80	O	C	1	36.00	2.40	36.00				
9 Type 3	WALL	12 4 7/8"	25	200	1 EA	3 5/8"	J Runner	200	LF	10 LF/PC	0.36	25	O	C	8	36.00	72.00	288.00	1.16	2.48	8,724.80	3.64
					1 EA	3 5/8"	J Runner	200	LF	10 LF/PC	0.36	50	O	C	4	36.00	72.00	144.00				
					1 EA	3 5/8"	CH STUD	150	PC	12 LF/PC	0.40	8	O	C	25	36.00	720.00	900.00				
					1 ea	1"	Shaftliner	200	LF	12 HT	0.32	4	O	C	50	36.00	64.00	1,800.00				
	to 10'				2 sd	5/8"	BASE	200	HT	10 HT	0.24	16	O	C	13	36.00	480.00	468.00				
	10 to 14				2 sd	5/8"	BASE	200	HT	2 HT	0.24	12	O	C	17	36.00	96.00	612.00				
	to 10'				2 sd	5/8"	FINISH	200	HT	10 HT	0.24	16	O	C	13	36.00	480.00	468.00				
	10 to 14				2 sd	5/8"	FINISH	200	HT	2 HT	0.24	12	O	C	17	36.00	96.00	612.00				
					1 ls		T & S	2,400	SF	12 HT	0.06	150	1	T	16	32.00	144.00	512.00				
							S.P.S.	7,200	SF		0.06						432.00	0.00				
					1 ln		CAULK	200	LF		0.11	200	O	C	1	36.00	22.00	36.00				
					1 ln		FIRE SAFE	200	LF		0.47	100	O	C	2	36.00	94.00	72.00				
					4 EA		CRNR BD	40	LF		0.12	80	O	C	1	36.00	4.80	36.00				
10 Type 3	WALL	14 4 7/8"	25	50	1 EA	3 5/8"	J Runner	50	LF	10 LF/PC	0.36	25	O	C	2	36.00	18.00	72.00	1.14	2.49	2,540.20	3.63
					1 EA	3 5/8"	J Runner	50	LF	10 LF/PC	0.36	50	O	C	1	36.00	18.00	36.00				
					1 EA	3 5/8"	CH STUD	38	PC	14 LF/PC	0.40	8	O	C	7	36.00	210.00	252.00				
					1 ea	1"	Shaftliner	50	LF	14 HT	0.32	4	O	C	13	36.00	468.00	468.00				
	to 10'	level			2 sd	5/8"	BASE	50	LF	10 HT	0.24	16	O	C	4	36.00	120.00	144.00				
	10 to 14	level			2 sd	5/8"	BASE	50	LF	4 HT	0.24	12	O	C	5	36.00	48.00	180.00				
	to 10'	level			2 sd	5/8"	FINISH	50	LF	10 HT	0.24	16	O	C	4	36.00	120.00	144.00				
	10 to 14	level			2 sd	5/8"	FINISH	50	LF	4 HT	0.24	12	O	C	5	36.00	48.00	180.00				
					1 ls		T & S	700	SF	14 HT	0.06	150	1	T	5	32.00	42.00	160.00				
							S.P.S.	2,100	SF		0.06						126.00	0.00				
					1 ln		CAULK	50	LF		0.11	200	O	C	1	36.00	5.50	36.00				
					1 ln		FIRE SAFE	50	LF		0.47	100	O	C	1	36.00	23.50	36.00				
					1 EA		CRNR BD	10	LF	10 LF/PC	0.12	80	O	C	1	36.00	1.20	36.00				

Part Two 09250 Gypsum Board

#	PART TYP	HT	THCK	GA	TOTAL LGTH					UNIT MATL		UNIT MATL COST	UNIT /HR	O/C	HRS	LABOR RATE	MATL COST	LABOR COST	MATL COST/SF	LABOR COST/SF	TOTAL COST	TOTAL COST/SF
11	Type 4 WALL	5 4 7/8"		25	2338	1 EA	3 5/8"	TOP TRACK	10 LF/PC	2338	LF	0.21	30	O/C	78	36.00	490.98	2,808.00	1.12	2.46	41,830.91	3.58
						1 EA	3 5/8"	BTM TRACK	10 LF/PC	2338	LF	0.21	60	O/C	39	36.00	490.98	1,404.00				
						1 EA	3 5/8"	STUD	5 LF/PC	1,754	PC	0.23	12	O/C	195	36.00	2,016.53	7,020.00				
		to 10'	level			2 sd	5/8"	FINISH	5 HT	4,676	LF	0.24	16	O/C	293	36.00	5,611.20	10,548.00				
						2 ls		T & S	5 HT	23,380	SF	0.06	150	1T	156	32.00	1,402.80	4,992.00				
								S.P.S.		23,380	SF						1,402.80	0.00				
						2 ln		CAULK		4,678	LF	0.11	200	O/C	24	36.00	514.36	864.00				
						1 ln		FIRE SAFE		2,338	LF	0.47	100	O/C	24	36.00	1,098.86	864.00				
						84 EA		CRNR BD	5 LF/PC	420	LF	0.12	60	O/C	7	36.00	50.40	252.00				
12	EXTERIOR WALL DRYWALL & INSUL	12	0	25	1770	0 EA	3 5/8"	TOP TRACK	10 LF/PC	0	LF	0.21	30	O/C	0	36.00	0.00	0.00	0.57	0.89	31,086.80	1.46
						0 EA	3 5/8"	BTM TRACK	10 LF/PC	0	LF	0.21	60	O/C	0	36.00	0.00	0.00				
						0 EA	3 5/8"	STUD	12 LF/PC	0	PC	0.23	12	O/C	0	36.00	0.00	0.00				
		to 10'	level			1 sd	5/8"	FINISH	10 HT	1,770	LF	0.24	16	O/C	111	36.00	4,248.00	3,996.00				
		10 to 14	level			1 sd	5/8"	FINISH	12 HT	1,770	LF	0.24	12	O/C	148	36.00	849.60	5,328.00				
						1 lyr	6"	INSUL	12 HT	21,240	SF	0.21	150	O/C	142	36.00	4,460.40	5,112.00				
						1 ls		T & S	12 HT	21,240	SF	0.06	150	1T	142	32.00	1,274.40	4,544.00				
								S.P.S.		21,240	SF						1,274.40	0.00				
13	EXTERIOR WALL DRYWALL & INSUL	14		25	443	0 EA	3 5/8"	TOP TRACK	10 LF/PC	0	LF	0.21	30	O/C	0	36.00	0.00	0.00	0.57	0.84	8,731.14	1.41
						0 EA	3 5/8"	BTM TRACK	10 LF/PC	0	LF	0.21	60	O/C	0	36.00	0.00	0.00				
						0 EA	3 5/8"	STUD	14 LF/PC	0	PC	0.23	12	O/C	0	36.00	0.00	0.00				
		to 10'	level			1 sd	5/8"	FINISH	10 HT	443	LF	0.24	16	O/C	28	36.00	1,063.20	1,008.00				
		10 to 14	level			1 sd	5/8"	FINISH	14 HT	443	LF	0.24	12	O/C	37	36.00	425.28	1,332.00				
						1 lyr	6"	INSUL	14 HT	6,202	SF	0.21	150	O/C	42	36.00	1,302.42	1,512.00				
						1 ls		T & S	14 HT	6,202	SF	0.06	150	1T	42	32.00	372.12	1,344.00				
								S.P.S.		6,202	SF						372.12	0.00				
14	TYPE C3 SOFFIT	3.5 3 1/8"		25	68	1 EA	3 5/8"	TOPTRK	10 LF/PC	68	LF	0.25	30	O/C	3	36.00	17.00	108.00	0.69	6.17	1,631.13	6.85
						1 EA	3 5/8"	BTM TRK	10 LF/PC	68	LF	0.25	30	O/C	3	36.00	17.00	108.00				
						1 EA	3 5/8"	STUD	3.5 LF/PC	54	PC	0.28	8	O/C	9	36.00	49.50	324.00				
		to 10'	level			2 sd	5/8"	FINISH	1 HT	136	LF	0.25	10	O/C	14	36.00	34.00	504.00				
		10 to 14	level			1 sd	5/8"	FINISH	0.33 HT	68	LF	0.25	10	O/C	7	36.00	8.50	252.00				
						1 ls		T & S	2.33 HT	158	SF	0.06	100	1T	2	32.00	9.51	64.00				
								S.P.S.		158	SF	0.06					9.51	0.00				
						14 EA		CRNR BD	10 LF/PC	140	LF	0.15	60	O/C	3	36.00	21.00	108.00				
15	TYPE C3 SOFFIT	5 3 1/8"		25	17	1 EA	3 5/8"	TOPTRK	10 LF/PC	17	LF	0.25	30	O/C	1	36.00	4.25	36.00	0.55	5.46	510.84	6.01
						1 EA	3 5/8"	BTM TRK	10 LF/PC	17	LF	0.25	30	O/C	1	36.00	4.25	36.00				
						1 EA	3 5/8"	STUD	5 LF/PC	14	PC	0.26	8	O/C	3	36.00	17.68	108.00				
		to 10'	level			2 sd	5/8"	FINISH	1 HT	34	LF	0.25	10	O/C	4	36.00	8.50	144.00				
		10 to 18	level			1 sd	5/8"	FINISH	0.33 HT	17	LF	0.25	10	O/C	2	36.00	1.40	72.00				
						1 ls		T & S	2.33 HT	40	SF	0.06	100	1T	1	32.00	2.38	32.00				
								S.P.S.		40	SF	0.06					2.38	0.00				
						4 EA		CRNR BD	10 LF/PC	40	LF	0.15	60	O/C	1	36.00	6.00	36.00				
16	TYPE C3 SOFFIT	3 3 1/8"		25	181	1 EA	3 5/8"	TOPTRK	10 LF/PC	181	LF	0.25	30	O/C	7	36.00	45.25	252.00	0.59	5.41	3,261.01	6.01
						1 EA	3 5/8"	BTM TRK	10 LF/PC	181	LF	0.25	30	O/C	7	36.00	45.25	252.00				
						1 EA	3 5/8"	STUD	3 LF/PC	145	PC	0.28	8	O/C	23	36.00	112.94	828.00				
		to 10'	level			2 sd	5/8"	FINISH	1 HT	181	LF	0.25	10	O/C	19	36.00	45.25	684.00				
		10 to 18	level			1 sd	5/8"	FINISH	0.33 HT	181	LF	0.25	10	O/C	19	36.00	14.93	684.00				
						1 ls		T & S	1.33 HT	241	SF	0.06	100	1T	3	32.00	14.44	96.00				
								S.P.S.		241	SF	0.06					14.44	0.00				
						19 EA		CRNR BD	10 LF/PC	190	LF	0.15	60	O/C	4	36.00	28.50	144.00				

Standard Estimating Practice

09250 Gypsum Board — Part Two

PART. TYP.		HT	THCK	GA	TOTAL LGTH				UNIT MATL		UNIT MATL COST	UNIT /HR		HRS	LABOR RATE	MATL COST	LABOR COST	MATL COST/SF	LABOR COST/SF	TOTAL COST	TOTAL COST/SF
17 TYPE C3	WALL	5	3 1/8"	25	45	1 EA	3 5/8"	TOPTRK	10 LF/PC	45 LF	0.25	30	O C	2	36.00	11.25	72.00				
						1 EA	3 5/8"	BTM TRK	10 LF/PC	45 LF	0.25	30	O C	2	36.00	11.25	72.00				
			14			1 EA	3 5/8"	STUD	5 LF/PC	36 PC	0.28	8	O C	6	36.00	46.80	216.00				
	to 10'					1 sd	5/8"	FINISH	1 HT	45 LF	0.25	10	O C	5	36.00	11.25	180.00				
	10 to 18		level			1 sd	5/8"	FINISH	0.33 HT	45 LF	0.25	10	O C	5	36.00	3.71	180.00				
						1 ls		T & S	2.33 HT	105 SF	0.06	100	1 T	2	32.00	6.29	64.00				
								S.P.S.		80 SF	0.06					3.59	0.00				
						14 EA		CRNR BD.	10 LF/PC	140 LF	0.15	60	O C	3	36.00	21.00	108.00	0.51	3.96	1,007.14	4.48
18 TYPE C3	SOFFIT	3.5	3 1/8"	25	1363	1 EA	3 5/8"	TOPTRK	10 LF/PC	1363 LF	0.25	30	O C	46	36.00	340.75	1,656.00				
						1 EA	3 5/8"	BTM TRK	10 LF/PC	1363 LF	0.25	30	O C	46	36.00	340.75	1,656.00				
			14			1 EA	3 5/8"	STUD	3.5 LF/PC	1,090 PC	0.28	8	O C	171	36.00	992.26	6,156.00				
	to 10'					1 sd	5/8"	FINISH	1 HT	1,363 LF	0.25	10	O C	137	36.00	340.75	4,932.00				
	10 to 18		level			1 sd	5/8"	FINISH	2 HT	1,363 LF	0.25	10	O C	137	36.00	681.50	4,932.00				
	18 to 26		level			1 sd	5/8"	FINISH	0.33 HT	1,363 LF	0.25	10	O C	137	36.00	112.45	4,932.00				
						1 ls		T & S	3.33 HT	4,539 SF	0.06	100	1 T	46	32.00	272.33	1,472.00				
								S.P.S.		4,539 SF	0.06					272.33	0.00				
						274 EA		CRNR BD.	10 LF/PC	2740 LF	0.15	60	O C	46	36.00	411.00	1,656.00	0.79	5.74	31,156.12	6.53
19 TYPE C3	SOFFIT	5.5	3 1/8"	25	341	1 EA	3 5/8"	TOPTRK	10 LF/PC	341 LF	0.25	30	O C	12	36.00	85.25	432.00				
						1 EA	3 5/8"	BTM TRK	10 LF/PC	341 LF	0.25	30	O C	12	36.00	85.25	432.00				
			14			1 EA	3 5/8"	STUD	5.5 LF/PC	273 PC	0.28	8	O C	43	36.00	390.10	1,548.00				
	to 10'					1 sd	5/8"	FINISH	1 HT	341 LF	0.25	10	O C	35	36.00	85.25	1,260.00				
	10 to 18		level			1 sd	5/8"	FINISH	2 HT	341 LF	0.25	10	O C	35	36.00	170.50	1,260.00				
	18 to 26		level			1 sd	5/8"	FINISH	0.33 HT	341 LF	0.25	10	O C	35	36.00	28.13	1,260.00				
						1 ls		T & S	3.33 HT	1,136 SF	0.06	100	1 T	12	32.00	68.13	384.00				
								S.P.S.		1,136 SF	0.06					68.13	0.00				
						69 EA		CRNR BD.	10 LF/PC	690 LF	0.15	60	O C	12	36.00	103.50	432.00	0.58	3.74	8,092.25	4.31
7 CEILING		1		20	5717	1 EA	1 5/8"	Grid System		5717 SF	0.38	100	O C	58	36.00	2,172.46	2,088.00				
	to 10'		level			1 sd	5/8"	DRYWALL	1 HT	5,717 LF	0.20	48	O C	120	38.00	1,143.40	4,320.00				
						1 ls		T & S	1 HT	5,717 SF	0.06	80	1 T	72	32.00	343.02	2,304.00				
								S.P.S.		5,717 SF	0.07					400.19	0.00	0.71	1.52	12,771.07	2.23
					10591 lf walls & soffits												107,286.55	252,144.00				
								supervision	6,236 hrs			312	0 CF	312	36.00	0.00	11,225.00				
								cleanup	210,415 sf	800	0.1	600	3 L	351	28.00	60.00	9,119.00				
								layout		10,591 lf		30	0 C	354	36.00	0.00	12,744.00				
															sub total					328,282.55	
																	11,225.00				
																	9,179.00				
																	12,744.00				
															sub total		252,144.00			359,430.55	
															CHECK	29.85%	70.15%			359,430.55	
																	10.00%	overhead			35,943.05	
																	5.00%	profit			17,971.53	
																		BID			413,345.13	

09510 – ACOUSTICAL CEILINGS

INTRODUCTION

Many types of ceiling systems are grouped under the heading "Acoustical Ceilings" and this discussion specifically addresses estimating the cost of lay-in ceiling tiles in a standard exposed grid system. Suggestions will be offered for other types of ceiling systems.

Many General Contractors estimate acoustical ceilings using the square foot cost method. An Itemized estimate is typically used by subcontractors and general contractors who self perform the installation of acoustical ceiling systems. An accurate takeoff and estimate can be used directly in ordering of materials and the budgeting of labor hours once awarded the given project. The takeoff and calculations are performed once, allowing the contractor to turn a project over to the field quickly, creating more time for management of the field staff.

The basic ceiling systems consist of inserts or anchoring devises for the attachment of hanger wires to the structure above. Wires are spaced four foot on center in both directions. The hanger wires are most often attached to clips which are attached to the bottom of the structure above with a power actuate fastener. In most commercial applications the deck above can be reached with a pole gun. The clip is slipped over the end of the long shaft of the pole gun. The end raised to the deck above. The power actuated fastener is fired by slight upward pressure from the bottom of the pole gun. It is important that the wires be sufficiently wrapped around itself once passed through the clip to assure against future pull out. The wire must be of sufficient length to allow the installer to run it through the grid and wrap it on itself to prevent future pull out from the weight of the system.

The grid system or acoustical suspension system generally consists of wall angles or molding fastened directly to the walls of the space or perimeter borders of the ceiling system. The grid is then snapped into place. Main runners, usually twelve feet in length, are run four foot on center in one direction. The mains are fastened to the hanger wires and leveled to the finished ceiling height as they are installed. A wall mounted laser level is very helpful in achieving a level ceiling. Four foot tees are then snapped into slots in the mains every two foot on center. If the system is for a twenty four inch by forty eight inch ceiling tile, the grid system is completed. Note Figure 09510-1 Room #101.

A system with a twenty four inch by twenty four inch tile requires another piece to complete the grid system. A two foot tee is snapped into slots in the center of the four foot tees. The two foot tees run parallel the main runners. Sometimes twenty four inch by twenty four inch tile systems are specified with main runners spaced at two foot centers with two foot tees in between. Note Figure 09510-1 Rooms 102 and 103.

Once the grid system is installed, leveled and locked into place, the ceiling mounted commodities of other building systems are installed. For example, ceiling diffusers, return air grilles, recessed light fixtures, etc. are laid into the grid (usually by other contractors). If required, the sprinkler contractor will cut back the drops to be level with the grid and install the sprinkler heads. While this work is done by others, coordination is required between the sprinkler and acoustical ceiling contractors.

The other trades now complete in the grid system, the ceiling installer can now install the ceiling tiles. The tiles are appropriately named as 'lay-in' tiles. They are laid into the grid system. The different edge types manufactured today require different work by the installer for a complete installation. There is inevitably *a* row of grid along the perimeter of the area in which nominal sized tiles will not fit. These smaller bays are called 'boarder cuts' by many tradesman. The installer not only has to cut the grid pieces along the walls, but he tiles must be cut to fit. If a reveal edge tile is used the tile must be not only cut to size but the reveal cut into the cut edge. Some specifications require these edges to be painted to match the tile. In most situations this is not required because the cut edge can not be seen due to its close proximity to the wall surface. The installer will have to make cutouts in tile for items such as sprinkler heads, high hat light fixtures and other items specified for installation in the acoustical ceiling system.

TYPES AND METHODS OF MEASUREMENT

Acoustical ceiling systems are generally measured in square foot of ceiling area. It is also important that the border or perimeter (lineal feet) of the ceiling be measured accurately.

Irregular areas of an 'L' type shape, comprising of two or more rectangular shapes, must be treated differently. The individual areas must be measured separately and their areas calculated. Add these areas together to find the total area of the space. The border or perimeter (lineal feet) must be measured directly from the design documents.

Areas with circular or triangular shapes or sides require more time to install. Production rates for complex shapes are reduced. The area and perimeter of a circle is easily calculated. Even the area of a rectangular space with a curved wall on one end can be broken into a triangle and an arched section. Calculate each area and add the calculations together to determine the total area. The perimeter must also be determined.

Since estimating the costs of the system will be done on a piece by piece basis, perform a takeoff of all the ceilings in the project before entering anything on the spreadsheet. The quantities for all rooms with the same ceiling type can be combined and the costs calculated at one time. The attached takeoff sheet is useful in compiling the total project quantity of a ceiling type. The total quantity will then be used in calculating the costs.

FACTORS AFFECTING TAKEOFF AND PRICING

1. Local labor force production rates. The skill and experience of the labor force can affect the cost of the system. Research actual historical data to establish production rates.

2. The size and shapes of the rooms to receive acoustical ceilings. The larger the rooms, the lower the overall percentage of border cuts and other cuts in the room. Rooms that are small in size will have a higher percentage of border cuts and light and sprinkler cutouts. These being time consuming work items, raise the labor hours per square foot of system, thus increasing the unit price per square foot.

3. Larger rooms will allow for larger work platforms to be built and moved around for easy access by the installers. Productivity will increase in large open rooms where installers can work off platforms up to twenty five foot by twenty five foot. Typical 'Baker' type scaffold is used when rooms are smaller and work may be performed in many rooms during the course of a day.

4. Odd shaped areas result in more border cuts and/or more waste. If a room is rectangular the border cuts may be used to fill bays along another wall. Odd shaped cuts typically result in unusable pieces, which increases the material costs.

5. The orientation of the ceiling in relation to the room will affect the cost of the system. Typically the ceiling grid is specified to parallel to the partitions and centered in the space in both directions. However, there are times that the system is skewed in the space for aesthetics. This will not only result in many odd shaped border cuts (producing more waste), but the layout and alignment of the grid system is more difficult and time consuming.

6. The height of the ceiling above the floor is another factor to take into consideration. Typical ceilings are around eight feet above the floor. The installer can easily work from a platform around three feet high. A ceiling system over twelve feet high will result in more elaborate scaffolding or working from powered lifts. The result is in more time spent building the scaffold, getting the lifts into or out of the building, lifting materials to the platform height, getting debris down to the floor level, and getting workers to the working height. Productivity is reduced when working from higher platforms.

7. The height of the structure above the floor level and the finished ceiling height will affect the cost. If the structure above is to high to reach from the floor, scaffolding is needed for attachment of hanger wires even if the ceiling is eight feet above the floor. The height of the structure above the ceiling system determines the length of the hanger wires. If the distance from ceiling to the structure above is bigger than the length of available wires, the cost of hangers will be increased.

8. The ceiling tile material and patterns have a great affect on the cost of the system. Obviously different materials and special patterned tiles will have a higher material purchase cost. They will also result in higher production costs. Metal or wooden ceiling tiles can not be scored and snapped like a mineral fiber board. Patterned tiles with scored lines and grooves will require special care and time to assure proper cuts. The patterned tiles also take more time to assure each tile is installed in the proper direction.

9. The type of grid system specified will result in different production rates. Standard 15/16 IN grid is the easiest and least time consuming to install. Fine line 9/16 IN grids or grids with slots and reveals in them take a little more time to install. They also tend to be a little less rigid before the tiles are set resulting in time spent realigning and adjusting the system if it is knocked or moved during the installation of other commodities in the ceiling system.

10. Concealed grid systems and metal pan systems with concealed systems have different installation procedures and unique productivity factors. The costs of such systems will not be discussed in this article. A similar approach to such systems can be used in determining their costs. If the pieces can be properly visualized and each one broken down to a single line item and the time to install each component properly analyzed; the estimator can arrive at a cost for any system, even if it has never been seen before.

11. Special local requirements such as 'hold down clips', seismic requirements and other specific requirements may add to the cost of performing the work.

It is important to be familiar with the many different ceiling systems incorporated in today's many diverse building systems. Nuances in the different systems can mean a great deal in the final cost of the system. To simplify the sample estimate, the standard 15/16 IN grid system with a simple lay-in reveal edge tile will be used.

PERFORMING THE TAKEOFF

Using the takeoff sheet provided, measure each room to arrive at both the square footage of ceiling area and the perimeter of the space. It is helpful on projects that may be subject to changes by addendum to list each room by room number, its length and width, the perimeter of the space. The heights of structure above and ceiling heights above finished floors should be noted during the takeoff process, along with bulkheads, coffers, drapery pockets and other variations in ceiling heights. The type of ceiling system must be noted also. Which is usually found on the finish schedule.

The takeoff sheet (Example 09510-1) is then used to calculate the values of the total square footage and total perimeter of each type of ceiling on the project. The quantities are then added directly to the spreadsheet for calculations.

THE SPREADSHEET AND CALCULATING THE COSTS

The spreadsheet (Example 09510-2) consists of a listing of the components of each ceiling type. The ceiling type and a brief description should be listed first. This assures an accurate listing of the ceiling system is available for future reference for ordering of materials or verifying the price of components from the supplier.

The ceiling system is then broken down into its basic components. The line items will be used to calculate labor and material costs for all the components. Acoustical tile, grid system, molding, mains, 4 FT tees, 2 FT tees (if needed), wires, and boarder cuts are the basic items for each ceiling system.

Acoustical tile is calculated from the square footage plus a percentage for waste. Material extensions are made by multiplying the square footage by the unit cost of the tile. The costs of the tile materials should be solicited from the supplier. Let them know the type, quantity and delivery date. The supplier should be able to give you an accurate quote that takes into all increases of prices and discounts for large shipment.

The labor for costs for acoustical tiles are extended by dividing the square footage by the unit production (amount per hour to be completed) to arrive at the total labor hours required to complete the installation of the tile. Assume the material is in close proximity to the installation site and that no loading or distribution of material will need to be done. A typical 24 IN × 24 IN lay-in tile can be has a production rate of 125 square feet per hour. Loading and distribution will be discussed later. The extensions are then entered into the appropriate spaces on the spreadsheet.

The grid system line item is used to calculate the labor to install the grid, including mains and tees. Moldings and wires are addressed elsewhere. A typical twenty-four inch by twenty-four inch grid system with mains, 4 FT tees, and 2 FT tees will have a production rate of around 125 SF per hour and; a twenty-four inch by forty-eight inch grid system will be installed at 150 SF per hour.

The molding line item will be used to extend both the labor and material for the wall angle or molding installed along the perimeter of the system. The lineal footage of boarder or perimeter is used for the quantity. The material is the unit price per foot of wall angle (around $0.21 per foot) times the linear footage. Wall angle is installed in the range of 50 to 60 lineal feet per hour dependant on the substrate it is to be attached to. Masonry applications will take more time.

The main, 4 FT tees and 2 FT tees line items are used to calculate the costs of materials only. The labor was calculated in the grid line item. The quantity for each line item is a factor of the total square footage of ceiling area. The lineal footage of mains is equal to twenty five percent of the area. The linear footage of 4 FT tees is equal to forty eight percent of the ceiling area and 2 FT tees are equal to twenty eight percent of the ceiling area. The linear footage is multiplied by the unit cost (around $0.21 per linear foot) for each to arrive at a total cost of each component.

Wires are typically installed on four foot centers in both directions. The total number of wires is then calculated by dividing the total square footage of ceiling area by sixteen. Wires average $0.25 each, including clips, and are installed at a rate of thirty per hour.

Boarder cuts on large projects will usually be a labor line item only. The discarded cut tiles from one area will be utilized in another in a well organized project. If the project is very small or a lot of waste will be incurred, the estimator will need to factor a cost for extra ceiling tile material to account for that waste. The labor for a typical lay-in tile is calculated at a rate of 30 lineal feet of boarder cuts per hour. The quantity being the lineal feet of the perimeter of ceilings.

The typical general condition items for an acoustical ceiling installation are: supervision, loading and distribution, and clean-up. Supervision can be calculated at a rate of five percent of the total labor to install the system. This labor will account for foreman's time to manage manpower and lay-out. Loading of materials can be accomplished at a rate of 600 to 800 SF per hour depending on site conditions. The loading is typically done with the help of dollies and carts. If a crane or lift must be used, that cost will need to be added to the estimate. Clean-up of debris resulting from discarded boxes, wrappers and cut-off materials can be completed at a rate of 600 to 800 square feet per hour.

The labor hours and material cost columns are then added to arrive at a subtotal for each. The total hours are multiplied by the burdened rate per hour ($36.00 per hour for this example) to arrive at the cost of labor. The material cost is added to the labor cost to arrive at the cost of work.

Mark-ups are then added to the cost to arrive at the bid or contract value. Take note that the general conditions factored into the cost do not account for overhead (or any home office costs for G&A, Project Management, Warehousing, Etc.), so those costs have to be factored into the estimate. Along with the desired profit from the project. The mark-ups for overhead and profit are often times a percentage of the total costs. Most companies know what percentage factor for overhead as a part of their accounting tracking systems (typically between 7.5% to 10%). The percentage factor for profit is developed by company management or owners. The mark-ups added to the costs is the total bid or contract value of the ceiling systems.

If the project has more than one type of system, then the costs of each system should be calculated like the one in the sample and then totaled for a complete estimate.

09510 Acoustical Ceilings

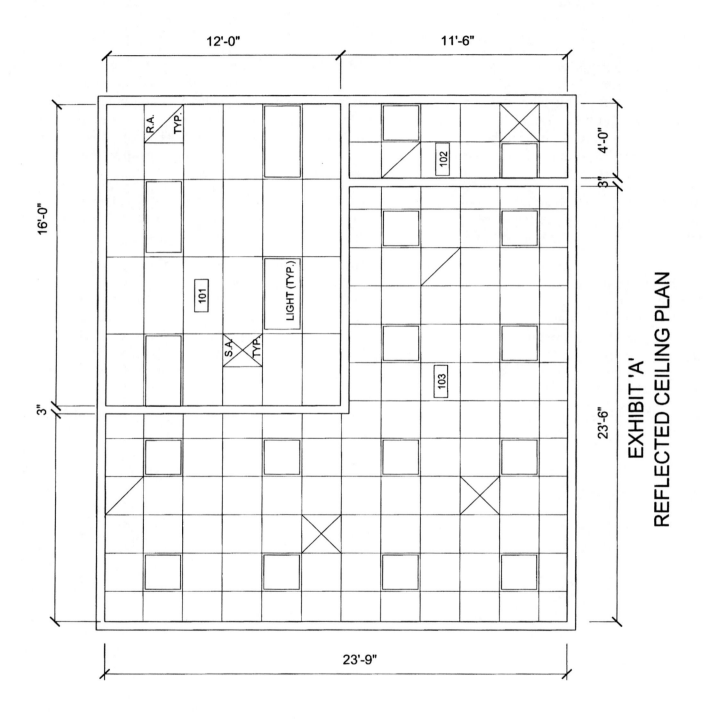

EXHIBIT 'A'
REFLECTED CEILING PLAN

ACOUSTICAL CEILING TAKE OFF SHEET

PROJECT ASPE SAMPLE
ESTIMATOR: MGG
DATE 21-May-97

ROOM OR NAME	ROOM SIZE	WALL ANGLE	CELING AREA - SF								
			TYPE 1 CEILINGS	TYPE 2 CEILINGS	TYPE 3 CEILINGS					
101	12-0 X 16-0		192							
102	4-0 X 11-6			46						
103	12 X 11-6			138						
	23-9 X 11-6			273.125						

Standard Estimating Practice — 423 — 09510 Acoustical Ceilings

TABLE NO. 1

SUSPENDED ACOUSTICAL CEILNGS

DESCRIPTION	MATERIAL COST/SF	MATERIAL % OF TOTAL COST	LABOR COST/SF	LABOR % OF TOTAL COST	EQUIP. COST/SF	EQUIP % OF TOTAL COST	TOTAL COST PER S.F.
SAMPLE #1 CEILING	$1.83	57.0%	$1.25	38.9%	$0.13	4.1%	$3.21
SAMPLE #2 CEILING	$1.86	45.2%	$1.81	43.9%	$0.45	10.9%	$4.12
2' x 2' 5/8" TEGULAR TILE IN 9/16" GRID (1)	$1.69	66.5%	$0.85	33.5%	$0.00	0.0%	$2.54
METAL PAN WITH ACOUSTIC PAD ON STEEL FRAME (1)	$2.93	51.0%	$2.82	49.0%	$0.00	0.0%	$5.75

(1) Taken from 1998 Edition of Means Building Construction Cost Data

ACOUSTIC CEILING ESTIMATE
PROJECT: ASPE SAMPLE
BID DATE: MAY 21, 1997 ESTIMATOR: MGG

DESCRIPTION	QUANTITY	UNIT MATERIAL	MATERIAL	UNIT PROD.	HOURS
ATC. TYPE 1:					
2'x4' LAY-IN					
STANDARD GRID					
TILE	1,920	0.75	1440.00	150	12.80
GRID	1,920			150	12.80
MODLING	560	0.21	117.60	50	11.20
MAINS	480	0.21	100.80		
4' TEES	922	0.21	193.54		
2' TEES	0		0.00		
BOARDER CUTS	0		0.00	30	0.00
WIRES	120		0.00	30	4.00
ATC. TYPE 2:					
2'x2' LAY-IN					
TILE	4,570	0.85	3884.50	125	36.56
GRID	4,570			125	36.56
MOLDINGS	1,260	0.21	264.60	50	25.20
MAINS	1,143	0.21	239.93		
4' TEES	2,194	0.21	460.66		
2' TEES	1,280	0.21	268.72		
BOARDER CUTS	315	0.42	132.30	30	10.50
WIRES	286	0.25	71.41	30	9.52
SUPERVISION	159			0.05	7.95
LOADING	6,490			600	10.82
CLEAN-UP	6,490			600	10.82
SCAFFOLD					
SUB-TOTALS		MAT.	$7,174.04	HOURS	188.72
		TAX	$430.44	X RATE	42.50
			$7,604.48		$8,020.78

TOTAL MATERIAL $7,604.48
TOTAL LABOR $8,020.78

TOTAL COST $15,625.26
OVERHEAD @ 10% $ 1,562.53
SUB-TOTAL $17,187.78
PROFIT @ 5% $ 859.39

TOTAL BID $18,047.17

09720 – WALL COVERING

Introduction

Refer to plans, specifications and schedules to determine each type of wall covering. List each type of wall covering by manufacturer, stock number, pattern, and color.

Survey wall coverings by either the area or strip method, or use both to prove quantities.

Determine locations of all wall coverings, and take off in a room-to-room sequence. List room number or name, wall orientation, and wall covering material type.

For walls, list height and length of each wall or total length of walls in each room. For ceilings, list length and width.

Quantity Survey

Base labor productivity rates of a SF, LY, SY or roll measure divided by an hourly or daily basis using a consistent method.

The estimator must use considerable judgment and experience when determining production rates. Rates vary based on pattern, texture, match, weight and size of wall covering. Production rates also vary with pre-trimmed, wall-cut or table-cut goods. There is volume work such as using pasting machines in corridors, and cut-up work such as small bathrooms using a pasting table. Whenever possible, develop production rates by accounting for these variables.

For each type of wall covering required, get the cost, length and width of bolt or roll, weight, type of pattern and match. Account for freight based on historical data and add to cost per roll, LY, or SY.

Account for all primer, paste and blades. On existing work, account for wall preparation such as stripping and skimming.

For area method, extend quantities of each type of wall covering and total. Divide by the net SF/LY or roll to determine total LY, or rolls of each wall covering type.

For strip method, is the material pre-trimmed or field-trimmed? Also, determine length of bolt or roll. Account for double or triple rolls or if an odd number of rolls requires a single roll.

To get the number of strips required, multiply total length of walls of each wall covering type by 12 in/LF. Then divide by the on-center dimension (OCD) of trimmed width in inches.

For bolt goods, divide height of wall by 3 LF/LY to determine strip height. Multiply number of strips required by LY/strip to determine total lineal yardage.

Example: A room to receive wall covering has a perimeter of 69 LF and is 9 FT high. What are the requirements for 54 IN wide material? After double cutting, the OCD is 51".

LY by strips = number of strips × lineal height in LY = (69 LF × 12 IN/LF ÷ 51") × (9 LF / 3LY/LF = 16.23 strips × 3 LY/strip = (round up) 17 strips × 3 LY/strip = 51 LY.

If this were a motel room and there were one hundred such rooms, the estimator would figure the number of full strips needed, and account for over door openings and around windows. For example, if there were two 3 FT swing doors and one 6 FT wide slider, the estimator might subtract 1 FT from door widths to allow for overlap and calculate as follows:

Full Strip LY = number of full strips - width of door openings × LY/strip = (69 LF = 2 LF - 2 LF - 5 LF) × 12 IN/LF / 51 IN OCD) × (9 LF/3 LF/LY) = 14.11 strips × 3 LY/strip = (round down) 14 strips × 3 LY/strip = 42 LY.

To this, add 1 LY over swing doors plus 1 LY over slider equals 4 LY. The lineal yardage required of one-hundred rooms is 400 LY. The estimator would then allow for waste with a percentage factor.

Deducting openings using the strip method on small quantities no allowance for waste. This is risky in consideration of possible flaws or damage.

For roll goods, divide number of strips required by number of net strips per roll. Add single rolls as required by double or triple roll packaging.

Add quantities for waste caused by material flaws, damage, or inexperience. Add materials specified for owner's future use as determined by the specifications.

09900 – PAINTS AND COATINGS

Introduction

The purpose of the quantity survey for painting is to define specific labor and material requirements in order to prepare an estimate for painting of a particular project. To this end, measure all items specified for painting in a practical, accurate and consistent method.

Quantity Survey

Survey all large surfaces using the area method. Measure walls in height times width. Measure ceiling and floor area in length times width. Make measurements in feet and inches as required for accuracy. Convert calculations to square feet (SF).

Measure all elements along a vertical or horizontal plane in lineal feet (LF) for determining labor. For determining material requirements, express the dimensions of the exposed surface area per lineal foot of each element in SF/LF. This also may be part of the description.

Examples: 10 IN Galvanized Fascia = 255 LF or Exposed 2 × 6 @ Ramada 2000 LF × 1.33 SF/LF.

Express other items, such as doors and windows in quantity (EA). For determining material quantities, enter the dimensions of the opening as part of the description.

Survey items of work in the most logical manner according to location. Take off according to substrate, finish, size and color. Whenever possible, state the location of work by room number or name, or elevation number, to the left of the description of work.

Separate new items of paint work from existing items of repaint work for determining different labor and material requirements.

When items of painting work vary and significantly impact labor or material requirements, develop a separate item listing.

The best method for exterior takeoff varies. On simple projects, the estimator may take off each group separately. On larger, more complex projects, take off all painted surfaces on each elevation, then move to the next elevation. In either instance, the goal is to be thorough and accurate.

Classify painting work into groups, developing a work sheet for class as required for larger projects. Carry forward all items to an estimate detail form, checking or highlighting each item and quantity entered.

Standards of Measurement

1. When determining labor requirements, consider an item which is measured along its length, such as conduit or small pipe, as at least 1 SF/LF. Logic: The difference between painting a 4 IN pipe and 1/2 IN conduit is the brush or spray fan size, not the amount of labor.

2. Consider an open solid form (such as chainlink, wire mesh or perforated metal) as no less than a solid with two sides.

3. Whenever the profile of a surface (such as corrugated siding or roofing) has greater surface area than can be measured in two dimensions, account for the additional surface area. Logic: A folded or formed surface doesn't diminish its true surface area, or the labor and material required to paint it.

4. Never deduct a small opening, under 100 SF in area, without sufficient reason. Logic: It's harder to paint a wall with an opening than a wall without one.

5. If a large opening is deducted, account for the wrap or return. Example: Measure a 10 FT × 10 FT overhead door opening in a masonry wall as a -70 SF deduct (-100 SF + 30 SF).

6. Never deduct the area of an obstruction (cabinets) or another surfacing material (ceramic tile). Logic: The labor increase is more significant than the material decrease.

7. If it isn't flat, it must be a lot more difficult to paint. Always account for ornamentation, intricacy, roughness, or other detail.

8. Account for any color change upon a surface (wainscot, stripe, reveal).

Accuracy should be proportionate to cost impact. For example, surface areas for wall coverings require closer measurement than flat wall paints.

Organization of the paint estimate is critical in getting a complete understanding of the work involved. Therefore, organize items of work into groups for determining cost impact of labor and material requirements for each group.

Taking off work in groups also reduces the likelihood of error, since the painting estimator develops mental checklists for each group. Grouping also enables the estimator to picture the proposed work. Visualization is as important as mathematics during takeoff procedures.

GROUP I: Walls, Ceilings and Floors

Review the material schedule, room finish schedule, sections, elevations, partition types and other pertinent information in detail to determine the extent of wall, ceiling and substrates, and different finishes required for these substrates.

Set up a particular method of takeoff for the type of work involved.

A room-to-room takeoff is the usual, preferred method when substrates, finishes and ceiling heights vary. Set up a columnar format of the wall, ceiling and floor survey. The leftmost column should contain the room number, or a descriptive reference for the room. To the right, provide three dimension columns for entering the height, length and width of each room. Provide additional columns as necessary to the right of the dimension columns for extension of quantities of each substrate and finish type. For example, a house with gypsum wallboard on all walls and ceilings might have four columns for quantity extensions. They could be "Flat Wall," "Flat Ceiling," "Semi-gloss Walls," and "Semi-gloss Ceilings."

On large jobs, to simplify organization of the wall and ceiling survey for painting, it may help to photocopy the room finish schedule. If possible, develop a system for identifying and highlighting different substrates and finishes. Example:

- ✓ Gypsum board wall or ceiling, Flat - no highlight
- ✓ Exposed structural ceiling, Enamel - Red
- ✓ Gypsum board wall or ceiling, Epoxy - Yellow
- ✓ Vinyl wall covering - Orange
- ✓ CMU wall, Flat - Blue
- ✓ No painted finish (CT, Marlite, etc.)-Purple
- ✓ CMU wall, Epoxy - Green

Use highlighters to organize the room Finish Schedule. Do not use this method on original drawings for return, or use by others.

GROUP II: Doors and Windows

Passage Doors and Door Frames

Review the door finish schedule and door elevations, if provided. Determine similarities and differences of doors and frames. Reference the door specification and note any shop-primed or prefinished doors or frames. Count each type of door and each type of frame or count each type of door/frame unit. Review opening widths and note pairs to assure an accounting of all doors. List carved, paneled, French, louvered and other specialty doors by count, species and finish. Count all doors and frames individually, then add the doors and frames.

Compare this to the total doors and frames on the schedule to assure correctness. Be sure that the door schedule uses a one-number-for-one-opening format. Make sure that one number does not represent a typical opening used in several locations. Distinguish between new and existing doors and frames if the project is a remodel. Note if there is a listing of existing doors on the door finish schedule. Judge whether or not any conditions warrant refinishing or repainting of existing doors and frames.

Large Doors

The estimator should cross-reference specifications to determine the amount of factory priming or finishing. Movable partitions and coiling counter doors may not appear on door schedule. Large doors may have milled doors and frames at pockets that do not appear on door schedule yet require field-finishing.

Take off each door that requires field-finishing by count, length and height and multiply by two for both sides. Reference sections and determine special requirements such as tracks, soffits, steel angles or wood moldings that require field-finishing.

HM Sidelites, Transoms, and Windows

When there are sidelites or transoms in door frames, or when HM windows occur, price each type of opening separately. Take off the sidelights, transom and windows separately, and show a total in lineal feet. Tally sidelights and transoms from the door frame schedule and windows from the floor plan. Be sure of the count by viewing elevations since clerestory windows may or may not show up on plan views.

Custom Wood Sidelites, Transoms, and Windows

Cross-reference millwork specifications to determine species and finish. Take off using either a unit count or show total lineal feet. Millwork also may include this item.

Steel Sash Windows

Take off each type of steel sash by count, width, height and number of lites.

"Averaging" Steel Sash Windows

Often, a paint estimator must face the task of pricing the painting of a multitude of different size windows with different number of lites. Rather than showing individual labor and production rates, determine an "average" window and show a unit production rate and material rate for this "average" window. "Averaging" aids in simpler visualization of the proposed work. How to "average" steel sash windows:

✓ Count the number of windows.

✓ Multiply the individual count times height and width of type to determine area of each type. Sum the area of all types. Divide by the total count to determine average area.

✓ Multiply the count times the width of each window type. Sum the widths of all types. Divide by the total count to determine average width.

✓ Divide average area by average width to get average height.

✓ Multiply the individual count times the number of lites for each type. Add the lites of all types, divide by the total count and round to determine average lites per window.

Example: Steel Sash

			Area	Width	Lites
30 EA	4' x 5'	8-lite	600 SF	120 LF	240
70 EA	5' x 7'	10-lite	2,450 SF	350 LF	700
10 EA	3' x 4'	6-lite	120 SF	30 LF	60
5 EA	4' x 15	24-lite	300 SF	20 LF	120
15 EA	2' x 3'	6-lite	90 SF	30 LF	90
130 Windows		TOTALS	3,560 SF	550 LF	1,210

Therefore,

The average area = 3,560 SF/130 Windows = 27.4 SF

The average width = 550 LF/130 Windows = 4 FT 3 IN

The average length = 3,560 SF/550 LF = 6 FT 6 IN

The average lites = 1,210 EA/130 windows = 9.3

The estimator now knows that the "average" window is very similar in nature to the 5'×7 FT 10-lite. Determine labor and material requirements for all 130 windows accordingly.

Wood Casement and French Windows

Carefully reference window specifications to determine the finish of casement windows (clad, primed, or stain.) What is the type of muntin (glazed or false) for French windows? Take off each type of window by count, width, height and number of lites. Casement and French window units also can be "averaged" like steel sash. List screens by description of frame, width and height.

GROUP III: Millwork and Interior Wood

Cabinets and Other Casework

Develop LF/SF multipliers for standard upper and lower cabinets and use them consistently. When cabinets or other casework vary from the norm, the estimator should develop a specific SF/LF for each type by examining elevations and sections.

To develop a casework multiplier, view the casework in section and determine SF/LF of exposed surfaces. Multiply by length of module. Determine SF of module sides and area of any drawers. Sum products and divide by module length to determine a developed LF/SF multiplier.

List all cabinets by room or elevation number. List species and finish. Determine lineal feet of each type of cabinet, the LF/SF multiplier, and the number of units provided. Cross-reference plans and elevations to assure the count of all cabinets.

Shelving

List all shelving by room number or name. Multiply number of shelves times length of shelving times two sides times depth of shelf for total surface area.

Wood Moldings

Determine amount of cornice, picture and other moldings, wood base, sills, lintels, box beams, and other wood elements. Develop this information from careful examination of plan notes, room finish schedule, room elevations, building and partition sections, and reflected ceiling plans.

Take off and list moldings by room number or name, type, species, dimensions and length listed in lineal feet. Breakdown built-up moldings into separate items if finishing before installation is a requirement.

Paneling

List paneling by room number or name, species and finish. Multiply length time height and express in square feet.

Wood Ceilings

List wood ceilings by room number or name. Determine flat area of deck. Figure additional surface area added by beams or open web trusses and add to surface area of deck. Developing a multiplier for this purpose may prove useful.

Example of developing a multiplier for beams:

A wood ceiling is 20 FT × 24 FT or 480 SF. The ceiling support is by 3 IN × 8 IN Beams, 30"OC. The actual height of an 8 IN beam is 7 ¼ IN. Therefore, add twice this amount, or 14 1/2 IN to the OC dimension of 30 IN. Therefore, the multiplier is 30 IN + 14 1/2 IN divided by 30 IN, or 1.48. The developed surface area of deck and beams is 480 SF times 1.48 equals 710SF.

Determine any attached mechanical, plumbing or electrical items on wood ceilings such as ductwork, sprinklers, or conduit and group with interior steel.

GROUP IV: Metals

Basic Methods

Distinguish between surfaces that are shop-primed or pre-finished, and those which require field priming and painting by referencing the specifications.

List and group galvanized and ferrous metal surfaces separately if surface preparation and priming requirements vary.

Depending on material finishes, colors and quantities, the estimator may wish to further group surface by interior or exterior exposure.

When attached elements such as ductwork, piping and conduit are to be painted the same as the adjacent surface, convert LF to SF. Treat this as a developed surface. When attached elements are a different color, take off item in LF for determining labor and SF for material requirements only.

Determine if color-coding is part of the work of this section or the responsibility of Division 15.

Miscellaneous Ferrous Metals

- ✓ List stairs by type, and number of flights. Consider railings as part of the assembly, to take off separately.
- ✓ List access and ships ladders by height in LF. List handrails in LF.
- ✓ List railings and fences by type, height and LF.
- ✓ List steel lintels, steel curbs, dock angles in LF.
- ✓ List bollards as striped or one color, in lineal feet.
- ✓ List security grilles by count, type, width, and height.
- ✓ List gates by count, type, width, and height.
- ✓ List columns by count, size and height.

Exposed Structural Steel Decking

Separate galvanized from prime-coated deck. Determine area of deck and use multiplier according to corrugation profile to determine actual surface area. Double for two side exposure.

Structural Steel Shapes

Refer to structural drawings. Group according to shape. Begin takeoff with largest members, and move in a consistent and logical manner until takeoff is both complete and accurate.

- ✓ List all steel members by count, shape, actual SF/LF, and length. Add 3 to 5 percent to plates and flanges.
- ✓ Figure open web trusses as solids or twice the length time depth.
- ✓ For grating, measure surface area of one side, and multiply by 4 to 6 according to type.

Piping

- ✓ For area of piping, multiply circumference times length.
- ✓ Figure normal hazard sprinkler pipe at 12 to 15 LF per 100 SF of ceiling area.
- ✓ Add for pipe hangers.

Galvanized Metals

✓ List conduit runs, such as those exposed as part of structure or which must paint to match an adjacent surface.

✓ Add for hangers.

Example: 1 IN conduit @ red brick - 255 LF

1 IN conduit pipe run with hangers - 110 LF.

List exposed ductwork in LF for defining labor and approximate surface area for material requirements by determining an average. If material is a special coating, or total exposed ductwork is large in nature, do a detailed takeoff. The takeoff includes each piece in lineal feet, determine SF/LF of each piece, and extend all quantities.

List reveals, fascias, downspouts, copings, and counterflashing assemblies for priming or finishing differently from the adjacent surface.

List each item separately, showing sum of exposed dimensions and express in lineal feet.

Example:

    18 IN galvanized fascia - 1250 LF

    14 IN coping - 5250 LF

    6 IN counterflashing - 4500 LF

    18 IN gutter - 750 LF

    14 IN downspout - 600 LF

GROUP V: Exterior Walls and Soffits

Exterior substrate and finish varies by geographic location and particular project requirements. The estimator should determine logical groups according to substrate and finish. The following are important considerations for exterior takeoffs:

View the roof plan and approximate sections. Determine requirements for painting roof-mounted equipment, flashings, and parapets.

Move in a logical and consistent manner around the building by determining a starting point and moving clockwise or counterclockwise back to that starting point. Be sure of wall heights by viewing section. Also, determine the finish of parapets and firewalls.

Check elevation lengths by floor plan dimensions. Be sure there are no hidden elevations. Check that all site work (site walls, gate, railings, bollards, etc.) is complete.

GROUP VI: Exterior Wood

✓ Determine siding in SF accounting for additional surface area of laps and grooves.

✓ Determine the amount of fascia, and other accented wood trim and list in LF.

✓ Determine amount of narrow overhangs and express in LF as well as SF/LF.

✓ Determine area of exposed decks or porches.

✓ Determine actual surface area by using a multiplier for exposed joists and beams.

Use of Spreadsheet Database Sorting

Doing a room-to-room takeoff of various finishes on many substrates may become extremely cumbersome. Computers offer a reasonable alternative when used with spreadsheet software. Set up necessary columns for doing a room-to-room takeoff, and soft the data into blocks.

Steps in spreadsheet database sorting:

1. Establish column designations. For example:

    ✓ Cell A2 = Room number

    ✓ Cell B2 = Orientation

    ✓ Cell C2 = Substrate

    ✓ Cell D2 = Finish

    ✓ Cell E2 = Length

    ✓ Cell F2 = Height (Width)

    ✓ Cell G2 = Area

2. In Cell F1 place the label "total area" for Cell G1.

Cell G1 would contain the formula @sum (G3.G202). In cell G3, enter the formula @sum(e3*f3). Copy the formula and paste to cells g3..g202.2.

Establish key abbreviations for column listings. Examples:

Orientation	Finish
✓ N (north wall)	✓ F (flat)
✓ A (all walls)	✓ E (epoxy)
✓ C (ceiling)	✓ VW (vinyl wall covering)
✓ F (floor)	

Similarly, F3 could be the 3rd color of flat or VW14 the fourteenth type, pattern and color of vinyl wall covering.

3. Lock titles on spreadsheets.

4. Beginning with the first room on the room finish schedule, take off all painted wall, ceiling and floor finishes according to substrate and finish. Then proceed in sequence until the survey is complete.

5. Save spreadsheet throughout takeoff procedure and before performing database sort.

6. Define block to sort and use substrate and finish columns as key columns for database sort.

7. In column H, @sum blocks and provide an appropriate label in column 1. Carry forward all quantities to an estimate detail form.

Waste Allowances

Waste allowances are an essential consideration in painting. Waste is any material that does not become part of the finished surface. In accounting for waste, the painting estimator should rely on historical data, experience and judgment.

Account for all waste items in the estimate. Some estimators approximate these items as a percentage of total materials. However, on large or complex painting jobs, such a method is inaccurate, and the result is an unsuccessful estimate. This is because the bid is too high, or reduced profit because the bid is too low.

The preferred method is to use a systematic approach defining each and all items required for each paint system for the particular job. For example, when painting wood door frames with lacquer undercoater and alkyd enamel finish, the estimator should account for all components. This includes lacquer thinner, paint thinner, sandpaper and masking tape as well as putty, caulking, undercoater and enamel.

The painting estimator also should determine the method of application for each finish. Account for paint waste due to spray loss, or fallout, in material consumption rates. When figuring dry film thickness (DFT) from theoretical coverage, the estimator should always add a percentage for waste. When spraying, waste factors from 20% to 50% are not uncommon depending on spray method used and profile of the surface.

Estimate solvents necessary for reduction and cleanup as a percentage of gallon of paint applied and converted into a developed cost.

Be sure to include costs for cartage and disposal of wastes such as dirty solvents or spent abrasives containing lead.

Example: An estimator determines that epoxy primer needed for 18,000 SF of structural steel is 100 gallons. The epoxy is $25 per gallon and the reducer is $15 per gallon. Expect that one quart of reducer is for reduction. Use daily cleanup material for reduction. However, 10 gallons of dirty solvent remain for disposal at the end of the job. The estimator expects 35% waste due to spray loss.

Total Material Cost for Epoxy Primer

100 gallons of primer	$25	2,500
25 gallons of reducer	$15	375
Subtotal		2,875
Add 10 gal waste disposal	$6	60
Add 35% spray loss		1,006
Total Material Cost		$3,931

In this example, the additional material cost for waste is almost 60% for spray loss, solvents, and hazardous waste.

Other types of paint waste which must be accounted for when estimating painting:

- ✓ Solvents used for reduction of material, solvent, cleaning, and cleanup of equipment
- ✓ Paint left for future use as part of contract closeout requirements
- ✓ Additional paint necessary due to number of color selections
- ✓ Additional paint necessary as a minimum amount of martial to use spray equipment
- ✓ Additional catalyzed materials used to fill equipment and lines, or otherwise mixed, unused material left from daily application
- ✓ Liquid strippers
- ✓ Wood bleaches
- ✓ Acids for etching and cleaning
- ✓ Amount of abrasives necessary for blasting
- ✓ Sandpaper, steel wool, wiping rags and other types of expendable preparation materials
- ✓ Masking paper, plastic and tape

Subtotal quantifies of all surfaces surveyed

Listed quality and types of materials in the schedule will determine actual material costs for each paint system specified for the project. For each paint system, define initial and intermediate preparation materials. Then define masking materials and number of gallons of specific material required for each cost of the paint system. Also include solvents necessary for reduction and cleanup. Insure sufficient gallonage as required by the color specification, closeout requirements and use of specific application equipment.

The solids by volume and the amount of waste govern gallons required by dry film thickness. Theoretically, one gallon of paint, consisting of 100% solids by volume, covers 1604SF of surface at 1 mil. When a gallon of paint is 50% solids by volume, one-half of the paint stays on the surface. The other half evaporates.

Example: Apply an enamel to 2,700 SF of primed metal at a total dry film thickness of 3 mils. What is the coverage rate per gallon if the enamel is 40% solids by volume with 25% spray loss using conventional equipment?

1,604 SF/GAL at 1 wet mil (wet film thickness)

.40 solids by volume

641.6 SF/GAL at 1 dry mil (dry film thickness)

641.6 SF/GAL at 1 dry rail / 3 mils = 214 SF/GAL

(100% usable-25% waste) = 75% applied material = 160.5 SF/GAL = 161 SF/GAL

2700 SF / 161 SF/GAL = 16.8 GAL

Note: There are two ways to look at waste percentages. In the above example, one quart of every gallon is spray loss, net SF/GAL is obtained, then total gallons required.

If another estimator figured net gallons first (2700 SF / 214 SF/GAL = 12.61 GAL) then added a 25% waste (3.15 GAL), the answer is the same. One way to account for waste is to think in "loss per gallon." The other way is to think "how much extra material does it take to account for waste." Either method is reliable, as long as an estimator is consistent in method.

Determine labor necessary for each paint system of the project by reviewing the surface preparation standards of the specification and the material schedule. At this point, picture the completed project, and consider the method and sequence of all paint systems applications.

Determining painting production rates is a function of defining labor necessary to perform each step of preparation and coat for each paint system. Production rates for a particular paint system may increase or decrease dramatically due to:

- ✓ color requirements
- ✓ access
- ✓ setup and cleanup necessary for the system
- ✓ weather

Proximity to other systems and finish construction material also is a factor.

09970 – COATINGS FOR STEEL

Introduction

The estimator should have or develop considerable skill, knowledge and experience in reading structural, mechanical, electrical and plumbing drawings. The estimator also should have extensive knowledge of geometry, and the ability to be thorough and accurate. Continuing education as well as membership in SSPC (Steel Structures Painting Council) and PDCA (Painting and Decorating Contractors of America) will help the beginning estimator in gaining such skills.

Do not use "determination of surface area by square feet/Ton, or gallonage of tank," when completed working drawings provide sufficient detail. Use such approximations only as the basis for preliminary estimates, or checks.

Degree of surface preparation suitability of primer and other coating requirements depend on accurately defining exposure and service requirements.

Quantity Survey

Read specification of Division 5 - Metals. Determine suitability of shop primer for coating system. In many instances, shop primer is temporary protection or not enough for final coating requirements. Shop-primer may need barrier priming or removed by abrasive blasting.

Refer to Division 15 and 16 to determine types of piping, conduit, and equipment. Refer specifically to Sections 15190 and 16195 for mechanical and electrical identification systems. Determine responsibility for color coding. Review Divisions 11, 13, and 14 as applicable for specific job requirements to determine accurate scope of work.

Structural drawings are usually diagrammatic and, therefore, steel members are not usually shown in section which is necessary to determine square feet/lineal feet. Refer to PDCA tables based on AISC shapes to determine square feet/lineal feet of each member. List structural members by AISC designation, length in lineal feet and square feet/lineal feet. Account for flanges, rivets and bolts.

To determine surface area of open forms, such as open-web trusses, wire mesh and chain link, figure as a solid surface, accounting for both sides. That is, multiply length times height times two.

Take off decking and grating in square feet. Determine actual surface area added by corrugation of decking, or nature of grate by studying details and sections. On decking, note whether exposure is one or both sides by reviewing insulating and roofing requirements.

Determine counts, areas or lineal feet of flights of stairs, railings, walkways, hatches, and ladders.

Account for piping and conduit by tracing flow or run. List separately exposed, concealed, or underground pipe and conduit for specific coating requirements. List piping and conduit by type, diameter and length. Account for hangers and valves. Specified sprinkler systems and other types of piping may have no specific details.

Determine requirements for equipment painting.

The general method of takeoff is to move from the largest to the smallest members. Therefore, takeoff procedures depend on type of structure. It is critical to be logical, thorough, and consistent.

For example, to take off a tank, determine the inside and outside areas of the tank, and the supporting structure, if elevated. Account for all hatches, ladders, catwalks, and stairs. On elevated tanks, account for the riser as well as other piping and conduit. Determine requirements, if any, for checkering or signage. Review mechanical, electrical, and plumbing drawings and cross reference with civil drawings for additional coating requirements.

To take off a structure, review drawings and develop a general idea of work involved. Devise a logical method of precedence and a point of origin. Move upward or downward from this point of origin, along logical vertical or horizontal axes as required.

Example (listed in precedence):

Structural Steel Ceiling -

Decking, major trusses and girders, minor trusses and bridging between major trusses and girders, columns, ductwork, sprinkler system, miscellaneous piping and conduit.

(Refer to 9900 Painting, Waste Allowances for additional information.)

Waste allowances for determining steel coating requirements for steel are extraordinary. Knowledge and experience of the estimator in this regard is critical to the success and accuracy of the estimate.

The estimator must be thoroughly familiar with methods of determining material waste allowances for paint and coatings.

Existing coatings containing lead and other heavy metals, or asbestos content of coating will require special procedures. Removal of such existing coatings will require extraordinary and costly procedures of testing, containment, handling and disposal due to governmental regulation of hazardous and toxic wastes.

Account for inclement weather that may affect loss of catalyzed materials. Be aware of certain environmental conditions, such as wind on mountain tops, which may affect loss of materials due to spray loss or crew shutdown.

Daily shutdown, onset of inclement weather, and rapid changes in temperature cause losses in catalyzed coatings.

Failure to coat within a specific period may require additional blasting due to formation of flash rust.

Failure to recoat certain materials within a specific period may require brush blasting for sufficient tooth for later coats.

Extension of Quantities

Review all plan sheets and applicable specifications to assure complete and accurate takeoff.

Visualization of the complete scope of work is a requirement of determining accurate material, labor and equipment requirements.

Group all surfaces by particular type of metal, size, degree of surface preparation, finish, location and access requirements.

Sum areas of large surfaces such as decking, siding, or steel plate. Account for additional surface area created by corrugation.

Group structural members. Determine areas of members by multiplying count times length times square feet/lineal feet. Account for flanges, rivets and bolts. Group duct, piping and conduit by size, type of metal, finish land access requirements. Account for hangars, valves and fasteners. Determine color coding requirements.

Determine actual areas of stairs, railings, and other access and safety steel.

When determining material requirements, use a systematic approach defining each requirement of surface preparation, priming and coating. Determine surface preparation requirements by exposure and service, according to specification, manufacturer's recommendations, and SSPC standards. Determine number of coats required to achieve minimum dry film thickness (DFT).

Account for all solvents for surface cleaning as well as material reduction and cleanup. Account for all blasting media.

Determine labor requirements by assessing type of structure, access, and coating system. For blasting, account for pot tender. Determine cleanup time for removal of spent media. Account for testing, containment and disposal of hazardous and toxic wastes. For coating applications, determine amount of unproductive time spent mixing materials.

For all operations, determine type of access. For specialized access and mechanized equipment, such as stages and scissor lifts, determine time necessary for each type of equipment. Then get current rental rates.

Determine all specialized equipment necessary for coating application, including mixing, breathing and safety apparatus.

If figuring work before erection, determine additional equipment and resources necessary for shagging and handling steel. Account for field touchup of shipping damage, welds and bolts.

(Refer to 09900 Painting for additional information.)

10800 TOILET, BATH, AND LAUNDRY ACCESSORIES

Toilet, Bath, and Laundry Accessories may be supplied under a variety of contract arrangements:

- ✓ Material is purchased directly from the supplier and installed by the general contractor.
- ✓ A specialty subcontractor may furnish and install the accessories.
- ✓ Material is owner supplied and installed by either the general contractor or a specialty subcontractor.

Toilet and Bath Accessories have three types of installations: 1) Surface Mounted 2) Semi-recessed and 3) Recessed. A surface mounted accessory is attached directly to the finished wall surface with various types of fasteners. There must be sufficient substrate behind the finished surface to accommodate the fasteners and the accessory's accompanying weight. A semi-recessed accessory has been designed to set into a framed opening exposing a few inches of the fixture. A recessed accessory is installed in a framed opening flush with the finished surface.

Standard items encountered in a typical estimate may include:

- ✓ Soap dispenser
- ✓ Soap dish (non-ceramic)
- ✓ Paper towel dispenser
- ✓ Waste receptacles
- ✓ Metal shelves
- ✓ Metal framed mirrors
- ✓ Toilet tissue dispenser
- ✓ Toilet seat cover dispenser
- ✓ Feminine napkin/tampon vendor/dispenser
- ✓ Facial tissue dispenser
- ✓ Ash tray/wall urn
- ✓ Tooth Brush/Tumbler/holder
- ✓ Paper cup dispenser
- ✓ Shower seats
- ✓ Towel Bar holder
- ✓ Mop holder
- ✓ Grab bars for handicap use
- ✓ Single/double robe/hat hooks
- ✓ Electric hand/hair dryers
- ✓ Diaper deck
- ✓ Specimen pass box
- ✓ Bed pan storage cabinet

10800 Toilet, Bath, and Laundry Accessories

- ✓ Automatic bed pan washer
- ✓ Lavatory stations
- ✓ Towel bar/ring
- ✓ Toilet paper holder
- ✓ Medicine cabinets
- ✓ Shower/tub door

Types of Measurement

The standard unit of measurement when performing a quantity survey for accessories should be designated by each (EA), number (NO), or occasionally other unit measures may be required.

Factors Affecting Takeoff and Pricing

Toilet, bath and laundry accessories are commonly shown on both the floor plan and interior elevation drawings and detailed in the specifications. It is necessary to cross reference the floor plans, enlarged rest room plans, interior elevations, and specifications to avoid discrepancies. If discrepancies are found, the architect should be notified and asked to clarify the differences. Sometimes the specifications will contain a quantity survey for the accessories. This information should only be used as a method for confirming the takeoff, not relied upon for the estimate.

The most efficient method to approach the quantity survey is to start at the lowest floor level taking off each rest room separately and noting the room number. The possibility for minor changes from floor to floor in "typical" rest room layouts must be assessed. For example, the accessories may remain the same from one floor to another, but the wall construction, material or dimensions may change.

Substrate or framing items should be estimated in accordance with their applicable divisions, 05400 Cold Formed Metal Framing, 09100 Metal Support Assemblies, or 06100 Rough Carpentry. The quantities should be included in their specific section divisions and should be noted that this is material to be used for substrate and/ or framing. It is good practice to estimate these items rather than assuming there may be waste to cover these materials. There is no need to estimate the accessory's fasteners; the manufacturer typically furnishes them. Specialty items such as lavatory stations and electric hand/hair dryers may require mechanical and electrical connections and are generally included in their respective sections. When performing the quantity survey, list the type of accessory, the primary and/or alternate manufacturers, model number, mounting configuration, room number, quantity and make any applicable notes. Refer to Exhibit "A" for sample quantity survey sheet. *Freight costs may significantly affect the total cost of materials. The estimator should confirm that freight cost is included in any material quotations.*

Overview of Labor, Equipment, and Material

The following are three accepted methods of pricing material, labor and other associated costs in the estimate. (The dollar amounts and labor hour production are examples only.) For the following examples no subcontractor's overhead, profit or

Part Two 10800 Toilet, Bath, and Laundry Accessories

other mark-ups were included to avoid confusion of methodology tabulations. These items should be addressed at the completion of this division or for the specific subcontractor. In the following examples sales tax and freight costs are included in the material unit price. The estimator should confirm their inclusion when preparing the estimate.

1) Material Priced & Labor Lumped

The material quantities will be listed separately and priced accordingly. The cost of the labor is then determined by the hours required to complete the work and the prevailing wages applied.

Item Description	Unit	Quantity	Unit Cost	Total Cost
A	EA	2	$80.00	$160.00
B	EA	1	$60.00	$60.00
C	EA	3	$120.00	$360.00
Total Material				$580.00
Labor 1 Carpenter		3Hrs.	$20.00	$60.00
			Total Material & Labor	$640.00

2) Material and Labor Priced and Shown Separately

In this method's example the unit costs and labor costs are displayed separately then combined to show a total cost for each quantity

Item Description	Unit	Quantity	Material Unit Cost	Labor Unit Cost	Total Cost
A	EA	2	$80.00	$8.60	$177.20
B	EA	1	$60.00	$7.40	$67.40
C	EA	3	$120.00	$11.80	$395.40
				Total Material & Labor	$640.00

Standard Estimating Practice

3) Unit Costs for Material and Labor Combined

This Unit Cost method is normally used for estimating where the labor hours involved are needed and their corresponding costs are not critical. This method combines the cost of material and the cost of labor for one total unit cost. The unit costs would be similar to those in Method #2.

Item Description	Unit	Quantity	Material & Labor Unit Cost	Total Cost
A	EA	2	$88.60	$177.20
B	EA	1	$67.40	$67.40
C	EA	3	$131.80	$395.40
			Total Material & Labor	$640.00

Ratios and Analysis Tools Used

A reliable rule for conceptual estimates as they relate to toilet accessories would be an allowance of one soap dispenser per lavatory, one toilet paper/seat cover dispenser per water closet, and one towel dispenser/waste receptacle per three lavatories. A reasonable ratio for bath accessories would be a toilet paper holder per separate water closet and/or bidet and one towel bar/ring per sink and one medicine cabinet, depending on the room configuration.

Method and Approach to Quick Budget Estimates

The following methods should only be used when performing preliminary or conceptual estimates and cannot replace detailed quantity survey and pricing.

1) Cost per square foot based upon the gross building area or rest room area

2) Cost allowance per rest room with possible variances for men verses women

3) An allowance based upon the total number of rest room fixtures or toilet compartments

4) A lump sum based upon previous projects of similar square footage, complexity, and/or dollar volume

5) An average material and labor cost is applied to each fixture

When any of these methods are utilized it must be remembered that these numbers should be updated to reflect the rise in costs since the previous estimate.

Post Project Analysis and Recording Historical Data

A record should be maintained from the conceptual phase of the estimate to the closeout of this specific division. This allows for the tracking of the accuracy for the entire phase of the project and will enable the estimator to compile accurate preliminary estimates in the future. With this information a data base can be created outlining the major areas of interest such as: project size/dollar value, number of rest rooms and accompanying fixtures, total hours for installation, date and location, etc.

Miscellaneous Pertinent Information

It is helpful to have access to the brochures or technical data sheets from the specific manufacturer. This enables the estimator to review the accessories dimensions and acts as an aid in determining the quantities required for the substrate, framing and the specific plumbing or electrical requirements.

Glossary

Backing - Material applied flush to the wall or partition to provide support for wall hung fixtures.

Blocking - A solid piece of material in the wall between the wall's framing members to provide support for wall hung fixtures.

Framing - Material applied between studs to box out a portion of the wall or partition to allow support of a recessed wall fixture.

Semi-recessed - A fixture set into a wall with exposed edges.

Recessed - A fixture set into a wall with the face of the fixture flush with the wall's finished surface.

Surface mounted - A fixture mounted to the finished surface of the wall.

Substrate - An underlying material that supports another material or fixture.

10800 Toilet, Bath, and Laundry Accessories

Exhibit A

QUANTITY SURVEY SHEET

Project Name		Project Size	SF
Estimate Date		No. of Stories	
Estimator			

DESCRIPTION	MANUFACTURER	MOUNTING	ROOM #	QUANTITY	UNIT	NOTES

Mounting: PR=Partial Recessed   R=Recessed   S=Surface

13280 - Hazardous Material Remediation

This standard is to create an awareness of the questions that arise in preparing and estimate for working with hazardous materials and should alert the estimator to possible hidden costs. The scope of work and specification general conditions may place a burden on the contractor, which may have not been included in the estimate. It is easy to identify the scope of work but the difficulty of performing some tasks may not be easily identified. A trip to the jobsite is not always possible and the estimator needs to ask all the possible questions about site and material conditions.

There are many hazardous materials that are toxic to plant and animal life when found in strong concentrations. This standard will deal separately with handling and disposal of materials contaminated with asbestos, mine tailings, and polychlorinated biphenyls (PCB's). Each of these materials may occur in varying degrees of hazardous concentrations. Federal regulations are consistent nationwide but state and local regulations may vary significantly in how these materials are to be handled. Often projects involving hazardous materials handling require experience and special training for workers and may require licensing and special insurance coverage.

Indirect Cost Items Affecting the Cost of Work

1) Variable site accessibility

    a) Types of site security

    b) Delays to personnel and material at access gates

    c) Monitoring while construction work is in progress

    d) Furnishing access to the site for inspections or monitoring of site conditions

2) Engineer's and owner's reputations

    a) Accuracy of drawings and specifications

    b) Timeliness in addressing change orders

    c) Prompt payment authorization for complete portions of work

3) Bidding environment

    a) Number of interested and qualified bidders

    b) Expected changes in wage rates for various trades

4) Ownership of Hazardous Material

5) Work Environment

    a) Union, non-union, or right-to-work area

    b) Extreme working conditions

6) Scheduling

   a) Requirements for phasing of work

   b) Other time restrictions such as "weekends only", etc.

   c) Lack of timely inspections

7) Contract Delays

   a) Possible delays identified in the contract documents

   b) Testing certain parts before beginning work on other parts of the project

Direct Cost Items

1) Types of contaminated material

2) Magnitude of safety precautions, special clothing and/or enclosures, and personnel safety equipment

3) Kinds of pre-construction, construction, and post construction health and physical costs

4) Type of testing, number of tests, and party responsible for testing at all stages of the project

5) Licensing and permit costs and restrictions

6) Requirements and locations of a licensed disposal facility for the contaminated material

7) Costs of transporting and disposing of all contaminated materials, including tracking records

8) Type of equipment used to reduce personnel exposure and increase labor productivity

9) Accessibility to project, e.g. security clearance or escort required

10) Responsibility to furnish safety clothing or equipment for the owner, engineer, or subcontractor personnel

11) Time constraints

   a) Short duration contract

   b) Specific time of day working hours or days of week

   c) Specific times work cannot progress in identified areas

Mine Tailings as a Hazardous Material

Mine tailings frequently hold concentrations of chemicals found near the site in their natural concentration. As the processing of the mined product and stockpiling of tailings continues, the concentration of the chemicals frequently reaches high levels. This chemical concentration can be hazardous to the environment. Considering neutralization of the tailing is imperative. This section addresses the stabilization of the waste for storage in a manner that should prevent it from oxidizing.

Conduct sufficient tests on the materials and the proposed storage location to determine the degree of difficulty of handling and compaction. Addition of a stabilizing agent to permit a safe storage of the materials is also a consideration.

The materials will often be exceptionally sticky and present a problem in drying, mixing, and transporting. The cohesiveness of the materials processed dictates the type of equipment used. The type of materials found also dictates the type and percent of additives for both neutralization and successful storage of the final product.

The successful preparation of an estimate requires plans that show the location(s) of the tailings, their depths, moisture content, and chemical composition. Test results should forecast volumes of any necessary additives and any special costs to deliver those additives to the mixing site.

Include in the estimate provisions for diverting water from outside sources and costs to prepare the designed storage and borrow facilities. Also provide for equipment cleanup.

Projects involving the cleanup of mine tailings are frequently at remote locations. This makes it necessary to include the cost of temporary quarters and utilities and the increased costs for delivery of construction materials.

The project specifications may require special handling of site surface water. They may also stipulate that the contractor is responsible for only treating the water to an acceptable level. The contractor must determine the costs to transport the contaminated water offsite for treatment, if permitted. If not permitted to transport, determine the costs to set up a temporary treatment site on location.

The contractor should be aware of hidden costs necessary to perform the work according to the contract. These may include:

Liners/covers for material hauling equipment

Dust control requirements

Types and number of soil tests

Contaminated liquid tests

Special local and state permits

Bonding costs to perform the work

Insurance coverage

Estimating the Disposal of Polychlorinated Biphenyls

List the specific quantities of all types of polychlorinated biphenyls for disposal in the estimate. Also define the type of disposal (burning or burial) and the degree of contamination.

The estimate should include costs of all construction equipment cleaning and disposal required by the work. Be sure to include special bonding costs for exposure of construction personnel to hazardous material.

The project may come to a halt during certain testing phases and personnel will be nonproductive while awaiting results of tests. This is true especially on small projects involving a single or few locations of the PCB contaminated sites. Recognize those costs in the estimate.

Estimating Removal of Asbestos Contaminated Material

Material which may contain asbestos (ACM) are usually thermal insulation, acoustical, and fireproofing construction materials. In addition to these basic materials, it my be present in:

Adhesives
9 IN × 9 IN vinyl floor tiles
12 IN × 12 IN floor tiles
Vinyl asbestos floor sheathing
Roofing material
Old electrical wire insulation
Older wall board
Paper wrap around air conditioning ducts
Ceiling tiles
Drywall compound

Because asbestos is a known carcinogen, controlled demolition is necessary during removal. In accordance with governing regulations, contain the area of asbestos removal. Devise special procedures to contain the release of asbestos fibers. Workers may be required to wear protective suits and respirators and change them regularly.

The owner should engage the services of an asbestos expert to determine the presence of asbestos and provide recommendations for the decontamination process. The proficient estimator must perform his own estimate of the work and ask questions to clarify the areas of work and quantities of asbestos. The estimator should read the specifications before the project inspection. Beware of the types of asbestos to be removed, requirements of the hygienists, and the type of protective gear needed.

At the project inspection the estimator must look at all areas in which ACM is present and other areas where ACM may be present. Consult available blueprints to understand the scope of the project. Then determine the level of difficulty in containing each area.

The estimator must identify:

    Which materials contain asbestos?
    How much of this material the owner wants to be removed?
    What is the needed level of protection for the workers and structure?

The next step is to measure the surface areas of the material and their respective depths. During the physical inspection, the estimator should note all obstacles in the way of removal including walls, doors, fixtures, and equipment that must remain. Determine waste allowances from historical data (when available).

Specific Demolition

Structural Beams

Measure all surfaces of the beam, including the length, height, width of the upper and lower horizontal sections. Include the point of connection (POC) with the other beams and deck. Estimate visually the average depth of the fireproofing in the area of removal. If there is a structural blueprint for the area, note the beam size and visually inspect it during the physical inspection.

Separate beams or concrete sections into groups by sizes of beams or section. Multiply length by height and width for all surfaces. Fireproofing and steel companies have charts that give the square footages of each beam size.

Add each group together and multiply by the thickness of the material sprayed on the beams to get the total cubic feet of material. Examples: the square foot sum is 20,000 SF of material and the average thickness is 2 IN. Two inches is one-sixth (1/6) of a LF. 20,000 SF * 1/6 equals 3,333.33 cubic feet of asbestos material for removal and disposal. Landfills usually charge by the cubic yard for disposal. To calculate the cubic yards of material to be disposed, divide the cubic feet by 27. 3333.33 27 equals 123.46 cubic yards asbestos waste.

Structural Deck

Measure the area of deck for removal of ACM. Note whether it is flat or corrugated, steel or concrete, and the depth of any corrugations. Note how many hangers are in the deck. Does a concrete deck have a square cube pattern with cube sides two to four feet in depth? How extensive is the fireproofing or acoustical spray?

For a flat deck multiply length by width. One method of calculating fireproofing or acoustical spray for corrugated deck is to multiply length by width by 1.6. Figure a concrete square pattern deck using the length of the sides of each square. Use the same formulas found in structural frames to get the amount of disposal material. Do not figure beams and deck together as the thickness of fireproofing is sometimes different.

Pipe Insulation

Define exact lengths and sizes of pipe for insulation removal. Identify the point of origin for the lines and follow them visually or with an accurate "as-built" drawing. Within the area of removal, not the following:

Diameter of the pipe

Thickness of pipe insulation

Location of structural beams

Location of mechanical equipment

Distance of pipe from walls and ceiling

Hardness of the insulation

Height of the pipe from the floor

Note whether or not there is any pipe insulation inaccessible within the walls. Controlled demolition and reinstallation of the walls containing ACM pipe insulation in will add to the amount of labor and removal required for removal of the insulation.

Separate lengths of pipe into groups by size, straight lengths, elbows, and connections. For disposal purposes, no other grouping is necessary. Multiply the length of pipe × the thickness of insulation × the diameter of the pipe × 3.14. Use the same formula found in structural frames to calculate the amount of disposal material.

Equipment Insulation

Define the exact equipment that is subject to insulation removal. Record the height, length, diameter, or width of each piece of equipment, extent of insulation, and the thickness. Note how secure the insulation is and whether it is in a state of disrepair. When looking at boilers and hot water storage tanks, check to see if block insulation is present. Is it secured with canvas only or tied on with wire mesh with canvas over the mesh? Note the height of the equipment and the difficulty in packaging the asbestos after the removal.

Spray-on Decorative Acoustical Ceilings

Define the square foot area of ceiling for acoustical material removal and the number of rooms involved. Note the walls and other fixtures in each area that could impede access and slow the work. Include in the square footage notes for each room the height of the ceiling and depth of the material. Note whether the substrate is wallboard, metal lath, or metal lath with plaster, or brown coat over the lath. ACM acoustics blown through the metal lath into a ceiling plenum can increase the amount of labor required for decontamination of the area. Note how large an area there is above a drop ceiling and whether other nearby areas may have contaminants. Make a separate line item for each ceiling for the labor section of the estimate. Use the formulas in structural deck to figure square footage and disposal yards.

Vinyl Tiles

Define the square foot area of floors walls for vinyl tile removal. Note whether the tiles are 9 IN × 9 IN or 12 IN × 12 IN. Physically inspect the condition of the tiles. Note their adherence to the substrate, whether the substrate is wood, concrete, or ceramic material. Also check for water damage.

Make a separate line item for each floor area for the labor section of the estimate and quantify in square feet. Floor tiles are cumbersome. For disposal, figure six floor tiles per plastic container, three plastic containers per bay of floor tiles, and one cubic foot per bag. As an alternate, estimate lining a dumpster with plastic and loading the material without bagging it first.

Vinyl Sheeting with Asbestos Backing

Calculate the square foot area of floors or walls for vinyl sheeting removal. Physically inspect the adhesive attachment of the vinyl sheet. Is the adhesive glue or rubber cement? Note the method of edge attachment. Determine the expected extent of damage in the removal process.

Make a separate line item for each floor or wall area for the labor section of the estimate and quantify in square feet. Since vinyl floor sheeting is lightweight, it may be removed in sections.

Duct Insulation

Measure the surface area of the ducts and depth of insulation. Note the location of the ducts and depth of insulation. Note the location of the ducts. Also note access for ACM such as:

Exposed
In an attic
In a shallow return air plenum
Height from floor
Obstructions between the floor and duct

Wallboard

Measure the square footage of ACM wallboard for removal and the manner of attachment to the substrate. Note whether there are successive layers of wallboard. Is there a wood frame backing or is there a direct application of ACM wallboard to a concrete or framed wall?

Miscellaneous Wire Insulation and Fascia Materials

Measure quantities of items containing ACM. Note whether the wire inside the insulation or the area behind the fascia materials will remain after removal of the ACM materials.

Containment Areas

A containment is an area surrounding items subject to removal of ACM. There is physical isolation with plastic and wood with negative pressure to contain the asbestos fibers. Ingress or egress is only through specially constructed decontamination chambers. These chambers impede the flow of air from the containment to outside areas. The specifications usually define the type and thickness of the containment materials. They also usually specify the containment areas for asbestos removal from mechanical equipment or structures. They may specify requirements for the negative pressure air system.

The requirements for negative pressure may include detailing air intakes and HEPA-filtered exhaust ports. The number and size of negative air units to achieve a minimum of four air changes per hour is an additional requirement.

Physically walk the area of removal as defined and record the height, length, and width of each area. These dimensions are necessary for determining material and equipment costs in the estimate. Locate an area for construction of the containment area(s) and their approximate size. Note any obstacles that will impede the smooth flow of the plastic sheeting and add labor and supervision time to seal an area. Be specific about what is to remain and what is to leave an area before containment. The use of additional wood, plastic, and fans may be necessary to protect items remaining in the area. This increases labor and material costs. Note the source of water and power and distance to the nearest connections. Determine who is going to make the connections into the building power when necessary. If it is the contractor's responsibility, include local trade costs in the estimate.

Multiply the floor area by each containment by two for the square footage of plastic, membrane, and paper required. Divide this figure by the total square footage in the rolls of plastic to determine how many rolls are required.

Multiply the wall area of each containment by two for the square footage of plastic required. Divide this amount by the total square footage of the rolls of plastic to determine how many rolls of plastic are required. Add at least 25% more plastic for the waste allowance.

Calculate the length of seams that require tape for securing all plastic sheeting together. Add 30% waste and divide by the number of feet in a roll of tape. Anticipate one case of spray glue for every two rolls of plastic.

Negative Air Equipment

Multiply the floor area of each containment by the height of the area to figure cubic feet of air for exhaustion from the room. Take 75% of the CFM listed on the proposed equipment and multiply this amount by 15 minutes. This will show the amount of air exhausted in 15 minutes. Divide the total cubic feet of air in the room by this amount to ascertain the number of pieces of equipment needed for the project. It is prudent to include the cost of standby equipment.

Decontamination Unit Materials and Allowances

A standard decontamination unit has a minimum of three chambers 4 FT × 4 FT and 6 FT in height. A large project will need at least twice that many chambers in a larger size. Anticipate enough framing lumber to make all the necessary chambers. Use 1/4 IN plywood to construct the walls and 1/2 IN and 3/4 IN plywood to construct the ceiling of the decontamination unit.

Allow a minimum of four hours for a plumber to hard plumb the supply and drain lines from the shower and/or decontamination units. Include this in the estimate as a subcontract cost. Determine the number of windows to be removed, then get a unit cost from the subcontractor to remove them. A connection into the building electrical system is usually necessary. Determine the location of the hook-up and estimate the cost.

Labor Required to Remove Asbestos Containing Material

The following are examples of labor requirements for asbestos removal. The amount of labor necessary to remove asbestos containing material can vary significantly with the thickness, hardness, and manner of attachment. Multiply the following labor hours by 1.5 if the fireproofing is over 2 IN in depth or difficult to remove. Multiply the time for removal of thermal insulation from equipment by a factor of 2 if there is heavy wire mesh covered by canvas.

Anticipated Unit Rates to Remove Fireproofing Material Working off a Scaffold:

Asbestos Removed (one worker)	15 SF floor area per hour
Detailer	25 SF floor area per hour
Bagger (two workers)	30 bags per hour
Set-up	Two workers per thousand SF per day
Load Dumpster	600 bags per day

Anticipated Unit Rates to Remove Pipe Insulation (2 IN pipe less than 6 FT in height):

Asbestos Removed (one worker)	120 LF per day
Bagging and Clean-up (one worker)	200 LF per day
Load Dumpster	600 bags per day

ACM equipment insulation should be estimated per job.

Anticipated Unit Rates to Remove Sprayed-on Acoustical Insulation:

Asbestos Removed (one worker)	600 SF floor area per day
Detailer (one worker)	900 SF floor area per day
Bagger (two workers)	30 bags per hour
Load Dumpster	600 bags per day

The Anatomy of a Mechanical Estimate

By John B. Stewart CPE

SYNOPSIS

"The Anatomy of a Mechanical Estimate" is a step-by-step guide in preparing a mechanical estimate. It is intended to be used as a model for the novice and professional alike, and to be modified as needed.

There are seven basic steps inherent to all mechanical bids. Each step is analyzed in detail without undertaking a specific discussion of the various trades, material, equipment and takeoff procedures involved with a mechanical estimate.

INTRODUCTION

Most estimators whose job it is to handle this type of work generally agree upon the basic principles of mechanical estimating. Although differences do exist in the forms and methods used, no one can argue the fact that the completed estimate must reflect all costs that are shown and called for in the project's drawings and specifications. These project costs include all labor, materials, equipment, direct and indirect costs, overhead and profit. With this in mind it is the intent of the author to generally describe the procedure for calculating these costs and to develop a complete mechanical estimate.

For the purpose of this discussion the estimate consists of the following.

1. Review of the project plans and specifications
2. Quantity takeoff of material and equipment
3. Pricing of the material and equipment
4. Estimating the labor
5. Calculating indirect project costs
6. Receiving subcontractor quotations
7. Arriving at a total project cost

REVIEW OF THE PROJECT PLANS AND SPECIFICATIONS

Upon receipt of an invitation to bid on a particular project, the estimator should make a general review of the project documents. The architect, engineer, type of building, approximate square footage, location, type of equipment, and any other information that would help in the evaluation should be noted. Once this has been done, a decision can be made as to whether or not to bid the project. This decision should be based on market conditions, type, size and location of job, and the estimator's bidding load.

Once the decision has been made to bid, the estimator should begin by reading very carefully, and checking every part of the drawings, specifications, request for bid and any other bid documents provided. It is most important to thoroughly understand these documents for two reasons: first, to avoid any mistakes or omissions in the bid and secondly the plans and specifications will eventually become part of the contract documents to which the contractor will be bound legally.

THE SPECIFICATIONS (CSI DIVISION 15)

The specifications are generally composed of an explanation of bid procedures, general and special conditions, from which the engineer's intent is derived. While reading the specifications, the estimator should pay particular attention to special procedures, equipment, material and construction specified by the engineer. These may include such items as AWWA rating for the valves, seamless pipe as opposed to ERW pipe, double wall duct construction with a perforated or solid metal liner. A listing should also be made of the manufacturers who the engineer has approved to supply the material and equipment for this particular project. A pre-printed form should be used to help organize this information for quick reference throughout the bidding process. This form should have space for the supplier/manufacturer of the item specified, along with space for the local representative, their phone number, and an area for notations where model numbers or some special type of construction should be listed.

Another area of the specifications the estimator should be aware of is the "related work by others." This related work could include, but is not limited to, such things as temporary heat, painting, concrete pads, excavation, structural steel, and cutting and patching. Other sections of the specifications may have to be reviewed in order to determine how much of this "related work" will affect the mechanical bid. Care should be taken not to duplicate these items.

While reading the specifications, note the particular sections, in division 15 that do not pertain to his trade's line of work. These sections will be needed to complete the estimate and should be covered by soliciting a subcontractor to provide a price for that particular section of the specification. This solicitation can be done in several ways; by use of a post card, formal letter FAX, or direct telephone calls. Whichever way is utilized, the estimator should follow up the request, prior to bid day, to verify which subcontractors will be quoting.

Depending on market area or company make-up, the subcontractors needed may vary, but may include any or all of the following:

- ✓ **SECTION 15080 — Mechanical Insulation**
- ✓ **SECTION 15300 - Fire Protection**
- ✓ **SECTION 15400 - Plumbing**
- ✓ **SECTION 15800 - Air Distribution (Sheetmetal)**
- ✓ **SECTION 15950 - Controls**
- ✓ **SECTION 15990 - Testing Adjusting and Balancing**
- ✓ **SECTION 16000 — Electrical Work pertaining to the controls**
- ✓ **Excavation & Backfill**

The subcontractors should be contacted well in advance of bid day in order to give them sufficient time to develop their estimates.

THE DRAWINGS

The first items that should he checked generally include the job location, scale, degree of completion, possible areas of interference with other trades, and missing items. It should also be noted if there are conflicts within the specifications. Generally speaking, if there should be a conflict between the various documents, the large details take precedence over the general plans and the specification takes precedence over all documents with the exception of the signed contract. If after referring back to the specifications, there is still any doubt, then the design engineer should be contacted to have any questions clarified. Normally, the engineer would then issue to all bidders, an addendum to the documents with the clarification.

After reviewing the plans and specifications, generate a scope of work. This scope should include all areas required to complete the work shown and specified. Itemize exceptions and clarifications within the scope of the estimate.

Develop scopes for the various subcontractors that will be required for a complete estimate. These scores of work will be used later in the bidding process to evaluate each subcontractor's quotation.

QUANTITY TAKEOFF

While reviewing the drawings organize the takeoff approach.

The quantity takeoff procedure is the measuring and counting of all materials and equipment indicated on a set of drawings. For an accurate takeoff, pre-printed forms should be used. These forms are used to record pipe, valves, fittings, ductwork, equipment and other materials needed. Only material quantities appear on these sheets, no labor or dollar amounts. In most cases, the front side of the paper or form should be used in order to minimize mistakes. The estimator should use given dimensions from the drawings, where shown, in lieu of scaled ones. Similar items should be grouped together in order to save time with pricing.

Items that aid the takeoff are:

- ✓ Colored pencils (use different colors for different systems in order to follow each system)
- ✓ Scale rules (architect's, engineer's and metric)
- ✓ Metallic tape (for quick takeoff of long pipe or duct runs)
- ✓ Automatic counter (a mechanical counter for single counts or a rotometer for long runs)
- ✓ Electronic takeoff machine (with the advent of computerized estimating these takeoff aids are becoming more popular)

Measure or record everything on the drawings that affects your scope of work. Note reduced size drawings, not-to-scale details, and changes in the scale from drawing to drawing and from detail to detail. Note missing or deleted items that may have to be included in order to make the job complete. An example would be steam trap stations that are needed to return the steam condensate to the system although they are not shown on the drawings.

Have a library of equipment catalogues, technical books, hook-up diagrams and installation manuals for reference when items are not clear and there is an application question.

A consistent method should be developed for performing the takeoff. For example the plan views should be taken off first, then the elevations and finally the details. This way the estimate can be traced back, if necessary, and checked for mistakes or omissions. If possible try to use repetition to speed things up, but use extreme caution not to duplicate an area.

PRICING OF MATERIAL AND EQUIPMENT

After the takeoff is complete develop costs for the bill material. This bill of materials can be priced in several different ways.

1. By computer, using the data the estimator has entered into the memory of the computer to create a database. This method is not only the quickest but also the most accurate.

2. Extension by hand, using pricing sheets provided by local suppliers. This method can be tedious and can lead to errors.

3. Supplier quotation is when the bill of material is sent to supplier(s) for pricing. In using this particular pricing method, make sure the supplier(s) have adequate time for pricing. The suppliers should be provided with the exact specifications, approved manufacturers, etc.

Equipment vendors should be contacted at the same time the subcontractors are and in the same manner. The equipment should be summarized on a spreadsheet for ease of reviewing. The quotations should be reviewed as soon as they are received, and should be checked for:

- ✓ Correct quantities
- ✓ Exceptions to the plans and specifications
- ✓ Special notes as to field installed options
- ✓ Number of pieces to be handled
- ✓ Shipping weight and availability

The vendor should correct any discrepancies before bid day. Normally material and equipment prices should include the freight cost of job site delivery and the sales tax (where applicable).

It is important to check and recheck the calculations. An uncomplicated system for doing this should be used. One such system is simply having someone else redo the calculations. Never trust the results until they have been checked. A simple mistake or forgotten items at this point may never be picked up until it's too late.

ESTIMATING THE LABOR COST

In order to arrive at a total job cost the estimator has to estimate the time required to install each individual item included on the estimate sheets. This is where the spreadsheets the estimator has used to list the material and equipment called for come in handy. From these sheets the estimator can tell how many feet of pipe, number of hangers, pieces of equipment, gallons of glycol, pounds of refrigerant, and so on, will have to be handled.

Estimating this labor is a difficult task at best, requiring both experience and knowledge of how the work is done in the field. To help with this task the estimator has pre-printed labor tables. These tables are based on average performance of workers under normal job conditions. The tables can be obtained from many different sources. Some of which are:

- ✓ MCA
- ✓ R.S. Means
- ✓ Page and Nations
- ✓ Octavianio
- ✓ SMACNA

Contractors may also have their own labor tables based on past historical job data.

The labor unit used in the majority of publications is the labor hour. Some estimators are more comfortable working with days. Because of this, the labor unit should be marked at the top of each sheet and care should be taken to make sure all labor units stay the same throughout the estimate. It can get to be very confusing if the labor unit changes from sheet to sheet.

The labor estimate must include the time for handling, erecting, and joining the material or equipment. This labor time can be easily calculated and totaled, but it is the rate of production that can alter this total. Many factors affect the rate of production including weather conditions, height of the building, access, existing conditions, and overtime. Analyze all job conditions carefully in order to determine if a correction factor, up or down, is needed.

Satisfied that everything has been covered and total labor time has been developed, the labor cost can be applied. The labor cost is the dollar equivalent of an estimated labor time. This labor cost is computed by adding the hourly base rate with the fringe benefits and insurances. Due to the almost certain increases in the cost of labor make sure that the most current rates are used. Be sure to anticipate any scheduled increases that may occur during the project.

CALCULATING INDIRECT JOB COSTS

Indirect costs are labor, material, and equipment not shown on the drawings but which will be needed to complete the job. These items will vary from job to job but may include temporary services, drinking water, toilets, storage sheds, field office, shop drawings, trucks, small tools, permits, warranty, guarantee, etc. Use a checklist in order not to overlook any of the indirect costs. Not all indirect costs are applicable to every project.

A few of the indirect costs are sometimes calculated as a percentage of the base labor or the job cost. These may include expendables, consumables, small tools and warranty.

Indirect costs may be a substantial part of the estimate and should not be overlooked.

RECEIVING SUBCONTRACTOR QUOTATIONS

The subcontractor bids are usually not received until the last few hours before the bid is due. Be prepared to evaluate each quotation as it is received. The scope of work generated during the specification review is used for this purpose. Review this scope with each bidder. Determine whether the subcontractors are bidding per the plans and specifications.

Special care should be taken in noting the subcontractor's exceptions and clarifications. These can and will often change the scope as it is listed. The most important thing to remember, when it comes down to the final decision on which subcontractor's price to use, is to make sure the quotations are equal and nothing is overlooked.

This subcontractor review process is very important. It not only protects the estimator, but it also protects the subcontractor as well. A missed note on the drawings, a forgotten piece of equipment, depending on the complexity of the project, can mean big losses for both parties, if not discovered prior to the bid.

DETERMINING THE BID PRICE

The bid price is normally determined by bringing all costs together on one final sheet. This sheet is called the recap or front sheet. This is where all costs for material, equipment, labor, indirect costs, and subcontractors are posted. Often a contingency factor is added to recognize risk and should reflect the additional cost for unusually high labor exposure or the completeness of the bid documents. Once all of these costs have been brought forward and the estimate has been checked and rechecked for completeness and accuracy, then the overhead and profit can be applied. The overhead costs are items incurred by every contractor in conducting business. These items include office rent, telephone, lighting, office staff, insurances, printing, automobiles and many more. Overhead is usually added as a percentage of the job cost and is set by management.

Profit should be applied as a separate percentage after all costs are included. This percentage is set by management and is the return on investment they feel is needed to operate the business.

Profit margin can vary, however, depending on the project cost. Normally smaller projects get a larger mark up while larger jobs will get a smaller markup applied.

CONCLUSION

In conclusion the author has only listed the basic steps, which should be used as an outline and modified as needed, and has not gotten into the specific and various trades, material, equipment, and take off procedures involved with a mechanical estimate.

The importance of a set and predetermined estimating procedure cannot be overly stressed. If the same procedure is followed on every job, large or small, the likelihood of errors or omissions will be greatly reduced and the estimator will feel confident that a complete and reliable estimate has been produced.

REFERENCES

Gladstone, John

Mechanical estimating guidebook for building construction, Rev. ed., McGraw-Hill, 1978

Burkholder, Lloyd K

Process & industrial pipe estimating, Craftsman Book Company, 1982

Khashab, A M

Heating, ventilating, and air-conditioning systems manual, McGraw-Hill, 1977

PART THREE

PROFESSIONAL ESTIMATING SERVICE GUIDELINES

Part Three

TABLE OF CONTENTS

Request for Proposal (RFP) for Professional Estimating Services 467

Contract For Professional Estimating Services between Owner and Estimator 473

Contract for Professional Estimating Services between Estimator and Client 483

Request for Proposal (RFP) for Professional Estimating Services

Sample #1 - Request for Competitive Sealed Proposals

The _____ extends an invitation to interested construction estimators to submit a competitive sealed proposal for estimating services relating to design and construction of various projects on _____
_____.

The individual projects will be varied in intent, scope, size and type. The number of projects initiated and assigned will be entirely at the _____ discretion. Aggregate maximum cost for fees for services under any such contract will be limited.

A request for proposal with instructions and form of contract providing additional information and the detail of services to be performed, is available at the office of _____
_____.

Eligible professionals are invited to seek the above RFP and associated information and make a proposal in accordance with criteria established. Proposals are due no later than _____ local time _____ at _____.

The _____ reserves the right to reject any or all proposals it may determine are non-responsive or unacceptable, and to waive or decline to wave any irregularities therein.

Sample #2 - Request for Competitive Sealed Proposals

Contents:

- ✓ **Request for Proposal**
- ✓ **Instructions to Proposers**
- ✓ **ASPE Contract for Professional Estimating Services between Owner and Estimator** (included in this section)
- ✓ **RFP Data Sheet**

Request for Proposal

The _____ requests proposals from qualified applicants to provide professional estimating services on open end contract basis for various types of remodeling, additions, alterations, and new capital projects in accordance with specifications and requirements on file in the office of _____. More than one such contract may be awarded.

Each contract awarded will be on the following basis:

- ✓ Each contract will not exceed $_____ in aggregate fees.

- ✓ The fee for individual projects will be negotiated on the basis of labor hours times contractually fixed rates.

- ✓ The _____ provides no guarantees as to the minimum or maximum number of projects to be awarded under each contract.

- ✓ Each estimate assignment (project) must be accepted and expedited to completion within the specified time constraints.

The _____ will receive sealed proposals until _____ local time _____ at the office of _____. Proposals received after that time will not be accepted.

The _____ reserves the right to reject any or all proposals or to withhold the award for any reason and to waive or decline to waive irregularities in any proposal. The _____ also reserves the right to hold all proposals for a period of 60 days after the opening date and the right to accept a proposal not withdrawn before the scheduled proposal opening date.

Instruction to Proposers

All proposals should follow the format and sequence described in the paragraphs below. This will allow a standard basis for evaluation. Failure to follow the instructions regarding format may result in rejection of the proposal.

Proposals shall be addressed and delivered to the _____ on or before the day and hour set for receipt of proposals. Each proposal shall be enclosed in a sealed envelope, marked with the name of the proposer and the date and hour the proposal is due.

No telegraphic proposals or telegraphic modification of proposals will be considered. Proposals received after the time fixed for receiving them will not be considered and will be returned to the sender. Proposers are solely responsible for the delivery of their proposals at the above location at or before the time and date specified.

Proposals may be withdrawn either personally or by written request any time before the scheduled opening date and time.

The technical proposal should display clearly and accurately the capability, knowledge, and capacity of the proposer to meet the technical requirements of this RFP. _____ copies of the proposal are required.

The technical proposal should be fully self-contained and should be divided into the following major categories:

- ✓ Description of the services to be provided
- ✓ Labor classification plan
- ✓ RFP data sheet
- ✓ Corporate resources and experience - GSA Form 254
- ✓ Other information as deemed appropriate by the applicant

Applicants may be requested to participate in an interview with the owner's representatives. Additional information may be required.

The _____ may contract with the selected firm at compensation that the _____ determines fair and reasonable, taking into account the evaluation criteria listed below, and may waive or decline to waive irregularities in any proposal.

A committee will evaluate proposals submitted in response to this RFP. The panel may request additional information through interviews, presentations, or correspondence.

The _____ will use the following criteria in the relative order of importance listed to evaluate the proposals:

- ✓ Demonstration of technical special expertise, qualifications, and experience, on similar types of projects successfully completed by the firm, and by personnel to be assigned to the project.

- ✓ Proposed consultants, their qualifications, expertise and experience, if any.

- ✓ The ability of the firm and its consultants to commit proper manpower (staff) and resources to the projects assigned.

- ✓ Demonstrated previous record regarding actual project costs vs. budgets and estimates prepared by the firm.

- ✓ Ability to meet the critical schedule requirements for assigned projects.

- ✓ Location of the firm's office where the work will be performed.

- ✓ Past performance in completing estimates within required scheduled.

- ✓ Client references.

A pro forma contract which the successful proposer will be required to execute is included in this RFP.

The firm to whom _____ awards the contract shall, within 15 days after notice of award, deliver to _____ signed copies of the contract as well as the required insurance certificates or policies. All policies of insurance shall be reviewed and approved by _____ before the successful proposer may proceed with the services. Failure or refusal to furnish required insurance policies or certificates in a form satisfactory to _____ shall result in rejection of the contract.

Applicants who desire clarification of the attached documents or who find discrepancies or omissions may request _____ in writing for an interpretation or additional information. All such inquires should be made to _____.

The following procedures apply: The applicant submitting the request shall be responsible for its prompt delivery. Interpretation or correction of RFP documents will be made only by addendum and will be mailed or delivered to each proposer of record. _____ is not responsible for any other explanations or interpretations of the RFP documents.

Each applicant shall visit _____ and become familiar with existing conditions and limitations under which the services are to be performed before submitting a proposal. Submission of a proposal will be construed as conclusive evidence that proposer has familiarized him/herself with the conditions involved with the services. Services may occur for work in any of the existing buildings _____ or for any proposed new building on _____.

Proposals in response to the RFP may contain data that the proposer does not want disclosed for any purpose other than evaluation of the proposal. If so, the proposer shall clearly identify those pages of the proposal that are to be restricted. _____ assumes no responsibility for disclosure or use of unmarked data. Unless identified, information submitted in response to this RFP may be disclosed pursuant to the freedom of information act and applicable revised statutes.

Should this RFP result in a contract, the terms, clauses, and conditions required by _____ procurement regulations shall apply to the contract.

This RFP does not obligate_____ to pay any costs incurred in the preparation and submission of proposals nor to enter into a contract with any of the applicants.

Request for Proposal (RFP) for Professional Estimating Services   Part Three

RFP Data Sheet

The RFP Data Sheet for Professional Estimating Services shall include the following:

✓ Firm name, address, telephone and fax (Office which will handle this project).

✓ Principal office location

✓ Proposed staff, staff qualifications and relevant estimating experience (on GSA Form 255). Specific project staff and staff responsibilities will be established for each assigned project and will be binding throughout the project unless modified by mutual consent.

✓ Proposed consultants and their experiences, expertise and qualifications with estimating for projects: (Name and describe on GSA Form 255 for any proposed consultants).

✓ Explain ability of you and your consultants to commit adequate labor and resources to assigned projects, and how many projects you can simultaneously handle.

✓ Confirm your willingness and ability to promptly initiate and execute necessary services to meet critical schedule requirements for each assigned project.

✓ Past performance in cost estimating accuracy versus actual project costs (list your estimate(s) and actual amounts for at least ten (10) projects ranging from renovations/ alterations to new capital facilities), $_____ value to $_____.

✓ Past performance regarding meeting schedules (list initial required dates and actual dates achieved for submitting estimates, for at least six (6) project estimates within the last three (3) years.

✓ Client References (include on Form GSA 255):

_____

Firm

by _____

Authorized Signature

RFP Data Sheet Supplement

_____

Firm

Contract For Professional Estimating Services between Owner and Estimator

This suggested document has important legal consequences. Consultation with an attorney is encouraged with respect to its completion or modification

This agreement, made and entered into this _____ day of _____, 20____, by and between the _____ on behalf of _____ hereinafter referred to as the "Owner", and _____ hereinafter referred to as "Estimator", shall be for the purpose of employing said estimator to perform the professional services hereinafter delineated.

The estimator certifies that the principals or officers of the firm, who will certify the work, represent themselves as being professionally experienced, competent and qualified to perform the services hereinafter described.

Now, therefore, for and in consideration of the mutual covenants, stipulations, and conditions hereinafter contained, the Owner and Estimator agree as follows:

Article 1 - Scope of Professional Services

A. Estimator agrees to perform those services described herein:

Project(s) Locations:

All projects will be in connection with _____.

Project Budgets:

Each individual project assigned for estimating will have a _____ cost budget and possibly an architect's estimate, which will be made available to the estimator.

Contract minimums:

The number of projects and individual project budget levels assigned is entirely at the discretion of _____. There is no guarantee as to the minimum or maximum number of projects to be assigned.

Contract maximum:

The total aggregate value of fees for all estimating services which may be assigned under each contract shall not exceed $ _____.

Number of contracts:

The _____ may enter into more than one separate contract.

Contract duration:

Each contract shall have a nominal term, or life, of _____ from date of contract. The _____ may elect to extend the contract up to _____. This is in the event that a project or services related thereto initiated within the nominal term cannot reasonably be completed before the close of the term, or for reasons deemed advantageous to the _____.

The actual duration or length of such extensions will be set by mutual agreement, at no increase of scheduled rates.

Nature and type of projects:

Projects may be alterations, remodeling, additions, or new major or minor capital facilities. Dollar value may range from a low of _____ to in excess of _____.

Initiation of project services:

A. Each project will be initiated by a work order request, which shall describe the project, its stage of development, include available drawings, specifications, and other descriptive material, the project schedule, budget, and estimate(s), if any, prepared by architects and/or engineers for the project.

B. Within three (3) working days of receipt of work order request, the estimator shall provide to _____ a cost proposal for the estimating services required thereby. Also included shall be a proposed time period to prepare and complete the estimate.

C. Upon review and acceptance of the written cost proposal and schedule as submitted, or as modified, the owner will issue a written work order, to proceed with the estimate.

Scope of services:

A. Cost estimates shall be prepared in accordance with the following scope and Part One - Section Two and Three - Levels of the Estimate and Scope of the Estimate as defined in the current Standard Estimating Practice Manual of the American Society of Professional Estimators, and included herein as exhibit "A".

B. The estimator warrants that the services performed by him (her) hereunder shall be in accordance with good estimating practice.

C. The size, nature and extent of any given project may vary. The contract has no guaranteed minimum amount nor number of estimating projects over the life of the contract.

D. The estimator shall in all cases coordinate closely with and keep owner's staff informed on the progress and results of each project that the estimator is preparing for the owner.

E. Upon completion of each estimate, the estimator shall submit her (her) dated and signed estimate complete with adequate back-up data.

F. Additional responsibilities of the estimator:

1. The estimator shall be responsible for the professional quality, technical accuracy, timely completion, and the coordination of all products of his (her) effort and other services furnished under this agreement. Without additional compensation, the estimator shall correct and/or revise any errors, omissions, or other deficiencies in all products of his (her) effort and other services he (she) provides.

2. Estimator shall perform such estimating services as may be necessary or required under the contract.

3. Approval by the owner of any services furnished shall not in any way relieve the estimator of responsibility for the technical accuracy of his work. Owner review, approval, acceptance, or payment for any of the estimator's services herein shall not be construed to operate as waiver of any rights under this contract or of any cause of action arising out of the performance of this contract, and the estimator shall be and remain liable in accordance with the terms of this contract and applicable law for all damages to the owner caused by the estimator's improper performance under this contract.

Article II - Changes

A. At any time the owner may make, by written order, changes within the general scope of this contract in the services or work to be performed. If such changes cause an increase or decrease in the estimator's cost of, or time required for, performance of any services under this contract, whether or not changed by an order, an equitable adjustment shall be made and this contract shall be modified in writing accordingly. The estimator must assert any claim for adjustment under this clause in writing within thirty days from the date of receipt by the estimator of the notification of change, unless the owner grants a further period of time before the date of final payment under the contract.

B. No services for which additional compensation will be charged by the estimator shall be furnished without the written authorization of the owner.

Article III - Length of Contract

This contract shall be valid for a period of _____ from the date signed by both parties. This contract may be extended for up to _____ if required due to projects in process but not completed, or for other valid purpose, based on mutual agreement of both parties. No other terms of the contract will change during such extended period.

Article IV - Compensation

In consideration of the services described owner agrees to pay estimator compensation in accordance with the following:

A. The estimator agrees to perform the work specified in the contract within the period(s) specified and within the cost allocation established in this article.

The work performed for each project, however, shall be as specified in the respective estimating services work order, and the approved manpower and cost projections.

B. Time is of the essence, and if at any time the estimator has reason to believe that the time for completion of a project will extend beyond or be substantially less than the projected time period, the estimator shall notify the owner in writing to that effect. The notification will state the revised estimate time and/or cost for completion of the project. Such notification will be submitted to the owner at the earliest possible date. Said revised estimated time and/or cost for completion shall be and remain without effect until agreed to by the owner in an amendment to the work order.

C. Compensation for each work order or task authorized on a cost reimbursement basis shall be approved personnel rates of the estimator times actual labor hours plus other direct costs, on a not-to-exceed cost basis.

D. Personnel rates, as a basis for computation of compensation, shall include all wages, salaries, insurance, benefits, overhead and profit or fees. Rates for various estimating personnel classification shall be as follows:

Title or Job Classification   Hourly Personnel Rate

1.

2.

3.

4.

5.

E. Other direct costs-other direct costs shall be other identifiable and pertinent costs of the estimator directly chargeable to performance if services under the contract, of prior approval has been granted in writing by the _____. Other direct costs shall be billed at cost and shall include the following: reproduction costs of plans, specifications and reports; commercial printing and binding; services of "special" consultants, if any; and living and travel expenses of employees when away from their home, or office or business while performing estimating services.

F. Payment - payment for each work order/project shall be made at the completion thereof, or on a predetermined schedule and shall include a breakdown of costs of personnel rates and other direct costs.

Article V - Termination

A. For cause - Owner's rights

1. The owner shall have the right to terminate this contract under the following circumstances:

a. If any or all work to be performed under the contract shall be abandoned by the estimator.

b. If the contract or any part thereof shall be assigned in violation of Article V, "assignability".

c. If the estimator shall become insolvent or unable to meet its payroll or other current obligations, or shall be adjudicated bankrupt, have an involuntary petition in bankruptcy filed against it, make an assignment for benefit of creditors, file a petition for an arraignment, composition or compromise with its creditors under the bankruptcy laws of any state, or shall have a trustee or other officer appointed to take charge of its assets.

d. If at any time it should appear to the owner that the schedule of work is not being maintained or that the estimator is violating any of the conditions or provisions of the contract, or if at any time the owner determines that the estimator is refusing or failing to properly perform the work or the estimator is performing the work under the contract in bad faith or not in accordance with the terms thereof.

2. Termination shall become effective only upon compliance with all the following conditions:

a. The owner shall give the estimator at least five days written notice by registered mail or hand delivered mail.

b. The notice shall specify the grounds for termination.

c. The notice shall designate a termination date. The estimator shall be given five calendar days after receipt of such notice to remedy default.

3. Once termination becomes effective, the estimator shall promptly discontinue all services affected, unless otherwise directed in writing. The estimator shall thereupon make available, and at the owner's option, deliver to the owner all finished and unfinished estimates, reports, and such other information and materials as may have been accumulated by the estimator in performing this contract.

4. The estimator shall be entitled to an equitable adjustment of any work completed and accepted. No amount shall be allowed as professional fee on unperformed services or other work not complete. In the event the owner terminates this contract for cause, the owner may withhold payments to the estimator for purpose of setoff until such time as an exact amount of additional costs occasioned to the owner by the estimator's default is determined. The estimator is responsible to the owner for additional costs occasioned by the estimator's failure to perform in accordance with this contract.

5. If the owner terminates this contract for cause, the owner shall have the right to complete such work by whatever method the owner may deem expedient including employing another estimator under such a form of agreement as the owner may deem advisable or the owner may provide all labor or materials and perform any part of such work that has been determined.

6. If the owner terminates this contract for cause and proceeds to complete the work thereunder, the expense of so completing, together with a reasonable charge for administering any agreement for such completion, shall be charged to the estimator. Such expense may be deducted by the owner out of such monies as may be due or may at any time thereafter become due to the estimator.

7. If the owner terminates this contract for cause and it is thereafter determined that none of the grounds for termination as outlined in paragraph III.A.1 of this appendix exists, termination shall be deemed to have been effected for the convenience of the owner under paragraph III.B.1 of this appendix.

B. For convenience of the owner

1. The owner shall have the right to terminate this contract at any time for its convenience. Termination shall become effective upon compliance with all of the following conditions:

a. The owner shall give the estimator at least five days written notice by registered or hand-delivered mail.

b. The notice shall specify that it is for the convenience of the owner.

c. The notice shall designate a termination date.

2. Upon receipt of notice, the estimator shall promptly discontinue all services affected unless otherwise directed in writing. The estimator shall thereupon make available, and at the owner's option, deliver to the owner all original finished and unfinished document, reports, estimates, and such other information and materials as may have been accumulated by the estimator in performing this contract.

3. If the contract is terminated by the owner as provided herein, the estimator shall receive an equitable adjustment for services accepted by the owner. The equitable adjustment shall include payment for contractual obligations reasonably incurred prior to termination. No amount shall be allowed for anticipated profit on unperformed services.

4. In the event the owner terminates this contract for its convenience, the owner may take over the work and prosecute the same to completion by agreement with another party or otherwise.

C. For cause - estimator's rights

1. The estimator shall have the right to terminate this contract under the following circumstances:

a. If the owner fails to provide compensation for previous services rendered within the time schedule agreed upon.

2. Termination shall become effective only upon compliance with all the following conditions.

a. The estimator shall give the owner at least five days written notice by registered mail or hand delivered mail.

b. The notice shall specify the grounds for termination and shall designate a termination date.

c. The owner shall be given five calendar days after receipt of such notice to remedy the default.

Article VI - Other Matters

A. Ownership of documents - All notes, reports, estimates, and technical data resulting from this agreement shall be the property of the owner. The estimator shall submit reproducible originals of the reports, estimates, and technical data to the owner.

B. Compliance - The owner and estimator agree that the work to be performed under this agreement shall be subject to and comply with all pertinent laws of _____. The owner and estimator agree that time is of the essence and that work to be performed shall not be delayed.

Article VII - Audit; Access to Records

A. The estimator shall maintain books, records, documents, and other evidence directly pertinent to performance under this contract in accordance with generally accepted accounting principles and practices consistently applied, including the financial information and data used in the preparation or support of the cost information of Article IV and a copy of the cost summary submitted to the owner. The owner shall have access to such books, records, documents, and other evidence for inspection, audit, and copying. The estimator will provide proper facilities for such access and inspection.

B. The estimator agrees to include all of this article in all his contracts and all subcontracts directly related to project performance.

C. Audits conducted under this provision shall be in accordance with generally accepted auditing standards and established procedures and guidelines of the reviewing or audit agencies.

D. The estimator agrees to the disclosure of all information and reports resulting from access to records under this section to the owner designated auditor, provided that the estimator is afforded the opportunity for an audit exit conference and an opportunity to comment.

The estimator may submit any supporting documentation on the pertinent portions of the draft audit report and that the final audit report will include written comments of reasonable length, if any.

E. The estimator shall maintain and make available records under this section during performance on work under this contract and for three years from the date of final payment of the project.

Article VIII - Subcontracts

A. All subcontractors and outside associates or consultants required by the estimator in connection with the services under this contract will be limited to such individuals or firms that were specifically identified and agreed to during negotiations, or that the owner specifically authorizes during the performance of this contract. The owner must give prior approval for any substitutions in or additions to such subcontractors, associates or consultants.

B. The estimator may not subcontract services in excess of $1,000.00 or thirty percent, whichever is less, of the price of a work order to subcontractors or consultants without the owner's prior written approval.

Article IX - Equal Employment Opportunity

The estimator agrees that he (she) will not discriminate against any employee or applicant for employment because of race, religion, color, sex, age, or national origin.

Article X - Insurance

A. The estimator shall procure and maintain during the life of the contract comprehensive general liability insurance in the amount of _____ for bodily injury and property damage. The owner shall be named on the comprehensive general liability insurance policies as a co-insured, and be shown on the insurance certificates provided to the owner by the estimator. Nothing herein shall be construed to limit the scope of indemnity set forth below.

B. The certificates shall provide that if the policies are canceled by the insurance company or the estimator during the term of the contract, written notice will be given to the owner not less than thirty days prior to the effective date of such cancellation.

C. Estimator shall indemnify, defend, and save harmless _____ for all claims, demands, suits, action, proceedings, loss, cost, and damages of every kind and description including any attorneys' fees and/or litigation expenses which may be brought or made against or incurred by _____ on account of loss of or damage to any property or for injuries to or death of any person, caused by, arising out, or contributed to, in whole or in part, by estimator, it's employees, agents or representatives during the performance of this agreement, or arising out of worker's compensation claims, unemployment compensation claims, or unemployment disability compensation claims of employees or estimator or claims under similar such laws or obligations. Estimator's obligation under this section shall not apply to any liability caused by the sole negligence of _____ or their employees, agents, or representatives.

D. The intent of the insurance requirements in Article X is that the estimator will provide comprehensive general liability insurance for bodily injury and property damage for the protection of the _____ at such times as the estimator may be on _____ property.

Whenever this type of insurance is available to the estimator only through the furnishing of professional liability insurance, the insurance requirements in Article X are considered as being null and void. Any requirements for the provision of professional liability insurance by the estimator for the purpose of furnishing estimating services is not within the scope or intent of those services.

The estimator shall provide statements from a minimum of two qualified insurance agencies that the general liability insurance is available to the estimator only after the securing of professional liability insurance.

The estimator and _____ mutually agree that the requirements of the article [are] [are not] to be included in this contract.

Article XI - Independent Contractor

The estimator will be deemed to be an independent contractor for the purposes of his contract, and neither the estimator nor any of its employees shall be considered an agent, employee or servant of the owner.

Article XII - Time of Essence

Time is of the essence of this contract; however, in the event the provisions of this contract require any act to be done or action to be taken on a date which is a Saturday, Sunday, or legal holiday, such act or action shall be deemed to have been validly done or taken if taken on the next succeeding day which is not a Saturday, Sunday, or legal holiday.

Article XIII - Assignability

This agreement shall not be assignable in whole or in part without the written consent of the owner.

Article XIV - Non-discrimination

The parties agree to be bound by applicable state and federal rules governing equal employment opportunity and non-discrimination.

Article XV - Arbitration

The parties agree that should a dispute arise between them, in any manner concerning the attached contract, and said dispute involves the sum of one hundred thousand dollars ($100,000.00) or less in money damages only, exclusive of interest, cost or attorneys' fees, the parties will submit the matter to binding arbitration pursuant to the construction industry arbitration rules for compulsory arbitration and the decision of the arbitrator(s) shall be final and binding upon the parties.

Article XVI - Obligation

The parties recognize that the performance by _____ for and on behalf of _____ may be dependent upon the appropriation of funds _____. Should the _____ fail to appropriate the necessary funds, the _____ may cancel this contract without further duty or obligation. The owner agrees to notify the estimator as soon as reasonably possible after the unavailability of said funds comes to the owner's attention.

The estimator shall not be obligated to perform any work under this contract until the owner notifies him in writing that the necessary funding has been appropriated.

Article XVII - Conflict of Interest

This agreement is subject to the provisions of _____ and _____ may cancel this contract if any person significantly involved in negotiating, drafting, securing or obtaining this contract for or on behalf of _____ becomes an employee in any capacity of any other party including a consultant to any other party with reference to the subject matter of this contract while the contract or any extension hereof is in effect.

Article XVIII - Law to Govern

This agreement is made under and shall be interpreted according to the laws of _____.

_____   _____
Owner   Estimator

_____   _____
Date   Date

Contract for Professional Estimating Services between Estimator and Client

This document has important legal consequences. Consultation with an attorney is encouraged with respect to its completion or modification.

This agreement, made and entered into as of _____ , 20 _____, by and between _____ whose address is _____, hereinafter called "estimator", and _____ whose address is _____, hereinafter called "client".

Witnesseth:

whereas, client desires certain professional construction estimating services in respect to projects within it's domain.

Whereas, estimator desires to perform such services for client.

Now, therefore, in consideration of the promises and mutual covenants contained herein, the parties do hereby agree as follows:

1. Term:

The term of this agreement shall commence on _____ , 20 ____, and shall continue for _____ unless terminated as hereinafter provided.

Upon expiration, this agreement may be renewed or extended on a month-to-month basis by mutual written agreement and consent of both estimator and client.

2. Scope:

Commencing with the effective date hereof, estimator agrees to render professional estimating services as defined in Appendix A, in a timely manner and consistent with the Code of Ethics and Standard Estimating Practice as defined by the American Society of Professional Estimators.

3. Compensation:

Client shall pay estimator for the work performed pursuant to this agreement, a fee or hourly personnel rate as defined in the attached Appendix B. This fee or hourly personnel rate includes all allowances for salaries and wages, direct labor burden (i.e. payroll taxes, insurance, vacation, sick leave, holidays, excused absences, fringe benefits, etc.), overheads, local travel costs, insurance, profits, fees, and all costs and expenses of whatever kind, except as otherwise specifically set forth in this agreement. Specific travel expenses for business trips made by estimator and other unusual expenses required by client and incurred by estimator will be reimbursed by client upon receipt of itemized application for payment.

4. Application and payment:

Estimator shall submit to client an application for payment on a periodic basis, as stated in Appendix B, for compensation earned hereunder and for expenses, detailing the nature of the services and the nature of the expenses. Estimator will furnish client supporting documentation of specific travel expenses and/or other unusual expenses. Client shall pay estimator within 10 calendar days following estimator's application for payment. There shall be no retainage on any of the amounts or withholding of funds for any reason. Late payment shall be subject to the legal rate of interest.

5. Conflict of interest:

During the term of this agreement, estimator shall not perform any services on behalf of any enterprise or any organization in direct competition with client nor engage in any business or activity in direct competition or direct conflict with client's best interest.

6. Confidentiality:

Recognizing the relationship of trust and confidence established between client and estimator by this agreement, both parties hereby agree not to use or disclose to others during or subsequent to the performance of the work (except as is necessary to perform the work and then only on a confidential basis satisfactory to both parties) any information (including any technical information, cost estimating data, operating experience or data) regarding estimator's and/or client's plans, programs, processes, products, costs, equipment, operations or customers which may come within the knowledge of, or which may be developed by, estimator or client in the performance of, or in connection with the work, without, in each instance, securing the prior written approval of both parties.

7. Exceptions:

Provisions of paragraph six above, however, shall not prevent estimator or client from disclosing to others or using in any manner information which estimator or client can show:

A. has been published or has become a part of the public domain other than by act or omission of estimator or client;

B. has been furnished or made known to estimator or client by third parties (other than those acting directly or indirectly for or on behalf of estimator or client) as a matter of right and without restriction on disclosure or use; or

C. was in estimator's or client's possession at the time it entered into this agreement and was not acquired by estimator or client directly or indirectly from either party or it's employees. For these purposes, no information obtained by estimator from client and client from estimator shall be deemed to be public domain or in the prior possession of estimator or client merely because it is embraced by more general information in the public domain or by more general information in the prior possession of estimator or client.

8. Ownership of materials:

All materials and information developed by estimator pursuant to this contract shall become and remain the exclusive property of estimator.

9. Indemnification:

Client hereby indemnifies and agrees to save estimator, it's directors, officers, and employees harmless from and against all claims, suits, demands, losses, costs and expenses brought by any person, firm or corporation for injuries to or the death of any person, or damage to or loss of property alleged to have arisen out of or in connection with estimator's services hereunder.

10. Taxes:

Estimator shall, where applicable, separately itemize and include in the application for payment, all valid use or sales taxes for services to be provided hereunder. This tax shall be in addition to the fee or hourly personnel rate specified in the attached Appendix B.

11. Independent Contractor:

The parties hereto agree that services rendered by estimator in the fulfillment of the terms and obligations of this contract shall be as an independent contractor and not as an employee, agent, or partner of client.

12. Notices:

All notices required or permitted by the terms hereof shall be sent to the following addresses:

Estimator _____

Client _____

13. Termination:

Estimator may terminate this agreement upon 15 days' written notice to client. In the event of any such termination by estimator, estimator shall be paid for services performed and expenses incurred to the date of termination for which estimator had not there to fore been paid by client. Client may terminate this agreement upon 15 days' written notice to estimator. In the event of any such termination by client, estimator shall be paid the outstanding balance of the fee, plus expenses incurred but not paid, at the time of termination.

14. Entirety:

This contract constitutes the entire agreement between the parties, and there are no oral promises, agreements, or warranties affecting it, and shall not be modified or rescinded except as herein provided.

This agreement shall be governed by the laws of the state of _____ and exclusive jurisdiction of any legal proceedings brought by either party to enforce this agreement shall rest with the _____ court.

15. Remedy:

The sole and exclusive remedies for estimator errors and/or omissions shall be (a) for estimator to correct his (her) errors and/or omissions without additional compensation from client, or (b) for client to have the errors and/or omissions corrected by others at estimator's expense.

16. Miscellaneous:

This agreement shall be binding on the heirs, successors and assignees of each of the parties hereto consistent with the terms and conditions of this agreement.

In witness whereof,

The parties hereto have executed this agreement as of the day and year first above written.

Estimator							Client

_____			_____

by								by

_____			_____

Title								Title

_____			_____

Attest								Attest

PART FOUR

ETHICS

Part Four

Table of Contents

The American Society of Professional Estimators Code of Ethics 489
Cannon # 1 489
Cannon # 2 490
Cannon # 3 490
Cannon # 4 491
Cannon # 5 491
Cannon # 6 492
Cannon # 7 492
Cannon # 8 492
Cannon # 9 492

The American Society of Professional Estimators

Code of Ethics

Introduction

Ethical principles are presented which are intended as a broad guideline for professional estimators and estimators in training. The philosophical foundation upon which the rules of conduct are based is not intended to impede independent thinking processes, but is a foundation upon which professional opinions may be based in theory and in practice.

Please recognize that membership in and certification by the American Society of Professional Estimators are not the sole claims to professional competence but support the canons of this code.

The distinguishing mark of a truly professional estimator is acceptance of the responsibility for the trust of client, employer and the public. Professionals with integrity have therefore deemed it essential to promulgate codes of ethics and to establish means of insuring their compliance.

Preamble

The objective of the American Society of Professional Estimators is to promote the development and application of education, professional judgment and skills within the industry we serve. Estimators must perform under the highest principles of ethical conduct as it relates to the protection of the public, clients, employers and others in this industry and in related professions.

The professional estimator must fully utilize education, years of experience, acquired kills and professional ethics in the preparation of a fully detailed and accurate estimate for work in a specific discipline. This is paramount to the development of credibility by estimators in our professional service.

Estimating is a highly technical and learned profession and the members of this society should know that the work is of vital importance to the clients and to the employers they serve. Accordingly, the service provided by the estimator should exhibit honesty, fairness, trust, impartiality and equity to all persons served.

CANON #1

Professional estimators shall perform services in areas of their discipline and competence.

1) Estimators shall to the best of their ability represent truthfully and clearly to a prospective client or employer their qualifications and capabilities to perform services.

2) The estimator shall undertake to perform estimating assignments only when qualified by education or years of experience in the technical field involved in any given assignment.

3) The estimator may accept assignments in other disciplines based on education or years of experience as long as a qualified associate, consultant or employer attests to the accuracy of their work in that assignment.

4) An estimator may be subjected to external pressures to perform work above or beyond qualifying education and experience. In fact, estimators must retain their integrity and professionalism by avoiding involvement in situations that may cause loss of independence and integrity as a professional estimator.

CANON #2

Professional estimators shall continue to expand their professional capabilities through continuing education programs to better enable them to serve clients, employers and the industry.

1) A member of the American Society of Professional Estimators will strive to gain the honored position of "Certified Professional Estimator" and encourage others in the society to obtain this honored position.

2) Members will lend personal and financial support, where feasible, to the schools and institutions engaged in the education and training of estimators.

3) Members will cooperate in extending the effectiveness of the profession by interchanging information and experience with other estimators and those in training to be estimators, subject to legal or proprietary restraints.

4) Members will endeavor to provide opportunity for the professional development and the advancement of estimators and those in training under their personal supervision.

CANON #3

Professional estimators shall conduct themselves in a manner which will promote cooperation and good relations among members of our profession and those directly related to our profession.

1) By treating all professional associates with integrity, fairness, tolerance and respect, regardless of national origin, race, religion, sex or age.

2) By extending fraternal consideration when giving testimony that may be damaging to a member of our society, as long as it does not violate this Code of Ethics and the laws governing the proceedings.

3) By accepting the obligation to assist associates in complying with the code of professional ethics. The professional character of our society is dependent upon continuing mutual cooperation with one another. It is the essential element of our continued success.

4) By recognizing the ethical standards set by other professionals, such as architects and engineers, directly related to our industry and by extending to them the common courtesies they deserve predicated upon the goodwill of all the obligation of the true professional to uphold the highest ethical standards in our free society.

5) By acting honorably, both in personal and professional life, by avoiding situations that may erode public respect. Ethical and personal character shall be paramount in estimators' life styles.

CANON #4

Professional estimators shall safeguard and keep in confidence all knowledge of the business affairs and technical procedures of an employer or client.

1) By not revealing privileged information or facts pertaining to methods used in estimating procedures prescribed by an employer, except as authorized or required by laws.

2) By holding in strict confidence all information concerning a client's affairs acquired during the fulfillment of an engagement and completion of an estimating procedure.

3) By serving clients and employers with professional concern for their best interests, provided however, this obligation must not endanger personal integrity or independence or a high degree of ethical conduct, as set forth in this Code of Ethics and related state and federal laws.

CANON #5

Professional estimators shall conduct themselves with integrity at all times and not knowingly or willingly enter into agreements that violate the laws of the United States of America or of the states in which they practice. They shall establish guidelines for setting forth prices receiving quotations that are fair and equitable to all parties.

1) By not participating in bid shopping as it is known in the building construction industry today. Bid shopping occurs when after the award of the contract, a contractor contacts several subcontractors of the same discipline in an effort to reduce this previously quoted prices. This practice is unethical, unfair and is in direct violation of the Code of Ethics as recognized by the American Society of Professional Estimators.

2) By not accepting quotations from unqualified companies or suppliers. Every effort should be made to pre-qualify any bidder to be used.

3) By not divulging privileged figures from subcontractors and suppliers to competitors prior to bid time in an effort to drive down prices of either. Should quotes be received from subcontractors or suppliers that are obviously low or appear to be in error, this person should be asked to review his price. When making this effort, quotes of others shall not be divulged.

4) By not padding or inflating quoted bid prices. An unethical practice for a professional estimator is to pad or inflate quotes when bidding with firms known for bid shopping. If not a violation of applicable laws, a professional estimator should not provide quotes to known bid shoppers.

5) Professional estimators shall not enter into the unethical practice of complimentary bids (comp bids). This practice is a violation of the Code of Ethics of the American Society of Professional Estimators.

CANON #6

Professional estimators shall utilize their education, years of experience and acquired skills in the preparation of each estimate or assignment with full commitment to make each estimate or assignment as detailed and accurate as their talents and abilities allow.

1) By not formulating estimates from a partial set of bid documents. This is in direct violation of the code of good estimating practices and is not acceptable. To formulate an accurate estimate in any discipline, a full review must be made of all related bid documents. Any other approach could cause errors or omissions that may endanger professional integrity and reliability. Exceptions to this rule should be considered only for the preparation of a conceptual estimate.

2) It is of paramount importance to a professional estimator to minimize the possibility of making mistakes or errors. The more detailed the estimate, the better the accuracy will be.

3) Each estimate shall be cross checked by means that will insure that it is technically and mechanically free from mistakes, oversight or errors. If possible and feasible, estimates should be checked by other professionals. If it is not feasible for someone else to cross check an estimate, the estimator should cross check their own in estimate by utilizing a different method, such as using the historical data or unit prices based on previous cost data on similar projects.

CANON #7

Professional estimators shall not engage in the practice of "bid peddling" as defined by this code. This is a breach of moral and ethical standards, and this practice shall not be entered into by a member of this society.

1) Bid peddling occurs when a subcontractor approaches a general contractor who has been awarded a project, with the intent of voluntarily lowering the original price below the price level established on bid day. This action implies that the subcontractor's original price was either padded or incorrect. This practice undermines the credibility of the professional estimator and is not acceptable.

2) The same procedure applies to a professional estimator engaged as a general contractor, as defined in the previous paragraph, when the estimator approaches an owner or client to voluntarily lower the original bid price.

3) When a proposal is presented, the professional estimator is stating the estimate has been prepared to the best of their ability using their education, expertise and recognized society standards. Entering into unethical practices such as "bid peddling" jeopardizes both personal and society professional credibility, while violating the trust of the clients.

CANON #8

Professional estimators and those in training to be estimators shall not enter into any agreement that may be considered acts of collusion or conspiracy (bid rigging) with the implied or express purpose of defrauding clients. Acts of this type are in direct violation of the Code of Ethics of the American Society of Professional Estimators.

1) Bid rigging, collusion and conspiracy, as defined by the American Society of Professional Estimators, may occur between two (2) or more contractors or two (2) or more subcontractors. Agreements are reached by companies or individuals in the act of conspiring to pre-set the price of a particular project (private or governmental) with the express purpose of predetermining the intended recipient of a contract to be awarded at a fixed price.

2) Professional estimators and those in training to be estimators shall not be associated with firms which are known to participate in the practice of bid rigging.

3) There are no conditions or social convention that will allow a professional estimator to enter into such fraudulent acts as those of bid rigging, knowing that they are held to be unlawful, immoral, unethical and unacceptable to this society.

CANON #9

Professional estimators and those in training to be estimators shall not participate in acts, such as the giving or receiving of gifts, that are intended to be or may be construed as being unlawful acts of bribery.

1) Professional estimators should not offer cash, securities, intangible property rights or any personal items in order to influence or that give the appearance of influencing the judgment or conduct of others that would place them in the position of violating existing laws or leave them with the feeling of obligation or indebtedness.

2) Professional estimators should not accept gifts, gratuities or entertainment that would place them in a position of breaking existing laws (municipal, state or federal) or that give the appearance of creating an inducement which would affect the estimator's professional credibility by placing them in a position of obligation.

PART FIVE

REFERENCE SOURCES

Acoustical Material Association, New York City, NY 10020
Air Conditioning and Refrigeration Institute, 1815 North Fort Meyer Drive, Arlington, VA 22209
Air Diffusion Council, 435 North Michigan Avenue, Chicago, IL 60611
Air Movement and Control Association, 30 West University Drive, Arlington Heights, IL 60004
Aluminum Association, 818 Connecticut Ave., N.W., Washington DC 20006
Aluminum Window Manufacturers Association, New York City, NY 10020
American Association of Highway Officials, 444 North Capitol, Washington DC 20006
American Concrete Institute, Box 19150 Redford Station, Detroit, MI 48219
American Gas Association, 1515 Wilson Boulevard, Arlington, VA 22209
American Institute of Architects, 1735 New York Avenue, N.W., Washington DC 20006
American Institute of Steel Construction, 1221 Avenue of the Americas, New York City, NY 10020
American Institute of Timber Construction, 333 West Hampden Avenue, Englewood, CO 80110
American Iron and Steel Institute, 1000 16th Street, N.W., Washington DC 20036
American National Standards Institute, 1430 Broadway, New York, NY 10018
American Society for Testing Materials, 1917 Race Street, Philadelphia, PA
American Society of Civil Engineers, 345 - 47th Street, New York City, NY 10017
American Society of Heating, Refrigerating & Conditioning Engineers, 345 - 47th Street, New York City, NY 10017
American Society of Heating & Ventilating Engineers, 1791 Tullie Circle, Atlanta, GA 30329
American Society of Mechanical Engineers, 345 East 47th Street, New York City, NY 10017
American Subcontractors Association, 1004 Duke Street, Alexandria, VA 22314
American Water Works Association, 6666 West Quincy Avenue, Denver CO 80235
American Welding Society, 2501 NW 7th Street, Miami, FL 33125
American Wood Institute, 2310 Walter Reed Drive, Arlington, VA22206
American Wood Preserver's Association 7735 Old Georgetown Road, Bethesda, MD 20014
Antifriction Bearing Manufacturers Association, 60 East 42nd Street, New York City, NY 10017
Architectural Wood Institute, 2310 South Walter Reed Drive, Arlington, VA 22206
Asphalt Institute, Asphalt Institute Building College, Park, MD 20740
Associated Air Balance Council, 1000 Vermont Avenue, N.W., Washington DC 20005
Associated Builders and Contractors, Inc., 6989 Washington Ave, Suite 200, Edina, MN 55435
Associated General Contractors of America, 1957 E Street, N.W., Washington DC 20006
Associated Specialty Contractors, Inc., 7315 Wisconsin Avenue, Bethesda, MD 20814
Building Stone Institute, 420 Lexington Avenue, New York, NY 10170
Cast Iron Soil Pipe Institute, 2020 K Street, N.W., Washington DC 20006
Chain Link Manufacturers Institute, 1101 Connecticut Avenue, Washington DC 20036
Commercial Standards (Bureau of Standards), Department of Commerce, Washington DC 20234
Concrete Reinforcing Steel Institute, 180 North LaSalle, Suite 2110, Chicago, IL 60601
Construction Specifications Institute, 601 Madison Street, Alexandria, VA 22314
Cooling Tower Institute, 9030 IH-45 North, Houston, TX 77037
Copper Development Association, 57th Floor, Chrysler Building, 405 Lexington Avenue, New York City, NY 10017
Edison Electric Institute, 90 Park Avenue, New York City, NY 10016
Factory Mutual System, 1151 Boston-Providence Turnpike, Norwood, MA 02062
Federal Specifications, General Services Administration, Specifications/Consumer Information, Distribution Sect (WFSIS), Washington Navy Yard, Building 197, Washington DC 20407
Gypsum Association, 1603 Orrington Avenue, Chicago, IL 60201

Heat Exchange Institute, 122 East 42nd Street, New York City, NY 10017
Illuminating Engineering Society, 345 East 47th Street, New York City, NY 10017
Institute of Electrical and Electronics Engineers, 345 East 42nd Street, New York City, NY 10017
International Conference of Building Officials, 5360 South Workman Mill Road, Whittier, CA 90601
Uniform Administrative Code
Uniform Building Code
Uniform Mechanical Code
Uniform Housing Code
Uniform Code for the Abatement of Dangerous Buildings
Training Manual in Field Inspection of Buildings and Structures
Uniform Building Code Standards
Uniform Fire Code
Uniform Fire Code Standards
Uniform Building Security Code
Analysis of Revisions
U.B.C. Supplements
Dwelling Construction Under the Uniform Building Code
Building Department Administration
Uniform Sign Code
Illustrated Mechanical Manual
Plan Review Manual
Concrete Inspection Manual
Marble Institute of America, 33505 State Street, Farmington, MI 48024
Metal Lath Manufacturers Association, 600 S. Federal Street, Suite 400, Chicago, IL 60601
Metal Lath/Steel Framing Association, 221 North LaSalle Street, Chicago, IL 60601
National Association of Architectural Metal Manufacturers, 221 North LaSalle Street, Chicago, IL 60601
National Association of Sheet Metal and Air Conditioning Contractors, P. O. Box 70, Merrifield, VA 22116
National Board of Fire Underwriters, National Electrical Code, 470 Atlantic Avenue, Boston, MA 02210
National Electrical Contractors Association, 7315 Wisconsin Avenue, Bethesda, MD 20814
National Electrical Manufacturers Association, 2101 L Street, N.W., Washington DC 20037
National Fire Protection Association, 470 Atlantic Avenue, Boston, MA 02210
National Lime Association, Washington DC
National Forest Products Association, 1250 Connecticut Avenue, N.W., Washington DC 20036
National Lumber Manufacturers
National Terrazzo and Mosaic Association, 3166 Des Plaines Ave., Suite 132, Des Plaines, IL 60018
Occupational Safety and Health Act, (Department of Labor), 200 Constitution Avenue, Washington DC 20210
Painting and Decorating Contractors of America, 7223 Lee Highway, Falls Church, VA 22046
Performance Handbook (Caterpillar Company), Peoria, IL
Portland Cement Association, 5420 Old Orchard Road, Skokie, IL 20076
Prestressed Concrete Institute, 20 North Wacker Drive, Chicago, IL 60606
Producers Council, Inc.
Rental Rate Blue Book, (Construction Equipment), 1290 Ridder Park Drive, San Jose, CA 95131
Sealed Insulating Glass Manufacturers Association, 111 East Wacker Drive, Chicago, IL 60601

Sheet Metal and Air Conditioning Contractors' National Association, 8224 Old Courthouse Road, Vienna, VA 22180
Southern Pine Association, P. O. Box 52468, New Orleans, LA
Steel Deck Institute, Box 3812, St. Louis, MO 63122
Steel Door Institute, 712 Lakewood Center North, Cleveland, OH 44107
Steel Joist Institute, 1703 Parham Road, Suite 204, Richmond, VA 23229
Structural Clay Products Institute, Washington, DC
Thermal Insulation Manufactures Association, 7 Kirby Place, Mt. Kisco, NY 10549
Tile Council of America, Box 326, Princeton, NJ 08340
Tile Manufacturers Association,
Tubular Manufacturers Association, 331 Madison Avenue, New York City, NY 10017
Underground Contractors Association
Underwriters' Laboratories, Inc., 207 East Ohio Street, Chicago, IL 60611
Uniform Building Code, 5360 S. Workman Hill Road, Whittier, CA 90601
Uniform Plumbing Code, 5032 Alhambra Avenue, Los Angeles, CA 90032
Vermiculite Institute, Chicago, IL
West Coast Lumbermen's Association, Portland, OR
Western Pine Association, Portland, OR
Wood Book, 157 Yesler Way, Suite 317, Seattle, WA 98104

Index

A

Abstracts	110
Accessories	258
Account Summary Form	97
Acoustic Ceiling Estimate	423
Acoustical Ceiling Take Off Sheet	421
Ceilings	415

B

Basic Masonry Materials and Methods	311
Standards	5
Bath Accessories	443
Beam an Girder Schedule	293
Beams	264, 291
Bid	3, 4, 9, 10, 13, 28, 30, 41, 53, 57, 63, 125, 152, 156
Analysis	125
Day Procedures	107
Documetns	37
Peddling	117
Price	463
Protest	157
Shopping	117
Process	152
Bored Piles	207, 213

C

Cabinets	431
CANON #1	489
CANON #2	490
CANON #3	490
CANON #4	491
CANON #5	491
CANON #6	492
CANON #7	492
CANON #8	493
CANON #9	493
Cashflow Analysis	183
Cast-In-Place Concrete	285
Change Order Aids	135
Change Orders	3, 4, 7
Checklists	3, 28, 39, 169
Clay Masonry Units	317
Coatings for Steel	439
Columns	264, 290
Company Policy	29
Conceptual Design	9
Concrete Accessories	251
Curbs	241
Footings	285
Forms	251
Materials	249
Walls	287
Construction Document	3, 10, 12
Equipment	175
Equipment Mobilization	177
Scheduling	56
Consumables Mobilization	177
Containment Areas	455
Contingencies	183
Contingency	103
Contract Knowledge	152
Performance	153
Contractor Overhead and Profit	123
Conversion Factors for Sloped Surfaces	342
Costs to the Owner	123
Custom Cabinets	357
Grade	353

D

Dampproofing and Waterproofing	361
Deferred Submittal	158
Demobilization	177
Design Contingency	103
Developing Crews	225
Differing Site Conditions	158
Direct Labor Burden	98
Documentation	3, 6
Door Grades and Applications	384
Door Hardware	395
Schedule	385, 395
Doors and Windows	429
Drill Pier Volume Charts	210
Drilled Caissons	207
Drywall Estimate	411
Duct Insulation	455

E

Economic Dimension	118
Economy Grade	353
Environmental Protection	175
Equipment	96
Escalation and Contingency	123
Estimate Basis	121
Format	91
Levels	91
Estimating Change Orders	131
Procedures	25, 169

Ethical Dimension 118
Ethics...3, 5, 21, 27, 487, 489
Excavation & Fill 203
    Support 185
Exterior Insulation and Finish Systems......................367
Exterior Walls 434
    Wood 435

F

Fascia Materials 455
Federal Claims 137
Final Cleaning 181
    Summary Form 98
Finance Expenses 182
Finish Carpentry 351
Fire Rating Requirements......................................386
Fixed Fees 99
Floor Framing 334
Forms 28, 169
French Windows 431

G

Galvanized Metals 434
General Assumptions 122
Grading..199
Grouting Pile Process 214
Gutters 241
Gypsum Board 401

H

Hardware Schedule 396
Hazardous Material Remediation............................449
HM Sidelites 430
Home office Staff 58
Horizontal Forming 253
Hot & Cold Weather Procedures.............................249

I

Installation of Manufactured Casework 354
    of Wood Base 353
Integrity..3, 5
Interpretation of Intent 87
Inventory Checklist 43

J

Job Site Security..179
Joint Sealants..381

Judgment..3, 5

L

Labor Hours..3, 7, 399
Large Doors 429
Laundry Accessories..443
Lawsuits 151
Legal Burden of Management 151
Levels of the Estimate 9
Lineal Feet Per Gallon 381
Litigation vs. Arbitration 155

M

Major Building Groups...255
Management and Engineering 173
    Support 182
Masonry Accessories..315
    Mortar 313
Master Checklist 41
Membrane Roofing 375
Mensuration 92
Metal Doors and Frames.......................................383
Metals 432
Millwork and Interior Wood..................................431
Mine Tailings..450
Miscellaneous Wire Insulation 455
Mobilization 176

N

Negative Air Equipment.......................................456
Notice of Ambiguity..158

O

Office Estimate 407
    Support 181
Order of Magnitude 3, 9, 10, 14-19, 33
Outside Services 178
Overhead...98
    Estimate Checklist 57
Owner's Construction Contingency.........................103
    Project Contingency 103

P

Paints and Coatings 427
Paneling...432
Performance Bonds..99

Personnel Mobilization	177
Pipe Insulation	453
Piping	433
Plan Review	89
Plant/Facility Mobilization	177
Polychlorinated Biphenyls	451
POST-Bid Procedures	125
Pre-Bid	152
Precast Structural Concrete	297
Prefabricated Structural Wood	339
Premium Grade	353
Presentation	121
Pricing	201
Pricing/Summaries	95
Prime Contractor Responsibilities	115
Production Process	223
Profit	99
Project Characteristics	257
Evaluation	29, 31
Project Office Expense	175
Site	171
Site Staffing	172
Protecting Impact Costs	159
Protest Reduction	159

Q

Quality Standards	353
Quantity Survey	91
Survey Sheet	448
Questions for Bidders	110
Quick Budget Estimates	446

R

Recommendations	122
Recording Historical Data	447
Reference Sources	495
Reinforced Unit Masonry Assemblies	321
Reinforcing Bars	207
Steel	271
Steel Placement	273
Removal of Asbestos	452
RFP Data Sheet	471
Rigid Pavement	237
Roof Decking	341
Maintenance and Repairs	379
Rough Carpentry	331

S

Sales or Use Taxes	99
Sanitary Sewerage	221

Scope Letters	110
of Estimate	21
Security and Electronics	386
Shelving	432
Sidewalks	245
Site Investigation Checklist	45
Slabs	267, 294
Small Tools	181
Soffits	434
Soil Conditions	208
Special Coatings and Material Types	386
Forms	39
Specific Demolition	452
Specification Review	85
Spray-on Decorative Acoustical Ceilings	454
Steel Sash Windows	430
Structural Beams	452
Deck	453
Steel	323
Steel Decking	433
Steel Shapes	433
Subcontracts	97
Supervision	173
Suspended Acoustical Ceilings	422
Systems Descriptions	92

T

Takeoff Procedures	93
Temporary Plant/Facilities	174
Utilities	174
Testing/Inspection	178
Tilt-Up Precast Concrete	303
Toilet Accessories	443
Transoms	430
Trenching	203
Truss Roof System	340, 344

U

Uniformity	3, 6

V

Value Engineering	3, 7, 31, 33, 36
Engineering Methodology	33
Engineering Team	34
Vendor Bidder Responsibilities	115
Vertical Forming	253
Vinyl Sheeting	454
Tiles	454

W

Wall Covering 425
    Framing 336
Walls 261
Walls, Ceilings and Floors 428
Waste Allowances 93, 285, 436
Water Repellents 363
Weather Protection 181
Windows 430
Withdrawal of Bid 157
Wood Casement 431
    Ceilings 432
    Moldings 432
    Sidelites 430